GENETICS OF INDUSTRIAL MICROORGANISMS

Volume II

ACTINOMYCETES AND FUNGI

Scientific Editors

Jan Nečásek, PhD
Professor of Genetics
Charles University
Prague

Františka Palečková, PhD
Research Institute
for Antibiotics and Biotransformations
Roztoky near Prague

Language Editor

Arnošt Kotyk, DSc
Institute of Microbiology
Czechoslovak Academy of Sciences
Prague

GENETICS OF INDUSTRIAL MICROORGANISMS
Volume II
ACTINOMYCETES AND FUNGI

Edited by: **ZDENKO VANĚK**
ZDENĚK HOŠŤÁLEK
JOSEF CUDLÍN

Institute of Microbiology

Czechoslovak Academy of Sciences

Prague

ELSEVIER PUBLISHING COMPANY

Amsterdam - London - New York 1973

Published in co-edition with
ACADEMIA, Publishing House of the Czechoslovak Academy of Sciences, Prague

Distribution of this book is being handled by the following publishers

for the U.S.A. and Canada
American Elsevier Publishing Company, Inc.
52 Vanderbilt Avenue
New York, New York 10017

for the East European Countries, China, North Korea, Cuba,
North Vietnam and Mongolia
Academia, Publishing House of the Czechoslovak Academy of Sciences, Prague

for all remaining areas
Elsevier Publishing Company
335 Jan van Galenstraat
P.O. Box 211, Amsterdam, The Netherlands

ISBN 0-444-40990-4
Library of Congress Card Number 75-180010
With 165 Illustrations and 101 Tables

Printed in Czechoslovakia

CONTENTS

Contributors

ALAČEVIĆ M., *Faculty of Technology, Zagreb, Yugoslavia*

ALIKHANIAN S. I., *Inst. of Genetics and Selection of Ind. Microorganisms, Moscow, U.S.S.R.*

BALL C., *Glaxo Laboratories Ltd., Ulverston, Great Britain*

VON BÉKÉSY N., *Research Institute for Medicinal Plants, Budapest, Hungary*

CALAM C. T., *I.C.I. Ltd., Pharmaceuticals Division, Alderley Park, Macclesfield, Great Britain*

CHEREST H., *Laboratoire d'Enzymologie du C.N.R.S., Gif-sur-Yvette, France*

CORCORAN J. W., *Northwestern University Medical School, Chicago, Illinois, U.S.A.*

DAGLISH L. B., *I.C.I. Ltd., Pharmaceuticals Division, Alderley Park, Macclesfield, Great Britain*

ELANDER R. P., *Wyeth Laboratories, Inc., West Chester, Pennsylvania, U.S.A.*

ENGEL P. P., *St. Benedict's College, Atchinson, Kansas, U.S.A.*

ESPENSHADE M. A., *Wyeth Laboratories, Inc., West Chester, Pennsylvania, U.S.A.*

ESSER K., *Institut für Allgemeine Botanik, Ruhr-Universität, Bochum, G.F.R.*

FINCHAM J. R. S., *University of Leeds, Great Britain*

GAITSKELL W. S., *I.C.I. Ltd., Pharmaceuticals Division, Alderley Park, Macclesfield, Great Britain*

GRENSON M., *Université Libre de Bruxelles and Institut de Recherches du C.E.R.I.A., Brussels, Belgium*

HASSALL C. H., *University College of Swansea, Great Britain*

HOPWOOD D. A., *John Innes Institute, Norwich, Great Britain*

HOŠŤÁLEK Z., *Inst. of Microbiology, Czechoslovak Academy of Sciences, Prague, Czechoslovakia*

HÜTTER R., *Eidgenössische Technische Hochschule, Zürich, Switzerland*

KAŠLÍK J., *Institute of Microbiology, Czechoslovak Academy of Sciences, Prague, Czechoslovakia*

KOBEL H., *Sandoz Ltd., Basle, Switzerland*

KOZOVÁ J., *Institute of Microbiology, Czechoslovak Academy of Sciences, Prague, Czechoslovakia*

LIERSCH M., *Research Labs of the Pharmaceutical Division of CIBA Ltd., Basle, Switzerland*

MACDONALD K. D., *Microbiological Research Establishment, Salisbury, Great Britain*

NÜESCH J., *Research Laboratories of the Pharmaceutical Division of CIBA Ltd., Basle, Switzerland*

PAN C. H., *Wyeth Laboratories, Inc., West Chester, Pennsylvania, U.S.A.*

PATHAK S. G., *Wyeth Laboratories, Inc., West Chester, Pennsylvania, U.S.A.*

ŘEHÁČEK Z., *Inst. of Microbiology, Czechoslovak Academy of Sciences, Prague, Czechoslovakia*

DE ROBICHON-SZULMAJSTER H., *Laboratoire d'Enzymologie du C.N.R.S., Gif-sur-Yvette, France*

ROMANO A. H., *University of Cincinnati, Ohio, U.S.A.*

ROPER J. A., *The University, Sheffield, Great Britain*

SAJDL P., *Institute of Microbiology, Czechoslovak Academy of Sciences, Prague, Czechoslovakia*

SANGLIER J. J., *Sandoz Ltd., Basle, Switzerland*

SERMONTI G., *Istituto di Genetica dell'Università, Palermo, Italy*

SPADA-SERMONTI I., *Laboratorio di Embriologia Molecolare, Napoli, Italy*

SPALLA C., *Farmitalia — Istituto Ricerche, Milano, Italy*

SUHADOLNIK R. J., *Albert Einstein Medical Center, Philadelphia, Pennsylvania, U.S.A.*

SURDIN-KERJAN Y., *Laboratoire d'Enzymologie du C.N.R.S., Gif-sur-Yvette, France*

TAMM C., *University of Basle, Switzerland*

TREICHLER H. J., *Research Labs of the Pharmaceutical Division of CIBA Ltd., Basle, Switzerland*

VANĚK Z., *Inst. of Microbiology, Czechoslovak Academy of Sciences, Prague, Czechoslovakia*

VINING L. C., *National Research Council of Canada, Halifax, Canada*

WHITEHOUSE H. L. K., *University of Cambridge, Great Britain*

WOODRUFF H. B., *Merck Sharp & Dohme Research Laboratories, Merck & Co., Inc., Rahway, New Jersey, U.S.A.*

1/ Introduction

1/1 A<small>LIKHANIAN</small> S.I.

Principal Results and Unsolved Problems in Microbial Selection

S. I. ALIKHANIAN

Institute of Genetics and Selection of Industrial Microorganisms, Moscow, U.S.S.R.

Four fundamental discoveries in microbial genetics (genetic transformation, explanation of microbial variability in mutational terms, revealing of sexual processes in bacteria and genetic transduction) played a decisive role in the development of applied branches of microbiology. On the other hand, microbiological and biochemical investigation of biosynthetic processes in microbial cells contributed very much to successful employment of genetic techniques in applied microbiology.

Development of the industry of microbiological synthesis, which dates back to mass penicillin production, brought applied aspects of microbial genetics to the level of paramount importance. After successful exploitation of genetic methods for selection of penicillin producers, the effectiveness of this approach was demonstrated in the selection of many other strains producing antibiotics, which resulted in the introduction of corresponding production processes. During recent years, the synthetic activity of microorganisms has been commonly used in different fields of industry.

Thus, it has been shown that commercial manufacture of a number of amino acids (*e.g.* lysine, glutamic acid, tryptophan), vitamins (*e.g.* B_{12} and riboflavin), nucleotides (*e.g.* inosinic acid), growth factors (gibberellins), enzymes and toxins employed as insecticides is possible only by using microbiological synthesis. The effective biosynthetic apparatus of microorganisms helped chemistry not only in the production of substances "discovered" by microorganisms, but also in carrying out some specific chemical reactions, *e.g.* certain steps in the chemical synthesis of cortisone and other steroids. Microbiological synthesis is also employed in industrial chemistry of polymers, for instance when microbiologically synthesized lysine is used for polylysine production.

The decisive points in organisation of all mentioned cases of industrial production based on microbial synthetic systems were firstly the increase of productivity of already known strains, for instance in the case of antibiotic producers, and secondly experimentally induced changes in microbial biosynthetic pathways for obtaining desired substances participating in a complicated metabolic pathway. An example of the latter approach is the creation of highly active amino-acid producers. In both cases the work is based on modern genetic techniques. Various genetic methods used for practical selection differed in their efficiency. Induced mutagenesis was the one most employed in microbial selection.

Selection by the use of various mutagens still continues to be commonly used for increasing the productivity of useful forms of microorganisms. Great experience has been obtained in breeding antibiotic producers using mutagens. The most important results were obtained with the penicillin producer *Penicillium chrysogenum* by STAUFFER, ALIKHANIAN and others and with the streptomycin producer *Streptomyces griseus* by DULANEY, STANLEY, ALIKHANIAN and others. It seems to me that these results allow one to make some important generalizations.

Firstly, the work of the mentioned authors makes it impossible to choose a single mutagen or a group of mutagens as the most advantageous for the selection of antibiotic producers. Positive results were obtained both for physical factors (UV, X-rays, γ-irradiation, fast neutrons) and for various chemical agents (nitrogen mustard, ethyleneimine, diethylsulphate, *etc.*). It has been shown that mutagens may differ from each other in the effectiveness of mutagenesis concerning different characters. As far as the effectiveness of different mutagens in increasing the productivity is concerned, no apparent superiority of any mutagen was observed. It seems that it is not the strength of a mutagen which determines its effectiveness but rather the properties of the producer, for instance origin of strain and particularly the number of previous mutation-selection procedures. A wild strain which had not been previously selected was shown to react effectively to any mutagenic treatment, while highly active strains isolated after a multistep selection (*e.g.* penicillin producers) reacted much more weakly.

Secondly, it is known that in the course of prolonged selection with the use of a mutagen the effectiveness of the latter progressively decreases. In this case, the effectiveness of selection may increase by the use of another mutagen. For instance, after repeated use of UV-light it is worthwhile to take a mutagen of another nature, *e.g.* a chemical one.

The third generalization concerns the choice of the effective dose. It has been observed that the highest yield of variants with increased activity was obtained at moderate doses which often do not correspond to the optimal doses for morphological or other alternative mutations. Moreover, the mutagen doses which induce maximal number of alternative, mostly morphological mutations result in the creation of variants with impaired ability to produce the antibiotic, *i.e.* increase the number of minus-variants.

The fourth conclusion may be drawn from the investigation of the correlation between morphological variability and changes in the antibiotic-producing capacity. Analysis of experimental data shows that the increase of strain activity as a rule is not connected with substantial changes of its morphology. On the contrary, drastic changes of colonial morphology are accompanied by partial or complete loss of producer activity. On the other hand, it is worth noting that pronounced shifts of activity usually lead to changes of strain morphology in respect to a number of microscopic characteristics, such as size and shape of colonies, extent of sporulation, structure of aerial or substrate mycelium, *etc.* As a rule, these morphological changes do not extend beyond individual variability of the given strain. Sometimes, these modifications are so slight, that they are hardly distinguishable from the unchanged variants in single-spore plating. Sometimes the changes may be revealed only when the studied strain is carefully compared with the one obtained in the course of further selection.

Fifth, it is well known that many selected strains are 30—50 times more efficient than the initial ones. Such strains as a rule result from multistep breeding with the use of mutagens, which consists of gradual accumulation of slight increases of activity over 10—20 generations (breeding steps). Such repeated treatment of the same genome with mutagens leads to its saturation with mutations and as a result decreases the probability of new mutations in this genome. In other words, we observe here the loss of effectivity of breeding with the use of mutagen which is expressed in a sharp decrease of the variation coefficient. These events resemble the situation characterized by specific types of dose-rate curves, when the mutation frequency rises with the dose up to the maximum, after which further increase of dose causes a decrease in mutation frequency. Decreased breeding effectiveness of highly active strains is connected with accumulation in the genome of numerous physiological mutations which affect viability. Further mutagen treatment of such a strain may create mutations fatal for the cells. In this case a kind of stabilization of the character in a population may be said to take place. The question arises if it is possible to impair such a stabilization of character experimentally. Some cases of destabilization of a character as a result of radical, mutagenically induced, rearrangement of the geno-type, known as "major" mutations, were described. The important effect of a "major" mutation is a sharp rise of the variation coefficient, which in turn restores selection effectiveness in highly active strains.

Great achievements in the selection of most of the industrially important pro-ducers are connected with the use of mutagens which have been employed for micro-bial selection for about twenty-five years. However, the use of mutagens in selection up to the present time poses many unsolved questions, the method being still in many respects empirical.

Numerous investigations of induced mutagenesis in microbial selection deal with the problem of specificity of mutagenesis.

A number of authors believe that the solution of this problem would lead to the possibility of obtaining directed mutations and therefore would markedly increase the effectiveness of selection with the use of mutagens. At the present time, the pro-blem of specificity of mutagenesis is far from being solved. We believe that the problem of mutagenic specificity does not exist at all in the state in which it is now formulated in the literature. It must be remembered that in general terms the con-ception of mutagenic specificity is usually considered in the literature as suggesting that certain types of mutations may be induced by certain mutagens and that it is possible to obtain a mutation with a desired phenotype by choosing proper mutagens. To evaluate this conception we need to bear two points in mind: the molecular nature of mutations and the multistep character of the mutational process. It is known that the molecular mechanism of mutagenesis is based on changes of nucleotide sequence in the DNA molecule which, in its turn, is connected with the substitution of the A—T pair by the G—C pair or *vice versa*. Therefore, possible modifications of genes are restricted to only two types of substitution (we deliberately do not mention here transversions and insertions and deletions of base pairs). Consequently, if a certain mutagen is able to induce substitution in both directions it may actually cause any kind of mutation at any site of any gene. If mutagens are used which are able to induce only unidirectional changes, induction of any type of base-pair substitution mutation may be achieved by using only two mutagens, provided that the direction of substitions caused by the two is opposite. These speculations show that potentialities of specificity and choice of "specific" mutagens are so limited, that it is hardly possible to speak about the reality of the idea about a set of mutagens which could ensure the obtaining of phenotypically determined mutations. The second aspect of mutagenesis, *i.e.* the multistep character of the process, makes the problem of specificity even more complicated. For instance, if we assume that one of the steps of mutagenesis consists in the penetration of the mutagen through the cell membrane, the mutagen once selected as specifically effective on this type of cells may be inactive as a result of change of cell permeability caused by a mutation.

Geneticists possess a number of tools for obtaining a certain microbial mutation. Thus, the method of selective media makes it possible to select certain biochemical mutations; using *amber* and *ochre* mutations allows to obtain reversions of certain triplets in the DNA molecule. Employment of these and other tricks of selection of mutants is in agreement with the conception of operated mutagenesis.

We believe that investigations of the increase of effectiveness of mutagens in the selection of microorganisms must develop in the direction of improvement of selective techniques of mutant screening, *i.e.* in the line of increasing the effectiveness of operated mutagenesis.

Recently, a new approach to microbial selection has developed, concerning the isolation of biochemical mutations. Successes in this field are based on detailed biochemical studies of the biosynthetic pathways of a number of metabolites.

Here I mean selection of producers of amino acids, nucleotides, and vitamins, *i.e.* substances composing a group of primary metabolites

The molecular structure of primary metabolites is much simpler than that of secondary metabolites which include antibiotics. Consequently, whole mechanisms of biosynthesis of most of the secondary metabolites are studied only in general. The main pathways of biosynthesis of amino acids, nucleotides and some of the vitamins are described in microorganisms in detail. It is known that the concentration of primary metabolites inside the cell is under the strict control of well-studied repression and feedback-inhibition mechanisms. As a result, primary metabolites normally do not accumulate in cells in large quantities. On the contrary, secondary metabolites are often produced in large amounts; investigation of mechanisms which regulate their biosynthesis is now only beginning.

Detailed knowledge of pathways of amino acid and nucleotide biosyntheis and of the corresponding control mechanisms, as well as the possibility of genetically blocking these processes at different stages determine the specificity of methods employed for selection of strains producing primary metabolites; these methods are based on using auxotrophic mutants with impaired control systems.

Auxotrophic mutants are characterized by their ability to accumulate intermediates of biosynthesis which precede a genetically blocked reaction; these products may undergo further conversions. At the same time, pathways of different metabolites are closely connected and consequently a block of one chemical reaction may result in the change of rate of other reactions coupled with the blocked one. Therefore, single biochemical mutation affecting the biosynthesis of a given metabolite may lead to a number of different changes. Nevertheless, isolation of auxotrophic mutants proved to be a very valuable method of selection of strains producing certain metabolites.

The advantage of this method is its relative rapidity and ease of the selection process. In fact, in this case the selection procedure is sometimes restricted to the induction of single or double auxotrophic mutants, characterized by high productivity, while breeding of highly active producers of secondary metabolites (e.g. antibiotics) requires multistep selection with the use of mutagens.

Availability of microorganisms with known pathways of synthesis of primary metabolites is very advantageous for microbial selection because in this case the result of a certain biochemical mutation may be predicted. The desired biochemical mutation may also be isolated spontaneously without mutagenic treatment of cells. Sometimes it is possible to obtain highly active spontaneous mutants from natural sources.

In 1956, a microorganism was selected from 500 examined soil bacterial strains, which was able to synthetize glutamic acid in large quantities and which was called *Micrococcus glutamicus*.

Since then, numerous bacterial species capable of synthesis of large amounts of glutamic acid were isolated. All of them were characterized by some common properties and, in particular, were biotin-deficient and produced up to 25—30 mg/ml glutamic acid under optimal conditions.

Another group of auxotrophic mutants of similar importance were obtained in several bacterial species: *Escherichia coli*, *Bacillus subtilis*, *M. glutamicus*. These strains were able to release large amounts of free lysine, and the most active were the mutants with hereditary changes in the aspartate metabolic pathway.

Aspartic acid is known to be the initial substrate in the biosynthesis of many amino acids: methionine, threonine, isoleucine and also lysine, the corresponding biosynthetic pathway being branched. The first of the branches is the one that leads to lysine formation. The other two lead correspondingly to methionine and to threonine and isoleucine. Thus, accumulation of excess of lysine takes place in all cases of genetic block at any step of methionine, threonine or isoleucine biosynthesis. It is quite clear that the most effective genetic block must occur at the earliest step of the pathway which precedes the formation of homoserine, the key amino acid in the synthesis of both methionine and threonine. And, in fact, homoserine-deficient auxotrophs were found to be most active in lysine production.

Simple calculations show that homoserine auxotrophs are 200—300 times more productive than the initial strain *M. glutamicus* which releases about 0.1 mg L-lysine per ml. This sharp increase of productivity as a result of a single biochemical mutation is one of the best practical achievements of current microbial genetics.

The key position in industrial amino-acid production is at the present time occupied by three amino acids: glutamic acid, methionine and lysine. Only two of them, glutamic acid and lysine, are produced microbiologically (methionine is synthesized chemically). Nevertheless, microbial geneticists have at their disposal highly productive mutants which make possible a large-scale production of other amino acids if necessary.

It is interesting to follow the procedure of breeding other strains of microorganisms producing purine compounds. Some microorganisms accumulate in the medium 0.3—0.5 mg/ml of these products, but no appreciable amounts of either inosinic or xanthosinic acids were found. Then the geneticists resorted to the procedure which gave brilliant results in the selection of amino-acid producers, *i.e.* to the isolation of auxotrophic mutants with a genetic block in the conversion chain of purines and their derivatives. The mutants synthesized up to 15 mg inosinic acid per ml. These results were another confirmation of the significance of auxotrophic mutants for breeding of primary metabolite producers.

The new possibilities of purine-deficient mutants are due to the purine derivatives as precursors in the riboflavin biosynthesis. Hence, the genetic block of purine biosynthesis at early steps may be used to breed mutants producing excess riboflavin.

The described method of microbial selection is rather simple and highly effective. However, it must be borne in mind that empirical isolation of biochemical mutants cannot yield any positive result, especially if we deal with secondary metabolites. Investigation of the biosynthetic pathways is essential for successful selection by the method of blocking different steps in the chain of biochemical conversions. Hence, this procedure of selection which seems to be very simple from the point of view of the breeder turns out to be rather complicated and difficult in its biochemical aspect

The work on the creation of superactive strains producing amino acids and purine compounds demonstrates that successful directed selection depends to a great extent on understanding the corresponding control mechanisms. Productivity of these strains which produce up to 30 mg lysine and more than 50 mg glutamic acid per ml markedly surpasses that of numerous antibiotic-producing strains. Productivity of the best of the latter when expressed in weight units has reached only 10—15 mg/ml (penicillin, streptomycin, terramycin) as a result of years of selection work, while the activity of others "younger" producers is as low as 2.5—3.0 mg/ml (oleandomycin, erythromycin).

The least developed technique is hybridization of useful forms of microorganisms, and because of that, the practical contribution of hybridization to microbial selection is very small. However, it is worth noting that genetic analysis in *E. coli* using conjugation and transduction, in *B. subtilis* based on transformation and in *Salmonella typhimurium* based on genetic transduction was brought to maximum perfection. Unfortunately, these microorganisms are but little employed in the microbiological industry and it does not seem possible to apply these achievements to practical purposes.

Genetic recombination in the classical object of microbial selection, the actino-mycetes, was discovered quite recently and its mechanism remained obscure up to the most recent time. It was initially believed that genetic recombination in actino-mycetes is in some respect similar to the parasexual cycle in lower fungi. However, it was found that it is in principle similar to that in bacteria. Exploitation of genetic methods, including genetic analysis, is essential for investigation of pathways of antibiotic biosynthesis and of the peculiarities of genetic mechanisms controlling antibiotic production. This in turn is necessary for successful application of hybridiza-tion for actinomycete selection. In this connection, it must be noted that genetic recombination is a widely distributed phenomenon amongst the members of the genus *Streptomyces*.

One of the main results of application of genetic analysis to actinomycetes was the discovery of circular linkage map in *Streptomyces coelicolor*, which most com-pletely defines the relative positions of different genetic loci. The characteristic of the genetic map of *S. coelicolor* is nonrandom distribution of loci; in some cases, loci which control related biochemical functions are closely linked.

The choice of *S. coelicolor* as an object for genetic studies of actinomycetes was successful in many respects, except the following: *S. coelicolor* does not produce an antibiotic and consequently the ability to form antibiotics which is characteristic of many actinomycetes was not included in the genetic analysis of this organism.

Our attention was drawn to the oxytetracycline producer *Streptomyces rimosus* which, in contrast to *S. coelicolor*, is only poorly genetically understood. It is of im-portance that the limited data concerning genetic recombination in *S. rimosus* agree with the results of genetic studies in *S. coelicolor*. Thus it was natural to try to study the genetic control of oxytetracycline synthesis in this organism. It was necessary to isolate mutations which affect the control of antibiotic formation, to localize and to estimate the number of genetic factors responsible for oxytetracycline biosynthesis.

For this purpose, mutants of active *S. rimosus* strains with impaired synthesis of antibiotics were obtained. In contrast to the initial strains which synthesized about 3000 µg oxytetracycline per ml, the amount of the drug formed by inactive mutants was about 20—70 µg/ml, and some of them excreted only traces of the antibiotic

When different combinations of these inactive mutants were simultaneously cultivated in mixed culture in a fermentation medium, production of large quantities of the antibiotic was observed in some cases.

Genetic analysis of these mutants revealed at least two sets of genetic loci responsible for oxytetracycline biosynthesis in *S. rimosus*. No active recombinants arose from crosses between different mutants of the same group which indicate a close linkage of the corresponding loci. Crosses between mutants of different groups yielded active recombinants. Thus, genetic investigation of inactive mutants has shown that the oxytetracycline biosynthesis is under the control of several genetic loci divided into two groups of closely linked loci.

Besides the investigation of recombination in actinomycetes and the well-known studies of the parasexual cycle in *Penicillium* species, a number of attempts were made to apply hybridization for selection of producers of penicillin, oxytetracycline and some other antibiotics.

First, hybrids of commercial microbial strains were described for penicillin producers. Both diploid and haploid recombinants were obtained in *P. chrysogenum*, but no apparent increase of activity in comparison with that of the initial strains was observed.

Later, a number of diploids were described obtained from crosses between two biochemical mutants of highly active strains. These hybrids surpassed the initial strains in their activity. It was suggested that genealogical relations of parent strains may influence the activity of hybrids. To test this possibility, a careful examination of a series of diploids, obtained between biochemical mutants of six *P. chrysogenum* strains, some of which were closely related while others markedly differed in their origin, was performed. The results obtained ruled out the above possibility.

A careful analysis of the penicillin-producing capacity of a number of diploid and haploid recombinants derived from the above-described diploids has shown that productivity of all types of recombinants was lower in comparison with that of the initial diploids.

At the same time, it has been shown that the extent of both spontaneous and UV-induced variability of penicillin-producing capacity was apparently higher in diploid strains than in the initial haploid strains.

In one case, two variants were selected after mutagenic treatment of conidia of a penicillin producer, the activity of which was 25—35 per cent higher than that of the diploid.

There are some literature data on the hybridization in another group of fungi, the *Aspergillus* species, which are used for production of a number of substances, particularly citric acid and some proteolytic enzymes. The practical effect of hybridization in this case was not higher than for penicillin producers.

In spite of the low efficiency of hybridization between fungal antibiotic producers these experiments are important for contributing to our understanding of the mechanisms of the biosynthesis of different metabolites and, which is more significant, our ideas about the nature of genetic changes occurring in the course of selection.

Even less is known about practical hybridization between commercial actinomycete strains.

Actinomycetes, just like bacteria, form haploid genetic recombinants and in this way differ from *Penicillium* and *Aspergillus* species which may produce diploid hybrids and thus demonstrate the heterosis phenomenon commonly used in higher plant hybrids. Therefore, in the case of actinomycetes, special ways of application of hybridization for selection must be sought. The following observations were made in crosses of active strains of actinomycetes: 1) prototrophic recombinants of a single cross markedly vary in their activity; 2) apparent differences of antibiotic activities of recombinants were observed in crosses between different biochemical mutants of two strains; 3) variable types of ratios of activities between initial strains and prototrophic recombinants were observed in different crosses, as well as in single crosses. Activity of prototrophic recombinants of actinomycetes may be characterized in either of the following ways: a) lower than in each of the initial strains; b) intermediate between the two initial strains; c) at the level of the more active strain; d) higher than the more active strain.

These data do not directly explain what determines the level of antibiotic activity of recombinants in different crosses and why recombinants obtained both in different crosses between various biochemical mutants of two strains and in each of these crosses differ in their activities.

Hybridization of actinomycetes, just like hybridization of lower fungi, need not have any detectable practical significance. However, the question about practical application of microbial hybridization cannot be completely ruled out, and the hybridization method must be included in the set of breeder's tools, especially in cases of strain variability of the producer.

Some cases of hybrid effects, more precisely the effect of genetic recombination in actinomycetes, have been described concerning the antibiotic activity and a number of other technological characteristics such as decrease of foam formation during fermentation in industrial conditions or development of resistance to high phosphorus concentrations in nutrient medium. An "indirect" hybridization effect was described for *Penicillium* species when mutagenic treatment of diploid strains detectably increased the activity and made it possible to obtain highly active strains.

Hybridization is the major tool of selection in higher forms where it is used for combination of characters from different organisms in a single organism, for elimination of unfavourable properties of hybridized forms and for obtaining the hybrid effect (heterosis). What are the perspectives of using microbial hybridization? *Firstly*, just as for higher forms, hybridization of microorganisms allows to obtain variants with the desired genotype. *Secondly*, hybridization may be used as a tool for destabilization of a character in population in cases when mutagenic selection of highly active strains fails. *Thirdly*, hybridization may be employed for obtaining diploids, necessary for the use of the phenomenon of polyploidy. *Fourthly*, hybridization allows one to create balanced heterozygous diploids for achieving the heterosis effect. *Fifthly*, hybridization may serve as a tool for elucidation of genetic systems which control the biosynthesis of metabolites, which may be important for selection purposes.

However, in spite of the apparent importance of microbial hybridization from the point of view of a breeder, the progress in this field is still insignificant. Hence, all the above-mentioned ways of application of hybridization must be extensively investigated.

We believe that beside hybridization, microbial selection must widely utilize phages, particularly the mutagenic action of phages, as well as their ability to transfer various genetic elements from one cell to another.

In conclusion, it must be noted that successes of theoretical exploration of microbial selection strengthen the power of genetic selection methods, increase their role in the improvement of applied microbiology and the development of industry based on microbial synthesis.

2/ Genetics of Streptomycetes

Developments in Actinomycete Genetics

DAVID A. HOPWOOD

John Innes Institute, Norwich, Great Britain

INTRODUCTION

In this communication I should like to review the recent developments in the genetics of the *Actinomycetes* and to present at least a brief treatment of the possible relationships between basic and applied research: how to profit from academic research in an industrial context.

I have taken account of the first point by selecting certain topics for detailed discussion, rather than giving a homogeneous review of the whole field, since the development of streptomycete genetics has been covered by a more balanced approach in previous reviews[1-5]. The topics chosen are primarily those to which new knowledge has been added since my 1967 review[3] or those where further comment may help to clarify differences of interpretation between my group and that of SERMONTI. These differences should not be overemphasized; they do not detract from the steady progress that has been made in placing *S. coelicolor* amongst the comparatively small group of microorganisms that are tolerably well-known from a genetical point of view. The treatment in this review is further unbalanced since two of the major recent growth points in *S. coelicolor* genetics, fertility variation[6] and genetic analysis of morphogenesis[7], have been reviewed separately by VIVIAN[8] and CHATER[9].

The second point is harder to cover. One can legitimately argue that basic research into the sexual biology and genetic analysis of the *Actinomycetes* is an essential prerequisite for any programme of strain improvement. How far could commercial strains of maize have been improved in the total absence of information on the ploidy of the organism, the occurrence of sexual reproduction, its morphological basis, including the sexual differentiation of different parts of the plant, to say nothing of more sophisticated knowledge involving such things as the induction and manipulation of male or female sterility, gene locations or the combining ability of different genotypes? Yet those engaged in improving strains of *Streptomyces* have been operating in a comparable state of ignorance.

It is now clear that genetic exchange is widespread if not ubiquitous in *Streptomyces*, and that at least some of its features are probably comparable with those in *S. coelicolor* A3(2), including even a similarity in linkage relationships (*see* the

21

contributions of ALAČEVIĆ[10] and COATS[11]); thus research on *S. coelicolor* A3(2) provides a model for the genus, and a review of this system has practical relevance. As for detailed suggestions for the solution of practical problems of strain improvement in the light of genetic knowledge, the literature is already rich with ideas (*see* particularly the Dubrovnik *Symposium on Genetics and Breeding of Streptomyces*[12] and SERMONTI's recent book on *Genetics of Industrial Microorganisms*[5].) Disappointingly slow progress with industrial strains does not reflect insuperable biological problems, but rather a lack of adequate commitment on the part of industry to this kind of research, which in many firms is attempted on far too small and tentative a scale. However, there are encouraging signs that some boards of directors of the fermentation industry are now prepared to consider a greater genetical content for the research effort of their companies, although still a negligible one by the standards of other firms likewise dependent on the quality of biological material: cereal and poultry producers are obvious examples, but there are many others.

THE SEXUAL BIOLOGY OF *S. coelicolor*

In my previous review[3], I pointed out our ignorance of basic questions concerning the sexual biology of *S. coelicolor*. Particular doubts concerned the morphological basis of conjugation; the nature of the sexual differentiation of strains; and the question of whether zygote incompleteness results from pre-zygotic or post-zygotic elimination, and if the former, whether zygote formation is preceded by a stage in which complete genomes from both parents occupy a common cytoplasm (heterokaryosis). Let us evaluate progress in illuminating these doubts.

THE MORPHOLOGICAL BASIS OF CONJUGATION

We are still totally ignorant of the means by which chromosomal material is transferred in a mating mixture of *S. coelicolor*, except that some kind of conjugation is involved[2]. Although providing no excuse for this ignorance, perhaps it is fair to remember that the nature of the union between mating *E. coli* cells is still unclear[13], in spite of all the elegant studies that have been made on chromosome transfer in that system.

SEXUAL DIFFERENTIATION OF STRAINS

This is an area in which a good deal of new information has been accumulated since 1967. Much of this has been reviewed by VIVIAN[8].

The strains of *S. coelicolor* used in our work have all been derived by mutational and recombinational steps from a single wild type, A3(2)[14], which had a considerable history of laboratory cultivation[15,16] before I began genetical work with it. Any sexual differentiation of our strains must therefore have occurred among the descendants of this strain. My early crosses at first involved exclusively mutant derivatives of A3(2) and subsequently also recombinants[17]; these crosses were characterised by a relatively low level of recombination, of the order of 1 in 10^5—10^6 of the total spores produced in a mixed culture being selectable recombinants. A second characteristic of these crosses was a statistically equal frequency of those complementary recombinant genotypes that could be selected. This finding pointed to equal contribution of genetic material by the two parents to the progeny, although not necessarily to each zygote[14]. (Indeed merozygosis was later shown to occur[3].) Thus in the early crosses, I was dealing with strains of uniform sexual properties. In later crosses the two features of the earliest matings no longer held true: recombination frequencies increased, and complementary genotypes differed in frequency[1,18].

It is now clear that the first sexual differentiation within the stock of strains (which by then had grown with the addition of further mutant and recombinant derivatives of the early stocks) is the explanation for the changed behaviour of the crosses. VIVIAN has described[8] our attempts to trace the evolution of sexual differentiation within the stock culture collection and the characteristics of the different fertility types encountered[6]. The sexuality displayed by the most striking class of strain, UF (ultra-fertile), has already been described[19]. In this paper I shall merely attempt to relate our conclusions on sexual differentiation within *S. coelicolor* A3(2) with those of SERMONTI and his associates[20–22].

We have deliberately designated the fertility types by terms that do not recall the plasmid-determined mating-types of Gram-negative bacteria (*E. coli*: F^+/F^- (Ref.[23]); R^+/R^- (Ref.[24]); Col^+/Col^- (Ref.[25]); *Pseudomonas aeruginosa*: FP^+/FP^- (Ref.[26]); *Vibrio cholerae*: P^+/P^- (Ref.[27])), since present indications are that the sexual differentiation we observe in *S. coelicolor* has no close parallel with that in eubacteria. The designations are not entirely arbitrary (IF — Initial Fertility — is that of the wild type; NF — Normal Fertility — is that of the majority of strains in the culture collection; UF — Ultra Fertility — is that of the strains required for maximum (approaching 100 percent) recombinant yield in a cross), but they could be regarded as arbitrary if further understanding of fertility variation in *Streptomyces* renders the present meanings of some of the terms inappropriate in relation to other fertility types that may remain to be recognised.

It seems clear that at least some of the strains designated R^- and R^+ by SERMONTI and CASCIANO[20] correspond with our IF and NF types respectively. This conclusion is reasonable since we can now be sure that the stocks of strains that I sent to SERMONTI in 1959—61 must have included both IF and NF types, and is indicated particularly by the finding that his strains numbered $8R^-$ and $109R^+$, which derived directly from my stock culture collection under the numbers 424 and 32, satisfy our

criteria for IF and NF strains respectively[28]. Moreover the patterns of marker segregation observed by SERMONTI and CASCIANO[20] in $R^+ \times R^-$ crosses, which were difficult
to interpret at the time in the absence of a knowledge of the relative positions of the
markers on the circular linkage map, which was not deduced until later[29-32], seem
compatible with those we observe in NF $\times$ IF crosses.

We feel that the nomenclature used by SERMONTI and CASCIANO[20] to describe
fertility variations in *S. coelicolor* A3(2) is inappropriate. The designation R^- (recombination-less)[33] for the IF strains seems particularly unfortunate (apart from
causing a possible confusion with the totally different *rec⁻* (recombination-less)
mutants of *E. coli*[34] and other microorganisms) because IF $\times$ IF mixtures are in fact
far from sterile under favourable cultural conditions: if they had been sterile, the
genetics of strain A3(2), involving as it did initially IF strains, could not have begun!
Perhaps because of a cultural difference, the levels of recombination in the crosses
of SERMONTI and CASCIANO[20] including those involving R^+ strains, were low compared with those we observe (of the order of a hundredfold lower), so that $R^- \times R^-$
crosses yielded recombinants at barely detectable frequencies, and were described
as sterile. However, this has always been a relative term [5,20,21].

The designation R^+ (recombination-positive) for the NF strains is likewise
misleading, since *all* classes of strain so far encountered (including the UF strains)
display a certain level of recombinant production with strains of homologous fertility
type. However, a more serious objection is that it deliberately implies analogy with
F^+ (or R^+) *E. coli* strains[5,22], whereas there is no close analogy; fertility type was
found to segregate amongst the progeny of some of the $R^+ \times R^-$ crosses of SERMONTI
and CASCIANO[20] and it certainly does so amongst the progeny of our NF $\times$ IF crosses,
leading to recognition of a chromosomal locus concerned in determining the difference in fertility[6], and providing no evidence for a "contagious extrachromosomal
factor" of the kind postulated by SPADA-SERMONTI and SERMONTI, although an
interaction between chromosomal alleles and one or more plasmids is by no means
excluded*. The historical origin of NF from IF strains[6], rather than the reverse,
in any case argues against a close analogy between R^+ and R^- *S. coelicolor* and
F^+ and F^- *E. coli* strains.

---

* The finding by SERMONTI and CASCIANO that a high proportion of the strains emerging
from an $R^+ \times R^-$ mixture with the markers of the R^- parent were R^+ is open to other explanations besides the one of infection of R^- by a plasmid from R^+. For example, an imbalance
in the parental input or growth rate could have resulted in the R^- parent being in the minority.
These strains could then have had a frequency comparable with that of the *selected* recombinants,
and have been recombinants which happenet to inherit a parental association of markers. Thus
genuine non-recombinants, converted by infection from R^- to R^+, may have been absent.

THE NATURE OF GENETIC TRANSFER

NF × UF crosses are *polarised* in that the genomic contributions of the two parents to the progeny are unequal: from the pattern of inheritance of markers by recombinant progeny we can deduce that all merozygotes contain a complete (or almost complete) genome from the UF strain and a genome fragment from the NF

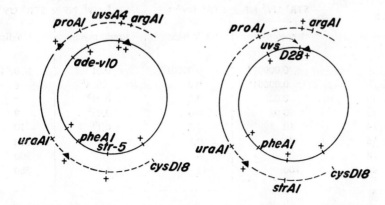

Fig. 1. Representation of the two crosses whose results are in Table 1. *Left:* cross I, *right:* cross II. The markers of the UF strain are on the *inside* circle and those of the NF fragment on the *outer* arc. Cross I is UV^S STR^S NF × UV^R STR^R UF; Cross II is UV^R STR^R NF × UV^S STR^S UF. *Triangles* indicate the alleles selected on the medium used to recover recombinants.

strain[19]. We attempted to use the NF × UF cross[35], with its high level of recombination, to throw light on the process of merozygote formation. The experiment involved resistance or sensitivity to streptomycin or UV-light (UV). In the case of streptomycin, the wild-type state is total sensitivity to 25 µg/ml streptomycin, while mutation in the strA locus confers total resistance to this concentration of the antibiotic (and indeed to much higher concentrations). In the case of UV the wild-type state is one of resistance, compared with the sensitivity conferred by mutation in one of a series of uvs loci, A—D[36,37]; however, the situation differs from that involving streptomycin in that resistance and sensitivity are statistical, not absolute, a particular dose of UV killing a much greater *proportion* of UV^s than of UV^r spores. Furthermore, streptomycin-resistance is recessive[1], and presumably involves the cytoplasmic ribosomes as in other bacteria, whereas UV-resistance is dominant[36,37], and presumably depends on enzymic repair of lesions in the chromosomal DNA.

Two pairs of strains were prepared, each consisting of one NF and one UF strain differing in several auxotrophic markers. In one pair the NF strain was resistant to both streptomycin and UV, while the UF strain was sensitive; in the other pair, resistance and sensitivity were reversed with respect to fertility type (Fig. 1). The experiment consisted of mixing dense spore suspensions of each pair of strains and spreading samples of the mixtures on a series of identical plates of non-selective

Table I

Effect on Recombinant Production of Treating a Mixed Culture with Streptomycin or UV-Light at Various Times[a]

Time before addition of strepto-mycin or UV-light h	Proportion of selected recombinants amongst total viable spores, $\times 10^3$			
	Cross I STR^S UV^S NF $\times$ STR^R UV^R UF		Cross II STR^R UV^R NF $\times$ STR^S UV^S UF	
	streptomycin	UV-light	streptomycin	UV-light
6	< 0.00001[b]	< 0.00001[b]	< 0.01[b]	< 0.01[b]
8	< 0.00001[b]	0.04	< 0.006[b]	6
10	0.0005	0.2	< 0.01[b]	5
12	0.02	0.6	< 0.02[b]	9
14	10	0.7	< 0.1[b]	10
16	30	0.8	< 0.1[b]	300
18	80	4	1	300
20	100	40	1	500
Control	200		400	

[a] For the marker arrangements *see* Fig. 1.
[b] No recombinants recovered.

medium. After varying periods of incubation, two plates of each cross were taken at random; one was treated with streptomycin (by spreading a drop of solution over the surface of the mixed culture), and one with UV (by irradiating the plate with approximately 1500 erg/mm^2 UV, a dose that should have reduced the survival of UV^s strains to 10^{-5} or less, but UV^r strains only to about 0.4 (Ref.[37]). The plates were then returned to the incubator. After four days' incubation, a spore suspension was prepared from each plate and spread at suitable dilutions on a non-selective medium and also on a medium selecting recombinants. Thus the experiment was designed to determine the time course of inhibitor sensitivity, during recombinant production.

The results of the experiment are in Table I.* The clearest conclusion is that recombinant production becomes insensitive to UV at about the same time in the two crosses, that is whether resistance is in the NF or in the UF parent, whereas recombinant production becomes insensitive to streptomycin much later when the UF parent is sensitive (Cross II) in comparison with the cross in which the UF parent

---

* The actual values for recombinant frequencies do not have very much meaning since they are the product of complex selective effects, different in the two crosses; they should be treated largely as a qualitative indication of the presence of recombinants.

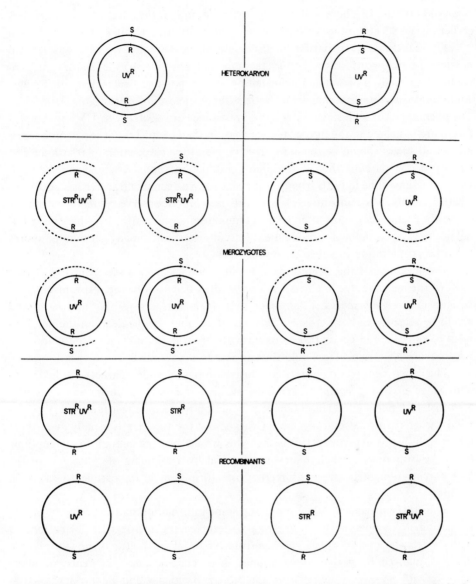

Fig. 2. Diagrams of the possible classes of heterokaryons, merozygotes and recombinants in the crosses whose results are in Table 1. *Left:* cross I, *right:* cross II. The full parental marker arrangements are in Fig. 1. In this Figure, only the constitutions (resistance or sensitivity) in respect of the *uvs* (*top* of the map) and *str* loci (*bottom* of the map) are indicated. The phenotype of each entity is specified in the centre of each circle, only resistance being indicated. The different possible merozygotes are those in which the NF fragment includes (1)   neither *uvs* nor *str* loci, (2)   *uvs* but not *str*, (3)   *str* but not *uvs*, or (4)   both *uvs* and *str*. Note that: in both crosses, at least some members of each class of entity are UV-resistant; heterokaryons in neither cross are streptomycin-resistant; some merozygotes are streptomycin-resistant in cross I but not in cross II, and some recombinants are streptomycin-resistant in both crosses.

is resistant (Cross I). There is also a slight suggestion that the effect of UV disappears earlier (at 6—8 h) than that of streptomycin (at 8—10 h) in Cross I.

Fig. 2 is designed to clarify the discussion of these results. We can consider two possible heterozygous stages before recombinant production: the merozygote immediately preceding recombinant formation, and a possible "heterokaryotic" stage, containing *complete* genomes from both parents, preceding merozygote formation. The phenotypes of each stage in respect of resistance or sensitivity to UV and streptomycin are indicated in the figure. We see that, in the case of UV, at least some members of all stages should be resistant, irrespective of the association of resistance and fertility in the parents; thus recombinant production should become insensitive to UV at the same time in both crosses (that is as soon as genome transfer has occurred, whether complete or incomplete), and this is what is observed. Genome transfer evidently begins very soon after spore germination since at 8 h, by which time UV did not prevent recombinant production, the cultures showed many germinated spores but little proliferation of substrate mycelium.

The meaning of the streptomycin result is less clear. We see from Fig. 2 that most of the merozygotes in Cross I are resistant, whereas none are resistant in Cross II. Resistant *recombinants* can, however, arise from both crosses. A "heterokaryon" would be sensitive in both crosses. This would account for the delay in the development of resistance in Cross II, compared with Cross I; recombinant production and release from phenotypic lag would thus appear to take some 8 h.

The experiment is not entirely conclusive on the question of whether a heterozygous stage involving *complete* genomes precedes merozygote formation. There is a consistent tendency for recombinant production to become resistant to UV somewhat earlier than to streptomycin in Cross I, but the time lag is usually quite short (about 2 h in this experiment). Whether this would be long enough for phenotypic lag to be overcome if a heterokaryon produced by a mingling of cytoplasm of the two parents obligately preceded merozygote formation is impossible to say with certainty.

Thus this experiment, while providing some information on the timing of zygote formation, is compatible with heterokaryosis preceding merozygote formation, or with the transfer of a complete genome from NF to UF, with little or no cytoplasmic mixing, followed by *post*-zygotic elimination of chromosomal segments, or alternatively with *pre* zygotic elimination. SERMONTI, BANDIERA and SPADA-SERMONTI[21], were likewise unable to exclude, in their experiment, a brief heterokaryotic stage preceding merozygote formation, as pointed out by SERMONTI[38].

PLASMIDS IN *Streptomyces*

As we have seen, there is presently no indication of the involvement of plasmids in sexuality in *S. coelicolor*, although it is not excluded. In view of the increasing evidence that plasmids occur in a very wide range of eubacteria, both Gram-negative and Gram-positive, associated with a broad spectrum of phenotypic properties (*see* recent reviews[39,40]), it would perhaps be surprising if the *Actinomycetes* were devoid of plasmids. There have indeed been reports of phenomena that could indicate the existence of plasmids in *Streptomycetes*, but unfortunately rigorous criteria, either genetic or physical, that would unequivocally have implicated plasmids were lacking. GREGORY and SHJU[41] described the origin, with a high spontaneous frequency, of tyrosinaseless variants (that is melanin-nonproducing) from a tyrosinase-positive strain of *S. scabies*. The frequency of such variants in this and in other *Streptomycetes* was greatly increased by growth in the presence of acriflavine[42]. More recently, OKANISHI, OHTA and UMEZAWA[43] found that prior incubation of *S. scabies* and *S. venezuelae* at high temperature in the absence of acriflavine, or at normal temperature in its presence, led to a high proportion of melanin-nonproducing colonies on subsequent incubation at the normal temperature. The same was true of nonproducers of aerial mycelium in *S. venezuelae* and nonproducers of antibiotics in *S. kasugaensis*.

In none of these cases was knowledge of chromosomal linkages adequate for unequivocal demonstration of extrachromosomal inheritance of the characters in question by genetic tests, although the genetic results with *S. scabies* were certainly suggestive. GREGORY and SHJU[41] found that unstable prototrophic colonies arising from mixed cultures of two auxotrophs, one tyrosinase-positive and the other tyrosinase-negative, and presumed to be heterokaryons containing the two parental genomes, segregated the two parental phenotypes in respect of nutritional markers and streptomycin resistance; however, almost all members of both classes were tyrosinase-positive. Likewise the few genuine recombinants obtained from the mixed cultures were nearly all tyrosinase-positive[44]. These observations led GREGORY and HUANG[44] to postulate carriage of the tyrosinase gene on a plasmid. Nevertheless, caution in interpreting such observations as evidence for plasmid involvement is desirable, particularly in view of past experience with other systems such as sporulation in *Bacillus* (*see* the discussion by SCHAEFFER[45] of the results of ROGOLSKY and SLEPECKY[46] and BOTT and DAVIDOFF-ABELSON[47]). Now that methods are available for the characterisation of plasmids by physical means, usually relying on the unique properties of small, closed circular DNA molecules[48-51], the melanin phenomena are obvious candidates for investigation and in our laboratory we have begun tests for plasmids in these species, and in *S. coelicolor*.

PHAGES

The genetic interactions between phages and their eubacterial hosts are legion: lysogeny, transduction, host-controlled modification and restriction, host-cell reactivation. While the diagnosis of these phenomena does not require a detailed knowledge of the genetics of the host bacterium, it is probably true to say that progress in their understanding would be severely hampered without such information. It is therefore not surprising that knowledge of phage — host interactions in the *Actinomycetes* is rudimentary[52], in comparison with those of some eubacteria, when we consider that no study of actinophages has been made in the only genetically well-known streptomycete, *S. coelicolor* A3(2). In an attempt to remedy this situation, we have recently begun the isolation and study of phages active on this strain. Contrary to the experience of others, we have found no difficulty in isolating such phages from a wide range of soil samples, and we can hope that their study will eventually fill some of the gaps in our knowledge of actinophage — host interactions. (*See* also the contribution of N. D. LOMOVSKAYA[53].)

GENETIC ANALYSIS

The basic requirements of genetic analysis in any organism are the means of performing long-range and fine-structure mapping, and tests of dominance and complementation. All these procedures are currently possible in *S. coelicolor*, but only long-range mapping is outstanding in the general context of microbial genetics. We must hope that future developments will improve the performance of the other tests.

LONG-RANGE MAPPING

The principles and methodology involved in the assignment of a new locus to an approximate position on the linkage map by selecting recombinants from a cross were described by HOPWOOD[3]. Perhaps it is worth emphasizing that a single cross of this type should not fail to give a positive result, since the whole linkage map can be covered by selection at two positions, unless a mutation at the locus concerned should lead to a severe reduction in viability; even then an approximate map position may be deducible from the distortion in segregation ratio at a nearby locus[54].

A variant of long-range mapping by selective means is nonselective analysis. The segregating heterozygous colonies (heteroclones) provided the first means of such analysis in *S. coelicolor*[1,2,55-57], and indeed in any bacterial system except for the pedigree analyses of *E. coli*[58,59]. The heteroclones themselves have to be selected but the spores produced on an individual heteroclone colony represent a population of sexual progeny, free of asexual background, that can be studied by plating on a nonselective medium. The limitations of the system arise from the biases introduced by the segregation of diverse recombinant subclones within an individual heteroclone colony[3,4].

A prerequisite for the selection of individual heteroclones is the availability of a pair of closely-linked selectable markers, one in each parent. To overcome this limitation, SERMONTI, BANDIERA and SPADA-SERMONTI[21] devised a "cellophane transfer technique" in which a mixture of the two parents was incubated for a period on cellophane over a nonselective medium, and later transferred to a medium selecting against the parents, but not necessarily for close markers. Minute "tufts" of aerial growth appeared on the surface of the mixed culture and these were analysed on the assumption that they were individual heteroclones. If this assumption is valid, the method certainly has possible technical advantages, although as pointed out by SERMONTI[60] the particular strains under investigation influence variables such as the optimal plating density, and the time of preincubation on nonselective medium, so that these have to be carefully determined in advance. The method does not, of course, avoid the biases due to heterogeneity of subclones within a single heteroclone; indeed such effects would be expected to be heightened since a proportion, at least, of the tufts no doubt originate from more than one mating event.*

More recently another nonselective system has become available in S. coelicolor[61]. This depends on the extremely efficient sexual reproduction occurring in

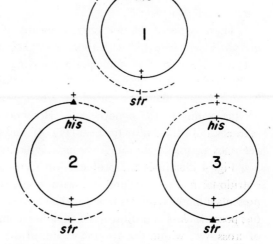

Fig. 3. Effect of semiselective analysis of an NF × UF cross. **1** no selection: in most of the merozygotes the NF fragment (outer arc) fails to include the hisA and strA loci; **2** selection of hisA⁺: crossing-over in the region clockwise of hisA occurs in all zygotes; **3** selection of strA: crossing-over in the region anticlockwise of strA occurs in all zygotes.

a mixed culture of strains of the two fertility types NF and UF. Spores produced on such a culture derive almost entirely from sexual reproduction, provided care is taken to achieve an approximately equal mixture of the two parental strains when they are inoculated. Thus the cross can be analysed by plating the spores on nonselective

---

* The possibility that tufts can contain the products of more than one round of mating would appear to invalidate SERMONTI's claim[33] to have disproved my hypothesis[3] that haploid recombinants can originate directly from a merozygote, when even numbers of crossovers occur within it. The claim was based on his statement that "no isolated haploid recombinant is ever directly formed on the basal mycelium"; however, in the results quoted[21], some of the tufts were in fact pure clones.

medium. Such crosses have allowed very efficient mapping of loci within the portion of the genome included in a reasonable percentage of the zygotes, which in practice means approximately the left-hand half of the map, from 6—12 o'clock[19]. For loci outside this region, it should be possible to increase the efficiency of mapping by making a semiselective plating; for example by selecting his^+ or str (Fig. 3) inherited from the NF strain. The selected sample of recombinants then includes a high proportion of those classes critical to the ordering of markers in the regions immediately clockwise of $hisA$ or anticlockwise of $strA$.

Even crosses of intermediate fertility, such as IF × NF or NF × NF (Ref.[6]) can be analysed nonselectively, since the total frequency of recombinants can range from about 0.5 to 10 percent or occasionally even higher. This means that many asexual progeny are handled for every recombinant identified, but the labour may well be justified in order to isolate a particular nonselectable recombinant class. The power of nonselective analysis should not be lost sight of in a commercial breeding programme.

SHORT-RANGE MAPPING

There is no doubt that current means of fine genetic analysis in S. coelicolor are not ideal compared with those available in some other microbial systems. The problem is not one of inadequate resolving power[62], but rather the difficulty of estimating the true frequencies of rare haploid recombinants: the problems arise because recombinants produced by crossing-over between a pair of close markers are (according to the hypothesis of HOPWOOD[3]) often still heterozygous for these markers and others near to them (Fig. 4), and they are accompanied by heteroclone genomes arising by crossing-over *outside* the interval between the two close markers.

Fig. 4 illustrates the problem for a cross between two closely-linked selective nutritional markers m_1 and m_2, with nonselective outside markers a and b. I represents a merozygote disomic for the chromosomal region concerned. IV represents one possible class of prototrophic haploid recombinant resulting from an even number of crossovers within the disomic region of the merozygote: an obligate crossover in interval 3 and the remaining crossover(s) in intervals to the left and/or right of interval 3. The patterns of such nonselected crossovers will determine the relative frequencies of the four classes of haploid recombinant with respect to the outside markers a and b; some of these frequencies will be the basis for deciding the order of the nutritional mutations m_1 and m_2: in the example of Fig. 4, an excess of class ab over a^+b^+ will indicate the order $(a)—m_1 — m_2—(b)$. Note that in the class of merozygote drawn in Fig. 4 a single crossover is required to the *left* of m_1, in intervals 1 *or* 2, whereas if additional crossovers occur to the *right* of m_2, they must occur in pairs. Thus nonselected crossovers will tend to occur preferentially to the *left* of the selected markers in this example, whereas nonselected crossovers to the *right* would have been favoured if the chromosome fragment had come from the other parent. This reasoning is the basis of my warning[3] that *polarisation of zygote formation* could have been a disturbing factor influencing the segregation data interpreted by PIPERNO, CARERE and SERMONTI[63] exclusively in terms of *polarisation in the mechanism of crossing-over*. The magnitude of any such disturbance would depend, apart of course from the extent of zygote polarisation, on the relative lengths of intervals **1** and **2** or **4** and **5**.

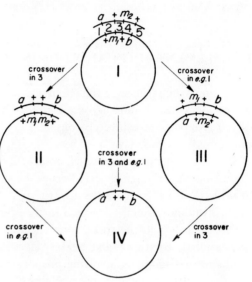

Fig. 4. m_1^+ and m_2^+ represent closely-linked selective markers; a and b represent outside, nonselected markers. **I** is a merozygote capable of yielding selected recombinants. **IV** represents a haploid recombinant produced by double crossing-over in the merozygote, an obligate crossover in interval 3 and a second crossover, for example in interval 1. **II** and **III** represent heteroclones arising by single crossovers within the merozygote; **II** is recombinant in respect of the closely-linked markers, and **III** is nonrecombinant in respect of these markers. Further single crossovers during the development of heteroclone colonies can convert **II** or **III** into **IV**, as shown.

Complications arise because some of the possible classes of recombinant are "heteroclones" (*e.g.* **II** and **III** in Fig. 4). **III** is the conventional type of heteroclone in which the coupling of the close selected markers is unchanged, while **II** represents the class in which crossing-over in interval **3** has led to a changed coupling of these markers. **II** will give rise to haploid segregants during the early stages of its growth into a colony on selective medium, by odd numbered crossovers to the left of interval **3**. However, provided a predominant genotype emerges in each colony, with a frequency unbiased by selective forces, then the classification of such colonies along with the "pure" haploids (**IV**) will not disturb the segregation of outside markers; however, such biases may not always be easy to avoid.

Heteroclones of the conventional kind (**III**), whose development depends on complementation between m_1 and m_2, may also segregate early in their development, in this case by a crossover in interval **3**, to give haploids; thus a proportion of the colonies finally classified as haploid will originate by this route also.

In summary, determination of the frequency of recombination between a pair of close markers, particularly complementing markers, by estimating the yield of recombinants between them, may be subject to considerable error caused by the presence of unrecognised heteroclones. Ordering close sites by selecting recombinants between them and observing the segregation of outside markers may also be ambiguous if segregation within heteroclones is nonrandom. Hence the importance of incorporating as many tests of internal consistency as possible, for example the use of reciprocal crosses with reversed coupling of outside markers (histidine[64] and tryptophan[65] mutants). The same precautions need to be taken when outside markers are selected and the frequency of recombinants of a particular class in respect of the close sites is used as a criterion of order (UV-sensitive mutants[36]).

COMPLEMENTATION AND DOMINANCE

Two systems have so far been used in the assessment of complementation: heteroclones and heterokaryons. Heteroclones have given a very clear indication of the grouping into three cistrons of a set of 20 closely-linked UV-sensitive (uvs) mutations[36,37] and have also been satisfactorily applied to the study of certain classes of auxotrophic mutations, notably those resulting in a histidine requirement: representatives of three clustered genes, hisA, B, C, were classified by HOPWOOD and SERMONTI[1] using heteroclones as criterion. Subsequent complementation analysis of the histidine system[66] in which further genes in the cluster were recognised, was done primarily with heterokaryons selected by the "cellophane transfer technique" of SERMONTI, BANDIERA and SPADA-SERMONTI[21]. This method gave self-consistent results in that mixtures involving phenotypically diverse mutations showed heterokaryotic "tufts", while mixtures involving phenotypically similar mutations, in general, did not. Exceptions to this latter generalisation were provided by all three G mutations, which complemented each other, one C mutation (his-14), which complemented four of the other C mutations, and one D mutation (his-15) which complemented three others (his-3, -4, -121); on this evidence, gene G was subdivided into three cistrons, C into two, and D into two.

The heterokaryon test of complementation appears to demand experience in its interpretation; in a brief attempt to confirm its usefulness, using mixtures of strains carrying hisA1 and hisC9, I found that "tufts" sometimes failed to develop in heterologous combinations, while growth in homologous combinations sometimes mimicked "tuft" formation. Moreover repeated attempts to demonstrate heteroclone formation in crosses of hisD15 with hisD3 and hisD4 proved negative, as originally found by HOPWOOD and SERMONTI[1], in contradiction to the conclusion of complementation by the heterokaryon test.

The findings of ENGEL[65] with tryptophan (trp) mutants highlight a further problem in testing complementation using heteroclones. He defined as heteroclones colonies that were heterogeneous, on subculture, in respect of markers close to the trp cluster and found that pairwise crosses of trp mutants yielded a more or less continuous spectrum of heteroclone frequencies, from (almost) zero to about 10 per cent. The result was that clear cistron boundaries were hard to draw. It is premature to assess the significance of low, but non-zero, heteroclone frequencies; it seems at least possible that heteroclones of class **II** (Fig. 4) may sometimes be involved, and these of course do not depend on complementation for their growth.

If current tests of complementation (and the same considerations refer to dominance tests) are less than ideal for auxotrophic mutations, the same is even more true of some other classes of mutation, for example those leading to morphological lesions[7]. As discussed by CHATER[9], progress in the functional analysis of such mutations, particularly those with lesions in the aerial mycelium, may depend on the development of a more stable heterozygous condition than that provided by the heteroclones or heterokaryons. The question of the autonomy of gene expression[67] is relevant here. Attempts are currently in progress in my laboratory to select a more stable heterozygous entity that might help to remedy deficiencies in present complementation tests[68].

RECOMBINATION IN OTHER ACTINOMYCETES

The scope and interest, from a comparative point of view, of actinomycete genetics have recently been widened by studies on two new systems. These already show features in common with streptomycetes, but there certainly will be interesting differences also.

RECOMBINATION IN *Thermoactinomyces vulgaris*

There is currently an upsurge in interest in the thermophilic actinomycetes with the realisation that they provide fascinating problems of physiological, ecological, taxonomic and medical relevance[69]. Within this group, *Thermoactinomyces vulgaris* appeared to us to be the best choice for a genetical investigation. Its technical advantages are its very rapid growth rate at temperatures above 50 °C (Ref.[70]), and its production of spores in great abundance. These spores provide a good contrast with those of the mesophilic streptomycetes[71,72] in being genuinely heat-resistant[69,70,73], like those of members of the eubacterial genera *Bacillus* and *Clostridium*; heat resistance is reflected in a multi-layered outer integument[74,75] and the presence of dipicolinic acid in the spores[74]. Moreover *T. vulgaris* is one of the two thermophilic actinomycetes implicated in the allergic condition known as Farmer's Lung[69,76]. The only obvious technical disadvantage of the organism is a currently poorly understood nutritional requirement satisfied by hydrolysed casein; however, this has not prevented the isolation and use as selective markers of auxotrophic mutations for vitamins, purines, pyrimidines and tryptophan as well as streptomycin resistance[77].

TABLE II

Results of Crossing Strains 1279 nic-1 thi-3$^+$ ura-1 str-1 and 1286 nic-1$^+$ thi-3 ura-1$^+$ str$^+$ of T. vulgaris

Genotypes[a] of progeny	Percentage of recombinants of each genotype on selective media supplemented as indicated[b]			
	nicotinamide thiamine streptomycin	uracil	thiamine uracil streptomycin	nicotinamide
nic + ura str	—	—	—	—
+ thi + +	—	—	—	—
nic + + str	79	—	—	82
+ thi ura +	—	—	—	—
+ + ura str	—	65	92	—
nic thi + +	—	—	—	—
+ thi + str	19	—	8	—
nic + ura +	—	—	—	—
+ + + +	—	33	—	16
nic thi ura str	—	—	—	—
+ + + str	0.5	2	0.5	0.5
nic thi ura +	—	—	—	—
nic thi + str	1.5	—	—	—
+ + ura +	—	0	—	—
+ thi ura str	—	—	0	—
nic + + +	—	—	—	1

[a] Arranged as complementary pairs.

[b] A dash indicates that a particular genotype could not grow on the medium in question.

Our first attempts to demonstrate recombination in T. vulgaris involved a well-known wild type, CUB76*. We failed to demonstrate recombination between two auxotrophic mutants of this strain or between an auxotrophic streptomycin-resistant mutant and the wild type. A few colonies appeared on selective media sown with spores from mixed cultures, but their frequency was not convincingly higher than could be explained by fresh mutations. We therefore tested for recombination in several other, newly isolated strains of T. vulgaris and were soon rewarded by a positive result in one particular strain, 1227. This strain also recombines with CUB76.

The frequency of recombination between mutant derivatives of strain 1227 is reasonably high: one selectable recombinant occurs for every 10^3—10^5 spores of parental phenotype derived from a slant culture inoculated with two parents bearing complementary markers. The results of the crosses indicate a process of true re-

* Kindly supplied by Dr. T. Cross, University of Bradford.

combination in which substitution of homologous segments of genetic material occurs, rather than some kind of repliconation[78] in which genetic material from one strain is merely added to the genetic complement of the other. This conclusion is indicated particularly by the finding of certain classes of progeny expressing auxotrophic markers of both parents.

Our information on recombination in *T. vulgaris* strain 1227 comes so far primarily from four-point crosses each analysed by recombinant selection on four

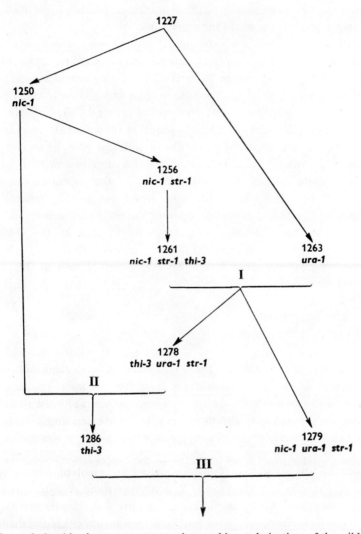

Fig. 5. Interrelationships between mutant and recombinant derivatives of the wild-type *Thermoactinomyces vulgaris* strain 1227. *I*, *II* and *III* are the crosses referred to in the text. *nic* = nicotinamide requirement; *thi* = thiamine requirement; *ura* = uracil requirement; *str* = streptomycin resistance.

different media. On each medium two markers are selected and two unselected, allowing the theoretical recovery of four classes of progeny on each medium, and a total of nine classes from the four media. This type of cross was first used by HOPWOOD[14] in *S. coelicolor* and further discussed by HOPWOOD and SERMONTI[1]. An example is given in Table II. This cross is one of a series involving the same four markers in three different coupling arrangements, the parents of successive crosses being related by mutational and recombinational steps (Fig. 5). The results of the three crosses (crosses **I, II, III** in Fig. 5) are similar in that the great majority of the selected recombinant progeny belong to the four classes differing from one or other parent by a single marker: these are the classes *nic thi$^+$ ura$^+$ str, nic$^+$thi$^+$ ura str, nic$^+$ thi ura$^+$ str*, and *nic$^+$ thi$^+$ ura$^+$ str$^+$* in all three crosses. Members of the remaining three classes, differing from both parents by two markers, represent a very small percentage (1.5% in the cross in Table II). This finding provides good evidence that we are dealing with a process of genetic transfer in which only small portions of the genome of one strain are transferred to the other; usually only the marker selected is transferred, although occasionally another (nonselected) marker is also inherited. The relative frequencies of cotransfer of different pairs of markers might provide the basis for linkage estimation, as in systems of genetic analysis based on transduction or transformation, but the frequencies of the critical classes are too small in the case of the four markers under consideration to give any reliable indication of linkage. Clearly no close linkage is apparent; indeed the frequencies of "double recombinant" classes are such that these classes might arise by the *independent* transfer of the two markers in question. Further progress with the analysis of this system, as of others, will probably depend on the finding of recognisable linkages involving further markers.

RECOMBINATION IN *Nocardia*

Just as members of the genus *Streptomyces* show extreme chemical versatility in the production of antibiotics, so members of the genus *Nocardia*[79] possess powers of chemical interconversion that may eventually lead to their wider use in industrial processes. *Nocardia* genetics may therefore have its commercial applications in the future. It could also be interesting from a comparative viewpoint, in relation to genetic phenomena in *Streptomyces* and other actinomycetes.

Recombination studies in the genus *Nocardia* have involved strains assigned to three different "species": *N. canicruria, N. erythropolis* and *N. restrictus*. ADAMS and BRADLEY[80] discovered that mixed cultures of auxotrophic mutants of a particular strain of *N. canicruria* failed to yield selectable prototrophic recombinants; the same was true of mixtures of mutants of *N. erythropolis*. However, mixtures consisting of one mutant derived from each "species" yielded recombinants. Similarly recombination was not observed between mutants derived from a strain of *N. re-*

strictus[81], whereas this strain recombined with derivatives of the *N. canicruria* strain of ADAMS and BRADLEY, although not with derivatives of their *N. erythropolis* strain.

Few results of genetic analysis of *N. restrictus*, which aromatises various 19-*nor* and 19-substituted Δ^7-steroids into equilin[81,82], have so far been published (but *see* the contribution of C.VÉZINA[83]). On the other hand, crosses between *N. canicruria* and *N. erythropolis* mutants have been the subject of numerous papers by ADAMS, BROWNELL and their associates[80,84-91].

The most striking feature of the *N. canicruria* × *N. erythropolis* crosses described in the earlier papers[80,84,85] was the predominant inheritance of nonselected markers of the *N. canicruria* parent: typically over 85 percent of the recombinants selected to contain one marker of each parent had all three or four nonselected markers of *N. canicruria*, although in one cross involving a particular nonselected tetracycline-resistance marker the great majority of recombinants inherited this marker from the *N. erythropolis* parent[85]; moreover the recombinants had the colonial morphology of *N. canicruria*[80]. These findings might have suggested that the zygotes arose by a donation of incomplete genomes from *N. erythropolis* to *N. canicruria*; the zygotes would thus obligately have included the selected allele of *N. erythropolis* (which was always the same marker, *purB2*$^+$, in these crosses), with inclusion of most non-selected markers of *N. erythropolis* at low frequencies. *A priori* a hypothesis of mero-zygosis was probably simplest, in view of the fact that all other bacteria that have been studied have been found to possess incomplete zygotes. However, the *Nocardia* results were analysed on the assumption that both *complete*, linear parental genomes were involved in zygote formation, and on this assumption all the markers were placed in a linear linkage group, with *purB2* of necessity occupying a terminal position[85].

Later[87], with the introduction of further auxotrophic markers into the parental strains, so that different combinations of selected alleles could be chosen, results incompatible with this hypothesis were obtained: a particular allele, *arg-1*, of *N. erythropolis* never appeared amongst the recombinants, no matter what pattern of selection was applied. The possibility of merozygosis was then considered, with particular regions of the *N. erythropolis* genome (including *arg-1*) being usually excluded from the zygotes, and so from the recombinant progeny; however, an alternative hypothesis, in which the zygotes were still considered to be complete, was also discussed.

More recently a further complication has arisen: selected recombinants have been found to segregate in respect of both parental alleles at particular loci[88], and this factor was suggested to have led to *"some erroneous information as to the segregation of unselected characters"* in previous work. The finding of segregating hetero-zygotes[90,91] may indicate the existence of entities analogous to the heteroclones of *S. coelicolor*, and conceivably these could eventually provide a hitherto lacking complementation test system.

Little appears to be known about the morphological basis of gene transfer in *Nocardia*, although the mechanism has been assumed to involve conjugation on the basis of negative results of attempts to select recombinants from mixtures of one viable and one heat-killed parent strain[84]. By studying the patterns of crossability of recombinants from the *N. erythropolis* × *N. canicruria* matings with the parent strains, which as we saw were self-sterile, a mating-type system was described in which the *N. erythropolis* and *N. canicruria* strains differed at two chromosomal loci; the alleles at these loci, *C/c* and *E/e*, segregated amongst the recombinant progeny. A phage, ΦC, was described as being "mating-type specific"[86,92] because it attacked *N. canicruria* but not *N. erythropolis*; however, the finding that some recombinants of each crossability pattern were resistant and others sensitive to the phage failed to support the notion that the phage was really specific for mating type rather than for some other chance difference between the parent strains.

As we have seen from the discussion of *S. coelicolor* genetics in this paper, an accurate linkage map is an invaluable aid to the elucidation of a variety of genetic phenomena, as well as to the experimental manipulation required in a strain-improvement programme. In view of the inherent possibility that the interspecific crosses that have provided the segregation data in *Nocardia* were complicated by chromosomal inhomologies of the kind that have led to inconsistent linkage data in eukaryotes when strains of diverse origin have been crossed, as well as the possible effects of persistent heterozygosis mentioned above, it would certainly seem wise to agree with ADAMS[87] that the linkage maps produced for *N. canicruria* and *N. erythropolis* are *"subject to future modification when further knowledge of nocardial hereditary phenomena is achieved."* It would seem particularly necessary to obtain critical evidence on the extent to which the zygotes are complete, since assumptions as to completeness or otherwise of the zygotes play a large part in determining the linkage relationships deduced from segregation data. It would also seem desirable to continue the search for recombination within a single isolate of a *Nocardia* since, if self-fertile strains exist, these could provide a simpler system for linkage analysis.* Our experience with *Thermoactinomyces vulgaris*[77], summarised above, is perhaps relevant. On the one hand, we see the power of crosses involving the same markers in different coupling arrangements, and each analysed by selection in several different ways, in demonstrating merozygosis; on the other the existence of self-fertile as well as self-sterile strains.

RELATIONSHIPS BETWEEN BASIC AND APPLIED RESEARCH

As I mentioned in the *Introduction*, I shall not aim to cover this topic in any detail. There are, however, a few points that have not been emphasized previously.

---

* A prediction from the conclusions of BROWNELL and ADAMS[86], is that their *CE* fertility type should be self-fertile. If so, a strain of this type might be a suitable starting point for further genetical work with a homogeneous group of strains.

MUTAGENESIS IN STRAIN IMPROVEMENT

There is a notoriously poor correlation between antibiotic yield as judged by performance in a plate test, or even in small-scale fermentation flasks, and yield under production conditions, although screening of mutants by plate tests is not entirely hopeless[5,93]. This means that any increase in the efficacy of mutagenesis can lead to a considerable saving of labour when random survivors of mutagenesis have to be evaluated by elaborate testing procedures. The definition by DELIĆ, HOPWOOD and FRIEND[94] of optimal conditions for mutagenesis in *Streptomyces* by the most potent mutagen so far discovered for other microbes, N-methyl-N'-nitro-N-nitrosoguanidine (NTG), has therefore paid an immediate dividend in several industrial laboratories. It turns out that *Streptomyces* is relatively resistant to the lethal and mutagenic effects of NTG, compared with, for example, vegetative eubacterial cells; hence for effective mutagenesis in *Streptomyces* the NTG concentration has to be high, the time of treatment long, and the rate of NTG decomposition to the effective mutagen (diazomethane) rapid, the latter being achieved by incubation at high pH. After mutagenesis under such potent conditions, a very high proportion of the survivors will harbour mutations affecting yield; many will in fact be multiple mutants and will be very "sick" compared with the wild type. However it is worth remembering that a strain that diverts much of its resources to antibiotic production must be very sick; the problem is to make it sick in the right way.

Although experiments of the kind performed by DELIĆ, HOPWOOD and FRIEND[94] could in principle have been carried out in a strain of *Streptomyces* in which information on linkage or the genetic system was totally lacking, as far as we are aware no such studies have been reported. We were in fact aided in our choice and understanding of test systems for mutagenesis by the ability to carry out genetic analysis, for example of prototrophic revertants to test for the involvement of suppressor loci; moreover our motive for starting the work was also academic. Therefore the outcome can legitimately be claimed as an example of academic research leading to something with commercial applications.

RECOMBINATION IN STRAIN IMPROVEMENT

The basic concept in the use of recombination in strain improvement is the coupling of two or more characters, originally present in diverse strains, to form a desirable combination within a single strain. For this basic manipulation, knowledge of the genetics of the strains concerned need not necessarily be very precise, although the more detailed it is, the more predictable the outcome of the crosses. The minimum requirements are normally at least one selectable marker in each strain, and the ability to select recombinants from mixed cultures of the strains. I referred to some of the problems of *recognising* recombinants in a previous paper[4]. As for the choice of selective markers, the familiar auxotrophic mutations are likely to prove the most useful.

It is worth remembering that most, although not all, auxotrophic mutations result in a pleiotropic depression of antibiotic yield, even in the presence of concentrations of required nutrients adequate for growth. Thus, either one has to operate exclusively with those mutations with negligible effects on yield, or else to arrange things so that only prototrophic progenitors and prototrophic recombinant progeny of auxotrophic strains are assayed for antibiotic yield. For example if two prototrophic strains are to be crossed, an auxotrophic derivative of each can be isolated, but the possibly low yield of these mutants should be disregarded. Prototrophic recombinant progeny selected from mixtures of the auxotrophs can then be assayed for antibiotic production (*see* the contribution of VLADIMIROV and MINDLIN[95]). In this way certain combinations of genes from the two parents will be obtained, but without knowledge of the linkage relations of the markers, there will be no way of knowing to what extent the range of possible combinations has been sampled. Only by having markers of known map position will it be possible to construct recombinants in a systematic way to include particular combinations of genome segments from the two parents.

Another possibility that may arise from advances in the fundamental genetics of a particular strain is that of performing nonselective genetic analysis, which, as discussed in this review, is now quite possible in *S. coelicolor* A3(2). Recognition and isolation of strains of particular fertility type will obviously depend on the availability of suitable markers of known map position, as we have seen from the discussion of fertility variation in *S. coelicolor*. However once suitable combinations of fertility types have been identified, the ability to dispense with genetic markers with possibly deleterious effects, for example on antibiotic yield, could clearly be advantageous.

We see, then, that something useful can be done on an empirical level, but a more sophisticated approach depends on a certain minimum academic knowledge of the genetics of the strain concerned.

PLASMIDS AND PHAGES IN STRAIN IMPROVEMENT

It is perhaps worth mentioning the potential interest of plasmids and phages in an industrial context. First let us remind ourselves of the tenuous nature of the dividing line between plasmids and phages, as indicated by several pieces of evidence from some elegant recent studies on eubacterial systems. It is now clear that, in physical terms, there is no important distinction between plasmids and many, at least, phage genomes. Plasmids are small circular DNA molecules, and so, at some stages in their life history, are most phage genomes[96]; the size ranges of the two classes of entity overlap. In terms of replication mechanisms, there are probably also close analogies, particularly perhaps between phages like P1 which lack a chromosomal location in the lysogenic cell[97], and plasmids which likewise fail to integrate into the host chromosome; and on the other hand between integrating plasmids, like F, and phages with chromosomal attachment sites. There is also the morphological similarity between the sex-factor-controlled pili of enteric bacteria[24] and the virions of the filamentous male-specific phages[98] suggesting a phylogenetic relationship between certain plasmids and phages, which is also strikingly demonstrated by the experimental evolution in the laboratory of a phage, by way of a defective phage, into a plasmid carrying bacterial genes[99]. The plasmids of eubacteria are characterised by two remarkable features: one is their frequently extremely broad host range (for example the sex factor, F, can exist in a wide range of Gram-negative bacteria[100,101]); the other is their capacity for taking part in genetic exchanges which result in pick-up by one plasmid of genes originally present on a different replicon. These properties are certainly powerful factors in contemporary enterobacterial evolution[102], and could be harnessed in any attempts to carry out "genetic engineering" in bacteria. Should plasmids with comparable properties to those of eubacteria be found in streptomycetes they could facilitate the build-up of a constellation of desirable attributes of diverse origin within a single industrially valuable strain, thus helping to overcome the limitations imposed by the circumscribed gene pool provided by the available strains of a particular "species"[103].

CONCLUSION

Topics in actinomycete genetics may be grouped under three main headings, which may perhaps be called: "applied", "comparative" and "fundamental". The first has an obvious meaning. The boundary between the second and third headings is somewhat arbitrary. By comparative questions I mean how do such things as genome topology[29,31], gene location[32], zygote structure and evolution[3], chromosome transfer, the control of fertility[6], and genetic interference[64] vary within the bacteria as a whole. Actinomycetes have a major role to play in this area because they represent a group of bacteria not closely related to the organisms which have provided most of the current ideas about bacterial genetics. The addition of *Thermoactinomyces* and *Nocardia* to *Streptomyces* should strengthen the value of actinomycetes in this comparative role. By fundamental topics I mean those in which an actinomycete is the organism of choice, because of some technical or biological advantage, for the study of a problem with general biological significance. The genetic control of morphogenesis in *S. coelicolor*[7], covered in detail by CHATER[9], is one example of such a topic. *Thermoactinomyces* could provide another good system for this approach. I think that, with the increasing interest in actinomycete genetics which is apparent at the present time, we can expect further advances in each of these three areas in the future.

It is a pleasure to thank Dr. Keith F. Chater, Mr. Richard J. Harold and Dr. Alan Vivian for helpful discussions during the preparation of this manuscript and Miss Helen M. Ferguson for her collaboration in the experiments summarised in the tables. I am grateful to Dr. James N. Adams, Dr. Paulinus P. Engel and Dr. Claude Vézina for letting me see manuscripts prior to publication.

REFERENCES

1. HOPWOOD D. A., SERMONTI G.: Adv. Genet. **11**, 273 (1962).
2. SERMONTI G., HOPWOOD D. A., in Gunsalus & Stanier (Eds.): *The Bacteria*, Vol. V. Academic Press, New York 1964.
3. HOPWOOD D. A.: Bacteriol. Rev. **31**, 373 (1967).
4. HOPWOOD D. A., p. 5 in Proc. Internat. Symp. Genetics and Breeding of Streptomyces (Dubrovnik). Yug. Acad. Sci. & Arts, Zagreb, 1969.
5. SERMONTI G., in Genetics of Antibiotic-Producing Microorganisms. Wiley-Interscience, London 1969.
6. VIVIAN A., HOPWOOD D. A.: J. Gen. Microbiol. **64**, 101 (1970).
7. HOPWOOD D. A., WILDERMUTH H., PALMER H. M.: J. Gen. Microbiol. **62**, 397 (1970).
8. VIVIAN A.: Paper at 1st Internat. Symp. Genetics of Industrial Microorganisms, Prague 1970. Abstract Book, p. 36.
9. CHATER K. F.: Paper at 1st Internat. Symp. Genetics of Industrial Microorganisms, Prague 1970. Abstract Book, p. 38.
10. ALAČEVIĆ M.: Paper at 1st Internat. Symp. Genetics of Industrial Microorganisms, Prague 1970. Abstract Book, p. 44.
11. COATS J. H., ROESER J.: Paper at 1st Internat. Symp. Genetics of Industrial Microorganisms, Prague 1970. Abstract Book, p. 48.
12. SERMONTI G., ALAČEVIĆ M., ŠUPUT J., PIGAC J., DELIĆ V. (Eds.): Proc. Internat. Symp. Genetics and Breeding of Streptomyces (Dubrovnik). Yug. Acad. Sci. & Arts, Zagreb 1969.
13. Ref.[23], p. 671.
14. HOPWOOD D. A.: Ann. N. Y. Acad. Sci. **81**, 887 (1969).
15. ERIKSON D.: J. Gen. Microbiol. **13**, 136 (1955).
16. STANIER R. Y.: J. Bacteriol. **44**, 555 (1942).
17. HOPWOOD D. A.: PhD. Thesis. University of Cambridge, Cambridge 1958.
18. HOPWOOD D. A.: Sci. Reports, Ist. Super. Sanità **1**, 463 (1961).
19. HOPWOOD D. A., HAROLD R. J., VIVIAN A., FERGUSON H. M.: Genetics **62**, 461 (1969).
20. SERMONTI G., CASCIANO S.: J. Gen. Microbiol. **33**, 293 (1963).
21. SERMONTI G., BANDIERA M., SPADA-SERMONTI I.: J. Bacteriol. **91**, 384 (1966).
22. SPADA-SERMONTI I., SERMONTI G.: Paper at 1st Internat. Symp. Genetics of Industrial Microorganisms, Prague 1970. Abstract Book, p. 34.
23. HAYES W., in *The Genetics of Bacteria and Their Viruses*. Blackwell, Oxford 1968.
24. MEYNELL E., MEYNELL G. G., DATTA N.: Bacteriol. Rev. **32**, 55 (1968).
25. NOMURA M.: Ann. Rev. Microbiol. **21**, 257 (1968).
26. HOLLOWAY B. W.: Bacteriol. Rev. **33**, 419 (1969).
27. BHASKHARAN K.: J. Gen. Microbiol. **23**, 47 (1960).
28. VIVIAN A., *personal communication*.
29. HOPWOOD D. A.: J. Mol. Biol. **12**, 514 (1965).
30. HOPWOOD D. A.: Genetics **54**, 1169 (1966).
31. HOPWOOD D. A.: Genetics **54**, 1177 (1966).
32. HOPWOOD D. A.: J. Cell. Physiol. 70, Suppl. **1**, 7 (1967).

33. Ref.[5], p. 301.
34. CLARK A. J., MARGULIES A. D.: Proc. Nat. Acad. Sci. U.S.A. 53, 451 (1965).
35. HOPWOOD D. A., FERGUSON H. M., *unpublished results* (1969).
36. HAROLD R. J.: p. 27 in Proc. Internat. Symp. Genetics and Breeding of Streptomyces (Dubrovnik). Yug. Acad. Sci. & Arts, Zagreb 1969.
37. HAROLD R. J., HOPWOOD D. A.: Mutation Res. 10, 427, 439 (1970).
38. Ref.[5], p. 300.
39. NOVICK R. P.: Bacteriol. Rev. 33, 210 (1969).
40. RICHMOND M. H.: Symp. Soc. Gen. Microbiol. 20, 249 (1970).
41. GREGORY K. F., SHJU W.: Nature 191, 465 (1961).
42. GREGORY K. F., HUANG J. C. C.: J. Bacteriol. 87, 1287 (1970).
43. OKANISHI M., OHTA T., UMEZAWA H.: J. Antibiotics 23, 45 (1970).
44. GREGORY K. F., HUANG J. C. C.: J. Bacteriol. 87, 1281 (1964).
45. SCHAEFFER P.: Bacteriol. Rev. 33, 48 (1969).
46. ROGOLSKY M., SLEPECKY R. A.: Biochem. Biophys. Res. Comm. 16, 204 (1964).
47. BOTT K. F., DAVIDOFF-ABELSON R.: J. Bacteriol 92, 229 (1966).
48. JANSZ H. S., POUWELS P. H., SCHIPHORST J.: Biochim. Biophys. Acta 123, 626 (1966).
49. RADLOFF R., BAUER W., VINOGRAD J.: Proc. Nat. Acad. Sci. U.S.A. 57, 1514 (1967).
50. RUSH M. R., GORDON C. N., WARNER C.: J. Bacteriol. 100, 803 (1969).
51. GORDON C. N., RUSH M. R., WARNER R. C.: J. Mol. Biol. 47, 495 (1970).
52. WELSCH M., p. 43 in Proc. Internat. Symp. Genetics and Breeding of Streptomyces (Dubrovnik). Yug. Acad. Sci. & Arts, Zagreb 1969.
53. LOMOVSKAYA N. D.: Paper at 1st Internat. Symp. Genetics of Industrial Microorganisms, Prague 1970. Abstract Book, p. 40.
54. VIVIAN A.: PhD. Thesis. University of Reading, England 1968.
55. SERMONTI G., MANCINELLI A., SPADA-SERMONTI I.: Genetics 45, 669 (1960).
56. HOPWOOD D. A., SERMONTI G., SPADA-SERMONTI I.: J. Gen. Microbiol. 30, 249 (1963).
57. HOPWOOD D. A.: Genet. Res. 6, 248 (1965).
58. ANDERSON T. F.: Cold Spring Harbor Symp. Quant. Biol. 23, 47 (1958).
59. LEDERBERG J.: Proc. Nat. Acad. Sci. U.S.A. 43, 1060 (1957).
60. Ref.[5], p. 276.
61. HOPWOOD D. A., HAROLD R. J., VIVIAN A., FERGUSON H. M.: Heredity 23, 628 (1968).
62. PONTOCORVO G., in *Trends in Genetic Analysis*. Columbia University Press, New York 1959.
63. PIPERNO R., CARERE A., SERMONTI G.: Ann. Ist. Super. Sanità 2, 393 (1966).
64. SERMONTI G., CARERE A.: Mol. Gen. Genet. 103, 141 (1968).
65. ENGEL P. P.: Paper at 1st Internat. Symp. Genetics of Industrial Microorganisms, Prague 1970. Abstract Book, p. 41.
66. RUSSI S., CARERE A., FRATELLO B., KHOUDOKORMOFF V.: Ann. Ist. Super. Sanità 2, 506 (1966).
67. PONTECORVO G.: Cold Spring Harbor Symp. Quant. Biol. 11, 193 (1946).
68. HAROLD R. J., *personal communication* (1970).
69. CROSS T.: J. Appl. Bacteriol. 31, 36 (1968).
70. TSILINSKY P.: Ann. Inst. Pasteur 13, 500 (1899).
71. WILDERMUTH H.: J. Gen. Microbiol. 60, 43 (1970).
72. WILDERMUTH H., HOPWOOD D. A.: J. Gen. Microbiol. 60, 51 (1970).
73. ERIKSON D.: J. Gen. Microbiol. 6, 286 (1952).
74. CROSS T., WALKER P. D., GOULD G. W.: Nature 220, 350 (1968).
75. DOROKHOVA L. A., AGRE N. S., KALAKOUTSII L. V., KRASSILNIKOV N. A.: J. Gen. Appl. Microbiol. 14, 295 (1968).

76. PEPYS J., JENKINS P. A., FESTENSTEIN G. N., GREGORY P. H., LACEY M. E., SKINNER F. A.: Lancet **2**, 607 (1963).
77. HOPWOOD D. A., FERGUSON H. M.: J. Gen. Microbiol. **63**, 133 (1970).
78. CLARK A. J.: J. Cell. Physiol. **70**, Suppl. 1, 165 (1967).
79. BREED R. S., MURRAY E. G. D., SMITH N. R., in *Bergey's Manual of Determinative Bacteriology*, 7th Edn. Baillière, Tindall & Cox, London 1957.
80. ADAMS J. N., BRADLEY S. G.: Science **140**, 1392 (1963).
81. VÉZINA C., *personal communication* (1970).
82. VÉZINA C., SINGH K., SEHGAL S. N.: Appl. Microbiol. **18**, 270 (1969).
83. VÉZINA C.: Paper at 1st Internat. Symp. Genetics of Industrial Microorganisms, Prague 1970. Abstract Book, p. 167.
84. ADAMS J. N.: J. Bacteriol. **88**, 865 (1964).
85. BROWNELL G. H., ADAMS J. N.: J. Bacteriol. **94**, 650 (1967).
86. BROWNELL G. H., ADAMS J. N.: Genetics **60**, 437 (1968).
87. ADAMS J. N.: J. Bacteriol. **96**, 1750 (1968).
88. BROWNELL G. H., KELLY K. L.: J. Bacteriol. **99**, 25 (1969).
89. WALSH R. S., BROWNELL G. H.: Bacteriol. Proc. 40 (1970).
90. BROWNELL G. H.: Bacteriol. Proc. 41 (1970).
91. FOLKENS A. T., ADAMS J. N.: Bacteriol. Proc. 41 (1970).
92. BROWNELL G. H., ADAMS J. N., BRADLEY S. G.: J. Gen. Microbiol. **47**, 247 (1967).
93. HOPWOOD D. A., p. 363 in J. R. Norris, D. W. Ribbons (Eds.): *Methods in Microbiology*, Vol. IIIA. Academic Press, London 1970.
94. DELIĆ V., HOPWOOD D. A., FRIEND E. J.: Mutation Res. **9**, 167 (1970).
95. VLADIMIROV A. V., MINDLIN S. Z.: Paper at 1st Internat. Symp. Genetics of Industrial Microorganisms, Prague 1970. Abstract Book, p. 46.
96. THOMAS C. A.: J. Cell. Physiol. **70**, Suppl. 1, 13 (1967).
97. IKEDA H., TOMIZAWA J.: Cold Spring Harbor Symp. Quant. Biol. **33**, 791 (1968).
98. MARVIN D. A., HOHN B.: Bacteriol. Rev. **33**, 172 (1969).
99. CHAKRABARTY A. M., GUNSALUS I. C.: Proc. Nat. Acad. Sci. U.S.A. **64**, 1217 (1969).
100. FALKOW S., CITARELLA R. V., WOHLHIETER J. A.: J. Mol. Biol. **17**, 102 (1966).
101. ROWND R., NAKAYA R., NAKAMURA A.: J. Mol. Biol. **17**, 376 (1966).
102. ANDERSON E. S.: Ann. Rev. Microbiol. **22**, 131 (1968).
103. MINDLIN S. Z., p. 147 in Proc. Internat. Symp. Genetics and Breeding of Streptomyces (Dubrovnik). Yug. Acad. Sci. & Arts, Zagreb 1969.

Fertilization in *Streptomyces coelicolor*

I. Spada-Sermonti, G. Sermonti

Laboratorio di Embriologia Molecolare, Napoli and Istituto di Genetica dell'Università, Palermo, Italy

The story of the fertility types in *Streptomyces coelicolor* followed in its general lines that of *Escherichia coli* K12. The strain employed, A3(2), turned out to be "homothallic", since recombinants were obtained between mutants obtained from one and a single strain[1]. The fertility of such crosses was low, though remarkably higher than that recorded in the first crosses performed in *E. coli* K12[2]. Later a self-sterile group of strains (R⁻) was reported[3] but the limit line between the selfsterile and the normal fertile strains (R⁺) was not sharp, and occasionally a strain changed, after repeated transfers, from the R⁺ to the R⁻ condition or *vice versa*, and intermediate types were observed. In this condition the infective nature of the fertility factor, claimed by Sermonti and Casciano[3], could not be safely stated.

A further step in the analysis of fertility variants in *S. coelicolor* A3(2), also paralleled in *E. coli* K12[4], was the discovery of a high-frequency recombination strain by Hopwood *et al.*[5]. This strain was called UF (ultrafertile), in contrast to the NF (normal fertile) strains, including all strains since recorded. On mixed cultures between UF and NF strains virtually all the spores were recombinant.

A strain 219 resembling the UF strains by Hopwood *et al.*[5], had been isolated by Sermonti *et al.*[6]. It turned out to behave quite differently when crossed with R⁺ or R⁻ strains. It showed very high fertility (approaching 100%) with some R⁺ strains and no fertility in crosses with R⁻ strains. Thus it belonged to the intersterile (R⁻) group of strains and it was named UR⁻ (ultra-R⁻) because of its high recombination rate in crosses with R⁺ strains. Thus it was not comparable with the Hfr strains of *E. coli*, which are highly fertile with any other strain. This provisional comparison between fertility variants in *E. coli* K12 and *S. coelicolor* A3(2) was suggested[7]:

E. coli Hfr   *S. coelicolor* R⁺   (Hopwood's NF)
E. coli F⁺   *S. coelicolor* (R⁺ to R⁻)
E. coli F⁻   *S. coelicolor* UR⁻   (Hopwood's UF)

TABLE I

Characteristics of Strains Used

Strain code N°	Fertility type[a]	Markers									
		strA1	*adeA3*	*pheA1*	*uraA1*	... *leuA1*	*metA2*	*hisA1*	*argA1*	... *thi(A)59*	*hisD3*
2	R⁺	*strA1*	+	+	+	+	+	+	+	+	+
44	R⁻	*strA1*	+	+	+	+	+	*hisA1*	+	+	+
4	R⁺	*strA1*	+	+	+	+	+	*hisA1*	+	+	+
39	R⁺	*strA1*	+	+	+	+	+	*hisA1*	+	+	+
310	R⁻	+	*adeA3*	+	+	+	+	+	+	*thi(A)59*	+
318	(R⁺)	+	*adeA3*	+	+	+	+	+	+	*thi(A)59*	+
79	(R⁺)	*strA1*	+	+	*uraA1*	+	+	+	*argA1*	+	+
118-4	R⁻	*strA1*	+	+	*uraA1*	+	+	+	*argA1*	+	+
311, 312, 313	UR⁻	*strA1*	+	*pheA1*	*uraA1*	+	+	+	*argA1*	+	+
165	(R⁺)	*strA1*	+	*pheA1*	+	+	*metA2*	+	+	+	*hisD3*
316	UR⁻	*strA1*	+	*pheA1*	+	+	*metA2*	+	+	+	*hisD3*
53	(R⁺)	*strA1*	+	*pheA1*	+	+	+	+	+	+	+
219	UR⁻	*strA1*	+	*pheA1*	*uraA1*	*leuA1*	*metA2*	+	+	+	+
80	R⁺	*strA1*	+	+	*uraA1*	+	*metA2*	*hisA1*	*argA1*	+	+
191	(R⁺)	*strA1*	+	*pheA1*	*uraA1*	+	*metA2*	+	*argA1*	+	*hisD3*
317	UR⁻	*strA1*	+	*pheA1*	+	*leuA1*	+	*hisA1*	*argA1*	+	*hisD3*
84	R⁺	+	+	+	*uraA1*	+	+	+	*argA1*	+	+

[a] R⁺ highly fertile, (R⁺) moderately fertile, R⁻ self-sterile, UR⁻ self-sterile, highly recipient.

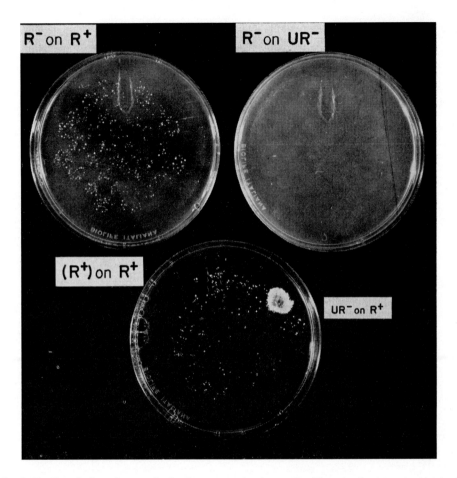

Fig. 1. Replica plating of spot-mixed cultures on selective media. The mixed spots were obtained by printing ordered spots from different colonies of the tested strain (surviving UV-treatment) on complete medium containing a dense background of spores of the tester strain, and incubating three days.

Top left: 310 R$^-$ (spots) × 39 R$^+$ (background); **top right:** 310 R$^-$ (spots) × 219 UR$^-$ (background); **bottom:** 165 (R$^+$) (spots) × 39 R$^+$ (background): in the two o'clock position, a thick spot of recombinants from a spot-mixed culture comprising on UR$^-$ variant of strain 165 is observed.

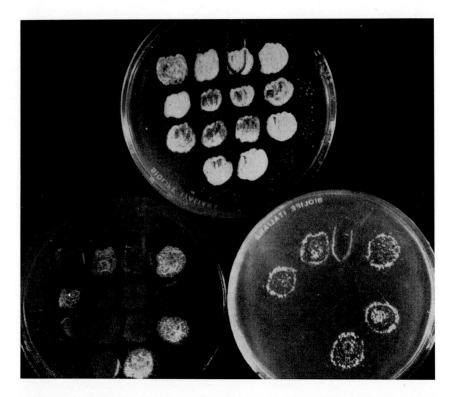

Fig. 2. **Top:** Spots from colonies of parental genotype 312 UR⁻ derived from a mixed culture with strain 53 (R⁺). **Bottom left:** Spot-mixed cultures obtained by printing colonies of 312 genotype on a dense background of the tester strain 219 UR⁻. **Bottom right:** Replica plating of spot-mixed cultures on selective medium: five of fourteen formerly UR⁻ genotypes have been converted to (R⁺).

ISOLATION OF OTHER ULTRA-RECIPIENT (UR⁻) VARIANTS

The first UR⁻ (UF) strain in our collection, 219 (as to its characteristics, *see* Table I), was isolated in a search for a high-frequency recombination (Hfr) strain in *S. coelicolor* by SERMONTI *et al.*[6]. It was isolated on a mixed culture (R⁺ × R⁻) grown for 24 h on a complete medium and then transferred by the cellophane technique[6] to minimal medium supplemented with streptomycin. It appeared as a distinct colony surrounded by a crowd of overgrowing tufts (heteroclones) in the zone of overlapping with the adjacent colonies. It was collected, but when it turned out to be a derivative of the R⁻ parent and still to be of the R⁻ type, it was not given further consideration. Only after the description of the UF strain by HOPWOOD *et al.*[5], the strain was reconsidered and employed in further studies.

Other UR⁻ strains were then looked for, using the replica procedure adopted by HOPWOOD *et al.*[5]. Spots of R⁻ colonies or R⁺ strains were arrayed on master plates and then replica-plated on to a complete medium on which a thick suspension of spores from an R⁺ (ultrafertile) strain had been plated. Spot-mixed cultures were thus obtained, which, after three or four days' growth, were replica-plated on to a selective medium (Fig. 1). In such conditions an R⁺ × R⁺ or an R⁻ × R⁺ mixed spot yielded a limited number of recombinant colonies, while a UR⁻ × R⁺ spot yielded hundreds of recombinant colonies, others displaying a continuous growth area[5].

As shown in Table II, UR⁻ variants were obtained only after ultraviolet treatment from a variety of strains, either R⁻ or R⁺, usually at a rate ranging from 0.15

TABLE II

Isolation of UR⁻ Variants

Original strain		Untreated sample			UV-treated sample		
Code N°	Fertility type	Number of colonies tested	UR⁻ detected Number	UR⁻ detected %	Number of colonies tested	UR⁻ detected Number	UR⁻ detected %
79	(R⁺)	286	0	0	153	20	13.1
310	R⁻	—	—	—	428	2	0.5
165	(R⁺)	150	0	0	611	3	0.5
191	(R⁺)	—	—	—	663	1	0.15
44	R⁻	120	0	0	1,590	3	0.2
80	R⁺	126	0	0	204	0	0
39	R⁺	120	0	0	180	0	0
4	R⁺	364	0	0	246	0	0
Total		1,166	0	0	4,075	29	0.4
Total (79 excluded)		880	0	0	3,922	9	0.23

to 0.5 per cent surviving colonies. Four strains did not produce any UR⁻ variant but the number of colonies tested was not high enough to warrant their inability to be converted into UR⁻. The R⁺ strains from which UR⁻ variants were obtained were only moderately fertile (R⁺). No spontaneous variant was detected. A peculiar strain 79 gave an unusually high rate of UR⁻ variants (about 13%) but no spontaneous variants. The pooled average rate of induced variation to UR⁻ (strain 79 excluded) was about 0.2% (9/4372), a rate comparable with 0.3%, observed by HOPWOOD et al.[5]. The spontaneous rate observed by HOPWOOD et al.[5] (0.03%) would have resulted in 0.3 isolates out of the 1226 untreated colonies tested by ourselves. We found none.

Repeated attempts to eliminate an assumed fertility factor from fertile strains with acridine orange (30 μg/ml) were unsuccessful but only 660 colonies from the spores treated were scored.

RECOMBINATION RATES BETWEEN DIFFERENT FERTILITY TYPES

The recombination rates between different combinations of strains was observed to range from practically zero out of hundreds of millions parental spores, to nearly 100% (Table III). An R⁻ (normal recipient) strain, crossed with an R⁺ strain yields 10^{-3} or more recombinants (Fig. 1, *top left*), crossed with an UR⁻ strain yields

TABLE III

Recombination Frequency between Pairs of Strains

Fertility type	Parents (code N°)	Recombinant frequency
R⁺ × UR⁻	39 × 316	0.78[a]
	39 × 219	0.51[a]
	84 × 219	0.77[a]
(R⁺) × UR⁻	53 × 312	2.5×10^{-2}
R⁺ × R⁻	84 × 310	200×10^{-6}
	4 × 118-4	230×10^{-6}
(R⁺) × R⁻	79 × 44	150×10^{-6}
	221 × 44	33×10^{-6}
-R⁺ × R⁺	39 × 80	25×10^{-6}
	39 × 191	35×10^{-6}
R⁻ × R⁻	44 × 118-4	0.02×10^{-6}
	118-4 × 310	0×10^{-6}
R⁻ × UR⁻	44 × 219	10×10^{-6}
	310 × 219	2×10^{-6}
UR⁻ × UR⁻	312 × 219	0×10^{-6}
	313 × 219	0.00×10^{-6}
	219 × 316	0×10^{-6}

[a] Unselective analysis on complete medium.

practically no recombinants (Fig. 1, *top right*). Crossed with an R^+ tester, an R^+ strain yields about 10^{-4} recombinants (Fig. 1, *bottom*), while an UR^- variant gives nearly 100 per cent recombinants (*see* also Ref.[5]).

Two progressions of fertility are thus evident. With an R^+ tester the three fertility types can be ordered as follows according to their ability to act as recipient: UR^-, R^-, R^+. With an UR^- tester, the gradient of donor ability is as follows: R^+, R^-, UR^-. Indeed there seems to be a series of intermediate strains between R^+ and R^-, while the step between R^- and UR^- appears to be clear-cut (using an R^+ donor as tester). Up to this point the comparison between R^+, R^- and UR^- of *S. coelicolor* A3(2) and Hfr, F^+ and F^- of *E. coli* K12 still holds.

INFECTIVITY OF THE FERTILITY FACTOR

The change of the fertility type in one parent, after its reisolation from a mixed culture with a strain of different type was usually observed, the other strain being unaffected. Invariably the changes occurred at a high rate and resulted in an increase of donor ability on the part of the recipient (less fertile) partner (Table IV).

In highly fertile crosses, $R^+ \times UR^-$, only a minority of parents could be recovered from the spores of the mixed culture. The R^+ exconjugant parental genotypes always retained their R^+ type, while one-half or more of the recipient parental genotypes turned out to have been converted in R^{+} [5,7] (Fig. 2). In experiments of interrupted mating (*see* later), hyphae of mixed $UR^- \times R^+$ cultures, collected after about one day showed a recombination rate ranging around 1% and a conversion rate ($UR^- \rightarrow R^+$) around 15% (176/1013). Moreover nearly all of the converted types (166/167) appeared only moderately fertile.

In normally fertile crosses, $R^+ \times R^-$, all the R^+ parents retained the R^+ type, while a variable fraction (from 10 to 50%) of the R^- parental genotype were converted to R^+. In an interrupted mating experiment the first converted exconjugants appeared after the first day of mixed culture, roughly at the time of the appearance of recombinants. The number of formerly recipient strains converted (at least five recombinants per mixed spot) increased with time — 1; 1.5; and 2 days — from 1.3 to 16.1 and to 40.3%. The degree of fertility obtained (calculated as the number of recombinants per fertile spot in mixed cultures with a UR^- tester) also increased with time from 2.4 to 5.1 and to 9.2%. Highly fertile strains appeared sporadically. They were more frequent among the spores of completely developed $R^+ \times R^-$ mixed cultures.

A mixed culture was also obtained between an R^- and a UR^- strain (44 × 316). Colonies derived from the spores of the mixed culture were tested against an R^+ strain. All the strains with the genotype of the R^- parent (150/150) retained the R^- type, while one-half of the strains with the genotype of the UR^- parent (90/165) had acquired the R^- fertility type. They were poorly fertile with the R^+ tester and virtually

TABLE IV

Fertility Type of Parental Genotypes after Mixed Heterotype Cultures

Fertility types		Cross		A parent		B parent	
A	B	A	B	R^+	UR^-	R^+	R^-
$UR^- \times R^+$		219×39^a		186^b	837	50	0
		219×39		30	6	59	0
		316×39		18^c	20	70	0
		311×53		14	28	—	—
				R^+	R^-	R^+	R^-
$R^- \times R^+$		44×80^a		25	242	—	—
		310×39		81	99	—	—
		314×53		14	28	—	—
		310×84		44	54	140	0
		$118\text{-}4 \times 39$		28	248	103	0
				R^-	UR^-	R^-	UR^-
$UR^- \times R^-$		316×44		90	75	150	0
		219×310		85	123	200	0

[a] Mating interrupted after about one day.
[b] 176/186 moderately fertile.
[c] 10/18 moderately fertile.

infertile when checked with a UR^- tester. Similar results were obtained in a cross $(R^+) \times UR^-$ (Fig. 2). In conclusion the donor ability (and the recipient inability) appeared to be transferable from the more fertile types to the less fertile types, from the less effective recipient strains to the more effective ones. The conversions observed showed the following pattern:

$$UR^- \to R^- \to (R^+) \to R^+$$

The transitions from R^- (or UR^-) to R^+ appeared indeed to be gradual. The "contagion" proceeded obviously in the opposite sense.

GRADUAL TRANSFER OF MARKERS IN $R^+ \times UR^-$ CROSSES

Mixed cultures between R^+ and UR^- strains were grown on cellophane discs lying on complete medium. At various intervals of time, usually ranging from 20 to

26 h, the mycelium was harvested by a loop, washed by centrifugation, ground in a Potter homogenizer, diluted and plated on different selective media[8]. In a cross (Fig. 3):

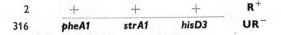

2	+	+	+	R⁺
316	*pheA1*	*strA1*	*hisD3*	UR⁻

the *strA* marker was selected for in each medium. The *pheA* donor marker first appeared in the sample taken after 20 h. The *hisD*⁺ donor marker made its first appearance in the 22 h sample and retained a very low rate, if compared to the *pheA*⁺, in the next two samples (24 and 26 h). It appears that the *pheA*⁺ marker is closer to the "required" region[5] contributed by the donor and that *hisD*⁺ is far removed from this region. The first *hisD*⁺ recombinants do not carry the *pheA*⁺ marker (Fig. 4 *middle*), while 80% of the *hisD*⁺ recombinants in the 26 h sample had received *pheA*⁺, (Fig. 4 *right*). This fact strongly suggests that the length of the contributed donor chromosome increases with time. *pheA*⁺ and *hisD*⁺ are linked and separated by the

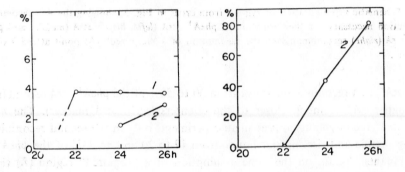

Fig. 3. Frequencies of recombinants bearing various donor markers among recombinants from cross:

316	*pheA1*	*strA1*	*hisD3*	UR⁻
2	+	+	+	R⁺

All recombinants are *strA* (resistant to streptomycin). **Left:** the frequencies are given as per cent of *pheA*⁺ recombinants; **right:** per cent of *hisD*⁺ recombinants. The recombinants have been obtained from samples of mixed cultures interrupted at various times. Notice the late appearance of the *hisD*⁺ marker (line *1*) and the even later appearance of recombinants bearing both *hisD*⁺ and *pheA*⁺ (line *2*).

counterselected marker *strA*⁺, so that recombinants bearing both genes must derive from zygotes which had received an almost complete donor chromosome (Fig. 4 *right*). On the assumption that the 9 o'clock region of the chromosome, which is responsible for the fertility of the R⁺ (NF) strain, must be always present in the R⁺ × UR⁻ (=NF × UF) zygotes[5], it turns out that the markers selected appear in the progeny the sooner, the shorter the donor chromosome contribution to the zygotes. They also attain a higher frequency (Fig. 4).

Similar results were obtained in crosses:

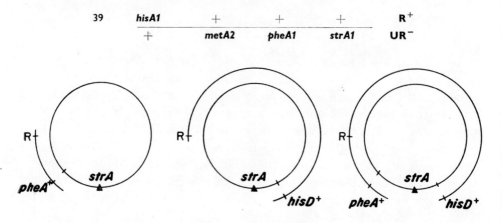

Fig. 4. Schematic representation of zygotes from cross in Fig. 3, showing the R$^+$ chromosome contribution necessary for the formation of *pheA*$^+$ *strA* (*left*), *hisD*$^+$ *strA* (*middle*) and *pheA*$^+$ *hisD*$^+$ *strA* (*right*) recombinants on the assumption of a "required" (R) point at the 9 o'clock position[5].

The *metA*$^+$ *strA* recombinants range from 30 to 50% of the *pheA*$^+$ *strA* recombinants (although *pheA*$^+$ is much closer to the counterselected *strA* marker). The *metA*$^+$ *pheA*$^+$ *strA*$^+$ recombinants appear in an experiment[9] only in the second recombinant-bearing sample and rise, in four hours, from 12 to 20 and to 33% of the *pheA*$^+$*strA* recombinants. Again, on the same assumption of a "required" region (R) the recombinants with a larger donor contribution appear later and attain a lower final level. If we compare the two experiments, *metA*$^+$ appears to be a much more "proximal" marker if compared to *hisD*$^+$, the latter being indeed the most distal marker from the 9 o'clock region, when the interruption of the donor circle at the *strA*$^+$ point is made compulsory by the selective conditions (Fig. 4).

GRADIENT OF MARKER FREQUENCY IN R$^+$ × UR$^-$ ZYGOTES

As was already noticed by HOPWOOD *et al.*[5], the alleles contributed by the R$^+$ (NF) parent show the maximum frequencies around the 9 o'clock region of the circular map, the so-called "required" region, which corresponds to the "silent" left arc of the map. This has been substantially confirmed by the unselective analysis of several of our crosses (Fig. 5). On the basis of the allele frequency gradient, the R factor, which is, as a rule, the most frequent donor marker, fits well at the 9 o'clock position, close to the *pheA* marker[5]. To check this assumption a number of genotypically recombinant *pheA* strains from a cross

39	*hisA1*	+	(R)	+	(R)	+	R⁺
219	+	*metA2*	*pheA1*		*strA1*		UR⁻

were selected and tested for their fertility.

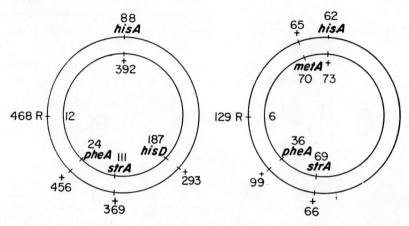

Fig. 5. Allele frequencies in two crosses R⁺ (*outer circle*) × UR⁻ (*inner circle*). R indicates the frequency of the R⁺ segregants in the whole samples of the progeny, as well as the "required" point[5] of the R⁺ genome. **Right:** cross 39 R⁺ × 316 UR⁻; **left:** cross 39 R⁺ × 219 UR⁻.

They were selected by plating an interrupted mixed culture (one day) on a selective medium for *metA*⁺ and *strA* (MM + phenylalanine + histidine + streptomycin), and isolated as unable to grow after replica-plating on a similar medium lacking phenylalanine. The *pheA* recombinants belonged to two classes, a larger (130) *pheA* UR⁻ and a smaller (67) *pheA* R⁺. This places R close to *pheA* on the *metA* side. Were the R factor on the right side of *pheA*, practically no *pheA* R⁺ recombinant would have appeared, since it requires two crossovers in the short area comprised between *pheA* and *strA*.

The occurrence in the cross (Fig. 3) of precocious *hisD*⁺ recombinants carrying the R factor and lacking the *pheA*⁺ markers (Fig. 4 *middle*) bears on the same direction, *i.e.* on the location of R to the left of *pheA*.

CONCLUSIONS

The comparison between the three fertility types in *E. coli* K12 and those in *S. coelicolor* A3(2) still holds at this point but for three substantial points of disagreement.

First, in $R^+ \times UR^-$ crosses (comparable to *E. coli* Hfr $\times$ F$^-$) virtually all the progeny was of the R^+ type, thus showing that the R factor has a role different from the F episome in *E. coli*. It appears to work more as a compatibility factor than as a chromosome mobilizer agent.

Second, although there is good evidence of an increasing contribution of the donor genome with time, it does not appear that there is a fixed origin and a determined sense of transfer in any single donor strain; even in crosses with UR^- recipients.

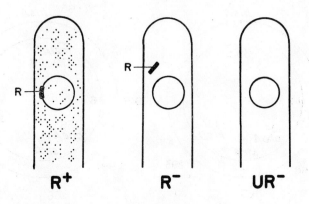

Fig. 6. A tentative model of the three fertility types discussed in this paper. The *outer line* represents the cell wall, the *inner circle* the circular chromosome. R indicates a fertility episome, the points in the R^+ cell suggest a cytoplasmic factor (possibly a repressor). Intermediate types between R^+ and R^- are interpreted as a mixture of the two types.

Third, in $R^+ \times R^-$ crosses (comparable to *E. coli* Hfr $\times$ F$^+$) we observe a gradual conversion of the R^- type to the R^+ type, which is not paralleled by the conversion of *E. coli* F$^+$ strains after contact with Hfr strains. It appears that some "contagious" agent is present in R^+ strains and lacking in R^- strains, although both of them carry the R fertility factor. One might advance the following hypothesis: the R particle is in a repressed condition in the R^+ strains (as a prophage in a lysogenic strain), and in an unrepressed condition in the R^- strains (as a F factor in a F$^+$ strain of *E. coli*). When R^+ and R^- strains are put in mixed culture the repressor (Fig. 6) migrates to the R^- strains, imposing upon the R particle a repressed condition and possibly promoting its connection with the chromosome. The "chromosomal" R factor might have a fixed abode (as λ in *E. coli*) in the "silent" 9 o'clock region of the map.

On this model we would expect a sort of zygotic induction of the R factor when it migrates to an UR^- cell from an R^+ mycelium.

As a matter of fact, in experiments of interrupted matings, in crosses $R^+ \times UR^-$ (39 $\times$ 219) after one day's mixed growth, most (166/176) of the formerly UR^- strains "fertilized" were only moderately fertile. One might assume that the transmitted R particle, reaching a UR⁻ cytoplasm devoid of repressor, is derepressed and the cell converted into the R⁻ condition, gradually reaching the R⁺ state, due to the spreading of the repressor agent.

The situation depicted in Fig. 6 although far from being satisfactorily documented, is compatible with all the data reported. The R factor appears as an intracellular particle, unable (?) to abandon the cell, though transferable at high frequency by intercellular contact either as a chromosomal factor (R⁺ × UR⁻ crosses) or as a plasmid (R⁻ × UR⁻ crosses). The R-repressor agent would have properties of gradual spreading from R⁺ to R⁻ (or UR⁻ cells). According to the model in Fig. 6 the R factor appears to be an episome, fixed to the chromosome in R⁺, free in the cytoplasm in R⁻ and absent in UR⁻. The symbols so far adopted, by ourselves (R⁺, R⁻, UR⁻) and by Hopwood *et al.*[5] (NF, UF) seem therefore inadequate and new ones should be discussed and agreed upon, possibly after the collection of new data.

The authors wish to thank Dr. Giuseppina Ficarra and Dr. Anna Maria Puglia for their collaboration in this work, particularly in the experiments of interrupted mating and in the search for new UR⁻ types, and for their permission to present some of their unpublished data.

REFERENCES

1. Hopwood D. A.: J. Gen. Microbiol. **16**, ii (1957).
2. Lederberg J., Tatum E. L.: Cold Spring Harbor Symp. Quant. Biol. **11**, 113 (1946).
3. Sermonti G., Casciano S.: J. Gen. Microbiol. **33**, 293 (1963).
4. Cavalli-Sforza L.L., Lederberg J., Lederberg E.M.: J. Gen. Microbiol. **8**, 89 (1953).
5. Hopwood D.A., Harold R.I., Vivian A., Ferguson H.M.: Genetics **62**, 461 (1969).
6. Sermonti G., Bandiera M., Spada-Sermonti I.: J. Bacteriol. **91**, 384 (1966).
7. Sermonti G., Spada-Sermonti I.: Atti A.G.I. (1970).
8. Ficarra G., Puglia A.M., *unpublished results.*
9. Puglia A.M., *unpublished results.*

Genetics of Tetracycline-Producing Streptomycetes

M. ALAČEVIĆ

Faculty of Technology, Zagreb, Yugoslavia

The interest in the genetics of *Streptomyces rimosus* was mostly centered on its ability to produce an important antibiotic. The other stimulus was to verify in another *Streptomyces* the main assumptions of the genetic system postulated for *S. coelicolor*[1-4], the only species of actinomycetes in which chromosome mapping had been widely carried out.

As shown by HOPWOOD[5], the normal genome of *S. coelicolor* A3(2) is a closed circular structure and the zygote genomes produced, after the genetic material is transferred from one parental strain to another, are only partially diploid. By quantitative evaluation of data it was possible in *S. coelicolor* to estimate the "distance" between many genes, as well as the length of two "empty" regions of opposite sites of the circle, each about one-fourth of the map.

HOPWOOD[1] and PIPERNO *et al.*[6] also show that the genes involved in the biosynthesis of histidine are partially clustered, with one cluster of 6 genes in one region and other genes scattered in other parts of the chromosome. The gene — enzyme relationship has been investigated in *S. coelicolor*[7] and compared with that of *Salmonella typhimurium*.

In our previous paper[8], the first attempt was made to clarify the genetic system in other streptomycetes, mainly *S. rimosus*, as well as to establish the order of some loci. By analysing trios of markers from many heteroclones, about ten markers were mapped on the chromosome of *S. rimosus*.

Several dozen of heteroclones of *S. aureofaciens* have been analyzed[9] and the pattern of recombination was comparable with that in *S. rimosus*. The unsuitability of *S. aureofaciens* for genetic analysis is due to the great instability of its auxotroph mutants and to the appearance of a great number of *arg* mutants, the only type of stable auxotrophs.

Interspecific recombinations were observed[8] in crosses between *S. rimosus* and *S. coelicolor*, as well as between either of these species and *S. aureofaciens*. The value of taxonomy of the genus *Streptomyces* was obviously questionable.

The genetics of antibiotic production has been studied so far in *S. rimosus* ATCC 10970, strain R7 low producer, and it is quite difficult because of interference of nutritional markers with antibiotic yield.

The present paper deals with a more refined analysis of the *S. rimosus* R7 map, and with its comparison with the map of *S. coelicolor* A3(2).

MATERIALS AND METHODS

MEDIA AND METHODS

The media and methods used as well as the rationale for the genetic analysis are almost the same as those developed for *S. coelicolor* A3(2)[2,10] which proved very effective for *S. rimosus*.

TABLE I

List of Mutant Alleles in *S. rimosus* R7

Alleles	Characteristics: requirement for	Mutagenic agent	Alleles	Characteristics: requirement for	Mutagenic agent
arg-1	arginine	UV	ilv-1	isoleucine and valine	UV
arg-2	arginine	UV	ilv-2	isoleucine and valine	UV
arg-30	ornithine or citrulline or arginine	UV	leu-2	leucine	NTG
arg-44	arginine	UV	lys-4	lysine	NTG
arg-49	arginine	NTG	met-95	methionine or homocysteine	UV
arg-50	arginine	NTG			
ade-2	adenine	UV	met-108	methionine or homocysteine	UV
ade-14	adenine	NTG	nic-1	nicotinic acid	UV
ade-15	adenine	NTG	nic-5	nicotinic acid	NTG
cym-102	cysteine or cystine and methionine	NTG	pan-1	pantothenic acid	NTG
			pan-9	pantothenic acid	NTG
cym-103	cystine and methionine	NTG	pdx-9	pyridoxine	NTG
			pro-1	proline	UV
cym-107	cysteine or cystine and methionine	NTG	pro-11	proline	UV
			rib-4	riboflavine	UV
gua-2	guanine	NTG	rib-5	riboflavine	UV
gua-3	guanine	NTG	thy-2	thymine	NTG
his-1	histidine or histidinol	UV	trp-3	tryptophan	UV
his-2	histidine or histidinol	UV	trp-5	tryptophan	UV
his-3	histidine	UV	trp-12	tryptophan	UV
his-4	histidine or histidinol	UV	trp-14	tryptophan	NTG
his-6	histidine or histidinol	UV	trp-15	tryptophan	NTG
his-16	histidine or histidinol	UV	tyr-1	tyrosine	UV
his-17	histidine	UV	tyr-6	tyrosine	NTG
his-21	histidine or histidinol	UV	tyr-9	tyrosine	NTG
his-24	histidine	UV	tyr-11	tyrosine	NTG
hom-1	homocysteine	NTG	ura-4	uracil	NTG

Our data have been obtained either from selective analysis or from analysis of heteroclones. The heteroclones were either produced by the selective technique on a minimal medium or by the cellophane method[3]. The latter method allows the growth of small, well distinguishable, semi-spherical tufts of sporophors on the surface of the cellophane. The mutants employed in re-combination start growth over the surface of cellophane discs on the plates with reproductive media and they stop growing on transferring the cellophane discs to proper selective media. These allow the development only of the merozygotes formed.

The isolation of heteroclones was not always possible, and tufts failed to appear from mixed cultures between asporogenous strains. In cases when heteroclone selection on cellophane was not possible, it turned out to be feasible after plating spores from mixed cultures on minimal media. Heteroclones appeared as small colonies not transferable by replica plating[2] on the same medium from which they derive.

In many cases the heteroclones, although detectable, had a very poor offspring and their analysis had to be confined to a few dozen segregants. This inconvenience has been partially overcome by analysing large series of sister heteroclones. Selective analysis was also employed in crosses involving large numbers of markers.

STRAINS USED IN CROSSES

Streptomyces rimosus ATCC 10970 R7 has been mainly used in this work. It has irregular colonies with rich sporulation on the reproductive medium. In submerged culture it produces 50—200 μg per ml of oxytetracycline. With different mutagens it gives stable auxotroph mutants for a variety of amino acids, purines and pyrimidines as well as vitamins. A list of mutations obtained by ultraviolet irradiation and N-methyl-N′-nitro-N-nitrosoguanidine (NTG) is given in Table I. The original strain is naturally resistant to streptomycin, at doses effective against *S. coelicolor* or *S. aureofaciens*.

Streptomyces coelicolor A3(2), the strain largely used in genetic analysis by Hopwood and Sermonti[2], has been used in interspecific crosses. This is the only strain among *Streptomycetes*

TABLE II

Circular Arrangement of Markers Based on Segregation from Four-point Heteroclones of *S. rimosus*[a]

Cross A: 113 *arg-1 pro-1* × 202 *his-1 tyr-1*[b]

Relevant genotypes	Number	Number of crossovers in excess of two required according to the marker arrangement		
		Likely arrangement	Unlikely arrangements	
		a t h p	*p t h a*	*t p h a*
+ + + +	87	0 × 87 = 0	0 × 87 = 0	2 × 87 = 174
arg + *his* +	58	0 × 58 = 0	2 × 58 = 116	0 × 58 = 0
Total quadruple crossovers		0	116	174

[a] Data from different heteroclones are pooled.
[b] *a arg-1, p pro-1, h his-1, t tyr-1.*

in which a linkage map has been constructed. In this work the mutant 219 *metA2, pheA1, strA1* UR⁻ has been employed in interspecific recombination with various mutants of *S. rimosus* R7*.

MAPPING METHODS

The procedure which has been almost exclusively adopted in constructing the map of *S. coelicolor* was the analysis of four, five, or six-point crosses. From the segregation data obtained either from heteroclones (Table II and III) or from selective

TABLE III

Location of a New Marker (*rib-4*) by the Analysis of Heteroclone of *S. rimosus*

$$\text{Cross: } \dots \frac{nic\text{-}1 \quad + \quad + \quad pan\text{-}1}{+ \quad his\text{-}4 \ trp\text{-}3 \quad +} \ \dots \ \frac{(\textbf{rib-4})}{+}$$

Segregants	Number of segregants
nic + + pan **rib**	0
+ his trp + +	106
nic + + + **rib**	174
+ his + + +	88
nic + + + +	83
+ + + + **rib**	16
+ + + + +	5
+ + trp + +	0

N°	Possible location of *rib-4*[a]	Number of quadruple crossovers required
1	**rib** nic + + pan + + his trp +	37
2	nic **rib** + + pan + + his trp +	83
3	nic + **rib** + pan + his + trp +	88
4	nic + + **rib** pan + his trp + +	0

[a] The location N° 4 minimizing the number of crossovers is preferred.

Allele ratios: $\dfrac{257 \quad 278 \quad 366 \quad \textbf{190} \quad 0}{215 \quad 194 \quad 106 \quad \textbf{282} \quad 472}$

* The strain was kindly provided by Prof. Sermonti, Palermo.

analysis (Table IV), the arrangement of markers was deduced by checking all the possible permutations of the circular arrangement of the involved markers and by choosing that arrangement from which the segregants obtained were formed by the minimum number of crosses, possibly only two. In four-point crosses the three possible permutations were checked (*see* Table II), *i.e.* abcd, abdc, acbd. In five-point crosses the number of possible permutations is twelve. When four markers had already been mapped (as was usually the case), these four were assumed to be in

TABLE IV

Location of a New Marker (*pdx-9*) in *S. rimosus* by Selective Analysis[a]

$$\text{Cross: } \ldots \quad \frac{nic\text{-}1 \quad + \quad + \quad + \quad rib\text{-}4}{+ \quad pro\text{-}1 \quad met\text{-}95 \quad his\text{-}4 \quad +} \quad \ldots \quad \frac{(\textbf{pdx-9})}{+}$$

Selected recombinant genotypes	Number of recombinants
his pro + +	15
+ + nic ***pdx***	2
his + + +	20
+ pro + +	12
+ + nic +	10
+ + + ***pdx***	21
Total	80

N°	Possible location of *pdx-9*[b]	Number of quadruple crossovers required
1	+ + pro met his + ***pdx*** nic + + + rib	21
2	+ + pro met his + nic ***pdx*** + + + rib	10
3	+ pro + met his + nic + ***pdx*** + + rib	30
4	+ pro met + his + nic + + ***pdx***	rib	22
5	+ pro met his + + nic + + + ***pdx*** rib	0

[a] By detection on selective media with histidine, proline, nicotinic acid and pyridoxine.
[b] The location N° 5 minimizing the number of crossovers is preferred.

$$\text{Allele ratios: } \frac{78 \quad 21 \quad 0 \quad \textbf{57} \quad 80}{2 \quad 53 \quad 80 \quad \textbf{23} \quad 0}$$

a known order (and this could easily be checked), and the four possible locations of
the fifth marker (**e**) among them were tested (Table III):

e a b c d,   a e b c d,   a b e c d,   a b c e d

The location of the sixth marker among five already located involved the testing
of five possible combinations (Table IV).

TABLE V

Representation of a Heteroclone of *S. rimosus*

	nic-1	+	+	+	pdx-9	rib-4
Cross:	+	pro-1	met-95	his-4	+	+

	nic-1	pro-1	met-95	his-4	pdx-9	rib-4
Allele ratios:	61	143	141	91	75	75
	102	20	22	72	88	88

Segregants	Number
his pro met + + +	8
+ + + nic rib pdx	49
+ + + + + +	1
his + + + + +	51
his pro + + + +	2
his + met + + +	1
+ + + nic + +	12
his pro met + rib pdx	10
+ + + + rib pdx	16
Total	163

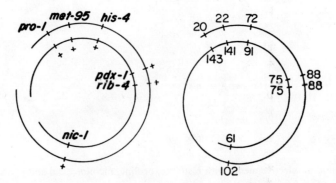

Fig. 1. Graphical representation of a heteroclone of *S. rimosus*. **Left:** assumed zygote composition.
Right: assumed heteroclone structure, with indication of allele ratio of genes in corresponding
positions.

This analytical procedure could be applied both to data obtained by segregation from heteroclones (Table II and III), and to data derived by selective analysis (Table IV). The latter situation was preferable because it was simpler, involving a reduced number of recombinant classes, and more informative since all recombinants derived from at least two crossovers in the two arcs of the map defined by the selective markers.

Once the arrangement of markers minimizing the number of crossovers was established, the gradient of allele frequencies along the map was observed. This was indicative of the structure of the zygotes in case of heteroclone analysis, and provided a control of the arrangement in case of selective analysis (Table V and Fig. 1). In the latter case the frequency of the alleles contributed by one parent, along the arc on which the new marker was located, regularly showed a continuous gradient going from zero to 100%. The latter type of analysis corresponds basically to that suggested by HOPWOOD[1], who first observed the allele frequency gradient, and then chose the arrangement of the new marker, along either arc of the map separated by the selected markers, minimizing the total number of crossovers.

THE MAP OF S. rimosus

The adopted procedures provided in some cases an unequivocal arrangement of the genes under study. In other cases the two best possible orders were quite indistinguishable or involved an almost equivalent number of quadruple crossovers. Some contradictory results were also occasionally met, which could usually be accounted for by an excess of prototrophs.

The circularity of the map was not exhaustively proved. In fact the adopted procedure should work as well also in the case of a linear arrangement of genes. Such an arrangement would be more committing, since it would involve the definition of the two ends of the map. These would correspond on the circular map to a point of high-frequency crossover or to a large "empty" region such as has not been observed. However, this is not a very valid argument, nor was the estimation of distances between pairs of adjacent markers accurate enough to warrant any safe conclusion. The circularity of the map was thus provisionally adopted by analogy with the situation recorded in S. coelicolor[1] and in other bacteria. The matter is still open to further investigation, involving the location of more markers.

The estimation of distances between a pair of markers was rather poor and often contradictory. Some pairs of markers appeared regularly linked, as for instance:

		distance
his-4	met-95	4,5
arg-1	met-95	11
trp-3	met-95	7
arg-1	rib-4	9
his-4	rib-4	13

Other pairs of markers turned out at different "distances" in different heteroclones, probably due to the large incompleteness of the zygotes. The distances resulting from the map representation (Fig. 2) are largely conjectural, and partially based on distances reported for similar pairs of markers in *S. coelicolor*[5]. The order of the markers, where unequivocally indicated, may be considered as reasonably certain. Some markers have been placed in a limited arc of the map, but their exact position is still

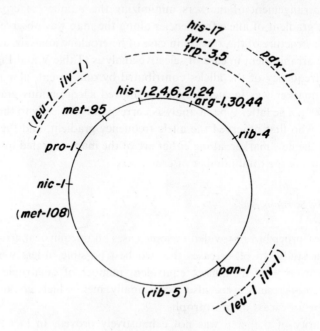

Fig. 2. Linkage map of *S. rimosus* R7 (explanation of symbols *see* Table I).

to be defined. This is also the case with the *tyr-1* marker, whose relative position with respect to *his-1* and *arg-1* appeared well established in a previous approach[8].

CLUSTERING OF GENES CONTROLLING THE BIOSYNTHESIS OF THE SAME METABOLITE

It is well-known that in bacteria the genes controlling various steps in the biosynthetic pathway of the same metabolite are often closely linked. In the example of the histidine cluster in *Salmonella typhimurium* this tendency is best evident. The way of clustering appears not to be constant in different groups of bacteria. In *S. coelicolor* according to PIPERNO *et al.*[6] a group of five genes (*hisB, I, A, G, C*) controlling the biosynthesis of histidine are in cluster, while three other genes are scattered on the map.

TABLE VI

Clustering of Histidine-Requiring Mutants of *S. rimosus*

Code of mutation	Accumulation products[a]	Growth on histidinol	Gene code[b]	Linkage
his-1	imidazolylglycerol-P	+	hisB	—
his-4	BBM	+	hisI, C or F	close to his-1
his-6	imidazolylglycerol-P	+	hisB	close to his-1
his-21	BBM	+	hisI, C or F	close to his-1
his-2	histidinol	—	hisA	close to his-1
his-24	histidinol	—	hisA	close to his-1
his-3	histidinol-P	+	hisD	allelic to his-16?
his-16	histidinol-P	+	hisD	distant from his-1
his-17	histidinol-P	+	hisD	distant from his-1

[a] Unpublished data kindly provided by S. Russi and S. Ferracin (Rome).
[b] As in *S. coelicolor*.

TABLE VII

Results of Crosses among Pairs of *his* Mutans of *S. rimosus*[a]

	his-3	his-16	his-17	his-1	his-2	his-4	his-6	his-21	his-24
his-3	0	—	—	(+)	(+)	+	+	+	(+)
his-16	—	0	—	+	+	+	+	+	(+)
his-17	—	—	0	+	(+)	+	+	+	+
his-1	(+)	+	+	0	—	—	—	—	—
his-2	(+)	+	+	—	0	—	—	—	—
his-4	+	+	+	—	—	0	—	—	—
his-6	+	+	+	—	—	—	0	—	—
his-21	+	+	+	—	—	—	—	0	(±)
his-24	(+)	(+)	+	—	—	—	—	—	0

[a] + recombinants formed, (+) only a few recombinants formed, — no recombinants, 0 not testable.

In *S. rimosus* nine mutants involved in histidine biosynthesis were examined (Tables VI, VII). Six mutations (*his-1, 2, 4, 6, 21, 24*) appear to be closely linked and remind one of the main cluster of *S. coelicolor*. Two mutations (*his-16, 17*) are distant from the main cluster as well as from each other.

COMPARISON OF THE MAP OF *S. rimosus* WITH THAT OF *S. coelicolor*

The arrangement of markers in *S. rimosus*, obtained by the methods discussed above, allowed a comparison with the order of similar markers in the map of *S. coelicolor* A3(2). It must be said, however, that in most cases the mutations were only defined according to the required final metabolite, and a corresponding location on the

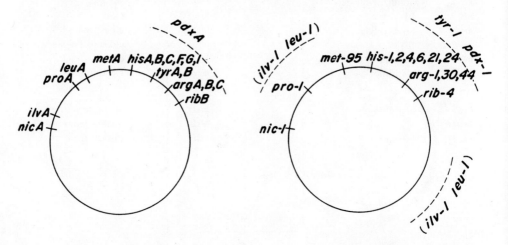

Fig. 3. Comparison of the maps of *S. coelicolor* (**left**) and *S. rimosus* (**right**).

two maps does not obviously mean that one is dealing with a gene coding in the two species for the same enzyme. Due to the tendency of genes involved in the same pathway (*i.e.* with the same requirement) to be either in a cluster or specularly located on the map[1], this was not such a serious bias as it would be, for instance, in comparing the map of two fungi, having mimic genes scattered on different chromosomes.

The correspondence of the order of the same markers in the map of the two species is illustrated in Fig. 3. This is quite good and possible disagreements could so far be accounted for by the fact that the same mimic genes may be located in more than one position of the map. Our analysis is not accurate enough as to warrant the conclusion that any gene is in a different location in *S. rimosus* with respect to the corresponding location in *S. coelicolor*. This matter which it is worth while to explore, could be investigated by more, extensive mapping in *S. rimosus* and a more complete phenotypic determination of the various genes, and by the analysis of interspecific crosses which should indicate possible regions of incomplete affinity.

The most striking result of this map comparison is that the genes of *S. rimosus* appear to correspond in their order to those located in a single arc of the map of *S. coelicolor*. The occurrence of similar genes specularly located in the opposite arc[1] was not noticed. It might well be that further marker location will reveal the occurrence of this series of specularly arranged genes, but on the basis of the number of located markers this seems rather unlikely. The hypothesis could then be put forward that the map of *S. rimosus* corresponds to a single arc, *i.e.* to one half of the map of *S. coelicolor*, the latter being (according to a suggestion by HOPWOOD[1]) a "duplicated" map. The situation, which would be of extreme interest on the basis of HOPWOOD's evolutionary hypothesis, is still in need of much more accurate testing, particularly by using interspecific crosses.

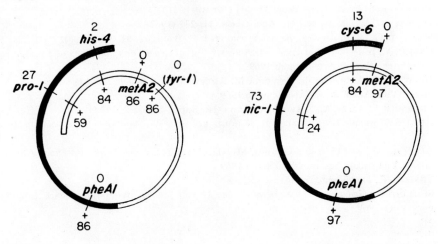

Fig. 4. Possible structure of two heteroclones from crosses of *S. rimosus* and *S. coelicolor*.

Left: cross

S. rimosus R[+]		+	pro-1 his-4	+	tyr-1
S. coelicolor UR[−]	pheA1	+	+	metA2	+

Right: cross

S. rimosus R[+]		+	nic-1 cys-6	+
S. coelicolor UR[−]	pheA1	+	+	metA2

The full line represents the *S. rimosus* (R[+]) contribution, the empty line the *S. coelicolor* (UR[−]) contribution. Notice the gap in the recipient genome at 9 o'clock position.

In crosses between an UR[−] (ultra-recipient) mutant of *S. coelicolor* 219 and a mutant of *S. rimosus* a relatively high frequency of recombinants was obtained, as well as many heteroclones (or heterokaryons) which turned out to be relatively stable. From the analysis of some of the latter, it turned out that the region contributed by the donor (*S. rimosus*) parent was limited, and regularly bridged the 9 o'clock (*pheA1*) position of the UR[−] (=UF) genome, as noticed by HOPWOOD in NF × UF crosses in *S. coelicolor* A3(2) (Fig. 4).

I wish to express my thanks and appreciation to G. Sermonti for his advice and sustained interest during this work, to S. Russi and S. Ferracin who provided their unpublished data and to Margareta Vešligaj for her assistance in the work reported in this paper.

This work was supported by "Pliva" Pharmaceutical and Chemical Works, Zagreb.

REFERENCES

1. HOPWOOD D. A.: Bacteriol. Rev. **31**, 373 (1967).
2. HOPWOOD D. A., SERMONTI G.: Adv. Genet. **11**, 273 (1962).
3. SERMONTI G., BANDIERA M., SPADA-SERMONTI I.: J. Bacteriol. **91**, 384 (1966).
4. CARERE A., p. 19 in G. Sermonti, M. Alačević (Eds.): *Genetics and Breeding of Streptomyces.* Yug. Acad. Sci. & Arts, Zagreb 1969.
5. HOPWOOD D. A., p. 5 in G. Sermonti, M. Alačević (Eds.): *Genetics and Breeding of Streptomyces.* Yug. Acad. Sci. & Arts, Zagreb 1969.
6. PIPERNO R., CARERE A., SERMONTI G.: Ann. Ist. Super. Sanità **2**, 393 (1966).
7. RUSSI S., CARERE A., FRATELLO B., KHOUDOKORMOFF V.: Ann. Ist. Super. Sanità **2**, 506 (1966).
8. ALAČEVIĆ M., p. 137 in G. Sermonti, M. Alačević (Eds.): *Genetics and Breeding of Streptomyces.* Yug. Acad. Sci. & Arts, Zagreb 1969.
9. EHRLICH T., *unpublished results.*
10. SERMONTI G.: *Genetics of Antibiotic-producing Microorganisms.* Wiley-Interscience, London— New York 1969.

3/ Genetics of Fungi

Breeding Systems and Recombination in Fungi*

KARL ESSER

Institut für Allgemeine Botanik, Ruhr-Universität, Bochum, G.F.R.

In general, genetics should not be restricted to a single organism, like *Drosophila, Neurospora*, maize, *etc.*, and not even to a group of organisms, like fungi, bacteria or viruses. This may be an objection to any kind of compilation on fungal genetics, be it a review, a book or a symposium. However, there are reasons which may invalidate this argument: Due to the numerous common properties by which any class of organisms used as a tool in genetics is characterized, it seems understandable that the particular group of organisms is subjected to a specialized genetic study which respects all the pecularities of these objects and which can use all the present specific theoretical knowledge and practical know-how. But apart from this there are many problems which can be studied solely with a certain class of living beings or for which the experimental use of a certain group is particularly to be recommended. With the fungi, problems of this kind are among others:

1. Mechanisms of recombination, due to the possibility of tetrad analysis.
2. Somatic recombination, due to the phenomenon of heterokaryosis unique in the fungi.
3. The whole complex of morphogenesis, since the fungi with their simply constructed organs are more accessible to both genetic and biochemical analysis than higher organisms.
4. Incompatibility and other control systems of sexual and vegetative propagation because they determine ultimately the mode of recombination of the genetic material and thereby, as breeding systems, the evolution of the group.
5. Extrachromosomal inheritance, the study of which has shown the existence of autonomous genetic information outside the nucleus.
6. Last, but not least, gene—enzyme relations are necessary for studying the function of the genetic material *per se* and in relation to the results obtained within the last few years with prokaryotic microbes.

The last notion leads to a general problem of significance for fungal genetics: This group of organisms is the most suitable for checking if some of the theories,

* With support of the Deutsche Forschungsgemeinschaft, Bad Godesberg, G.F.R.

models and hypotheses found from the study of microbes may be used to explain the functions of life within the chromosome-containing organisms.

The main aim of this paper is to introduce the topic of fungal genetics and to summarize information about the general significance of breeding systems with respect to their control of recombination and about the necessity of considering these factors in fungi for industrial purposes.

Since the days of Mendel it has been realized that recombination is one of the basic attributes of the genetic material. The formal genetics, done in the first half of this century mostly with eukaryotic organisms, has certainly achieved a detailed description of this phenomenon, but as regards comprehending the mechanism of recombination it has left us at a dead end. This gap has been partially filled within the last two decades by ideas originating from research performed with prokaryotic microbes. In spite of this success of molecular genetics our understanding of recombination is far from being complete. This lack of knowledge concerns not only the mechanism of the recombination process but also its control by external and internal genetic and non-genetic factors. Furthermore, due to the rather complicated structure of the chromosomes it has still to be proved to what extent the models of recombination arising from study of bacteria and bacteriophages can be extrapolated to eukaryotic organisms.

As briefly mentioned above, fungi are the most suitable eukaryotic organisms for research of this kind. It is trivial to mention that this has been realized by numerous scientists because, parallel to research on microbes, recombination in fungi has been studied within the last few years from numerous aspects.

The understanding of recombination is a prerequisite for successful handling of microorganisms in industrial microbiology. The significance of breeding systems for recombination and also its theoretical value for understanding evolution and its practical value for the use of fungi in industry is generally acknowledged.

Therefore we felt it worthwhile to summarize the actions and the interactions of the different breeding systems in a general model which might help, on the one hand, to overcome the confusion in terminology which exists in the literature and, on the other hand, to aid the mycologist in recognizing the breeding pecularities of these fungi.*

"The term breeding system is used to cover all those variables apart from mutation which affect the genetic relations of the gametes that fuse in reproduction. The variable components of the breeding systems are numerous and not easy to estimate." This definition given in the *"Glossary of Genetics"*[2] is, on the one hand, rather imprecise, due to the great diversity of genetic and physiological determinants of the breeding systems in different plants and animals. On the other hand, especially where the fungi are concerned, it is too narrow since it considers only sexual processes as basic events for breeding systems. Any definition of breeding systems valid for fungi must take

---

* Since the content of the following part of this lecture will be published elsewhere in detail, we have omitted quotations of literature and refer to the publication[1].

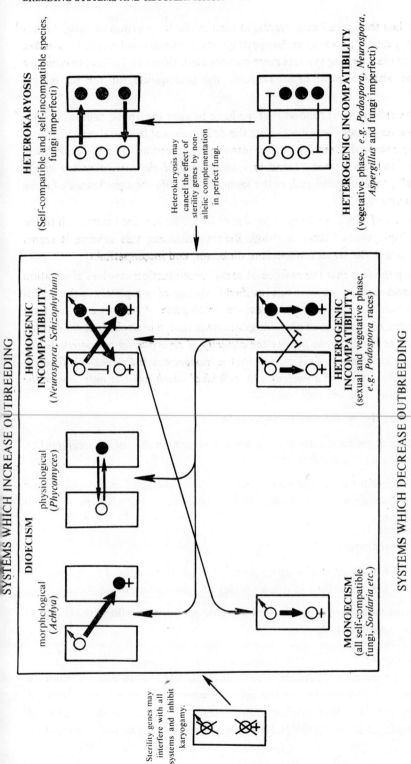

SYSTEMS WHICH INCREASE OUTBREEDING

HETEROKARYOSIS
(Self-compatible and self-incompatible species, fungi imperfecti)

Heterokaryosis may cancel the effect of sterility genes by non-allelic complementation in perfect fungi.

HOMOGENIC INCOMPATIBILITY
(*Neurospora, Schizophyllum*)

DIOECISM

morphological
(*Achlya*)

physiological
(*Phycomyces*)

Sterility genes may interfere with all systems and inhibit karyogamy.

HETEROGENIC INCOMPATIBILITY
(vegetative phase, *e.g. Podospora, Neurospora, Aspergillus* and fungi imperfecti)

HETEROGENIC INCOMPATIBILITY
(sexual and vegetative phase, *e.g. Podospora* races)

MONOECISM
(all self-compatible fungi, *Sordaria etc.*)

SYSTEMS WHICH DECREASE OUTBREEDING

Fig. 1. Action and interaction of breeding system in fungi. The *large square* in the center displays the main systems. Heterokaryosis occupies the *right* side of the figure. Representative organisms for each system are given in *parentheses*. The *rectangles* represent single individuals (except for heterogenic incompatibility in the central part, where the rectangles represent single *Podospora* races). The *male and female symbols* represent nuclei of different sex. Differences of the genetic equipment of nuclei are characterized by *white* and *black*. In the case of physiological dioecism and heterokaryosis, where sexual differentiation of the nuclei cannot be proved, they are symbolized by *white and black circles*. The *thick arrows* indicate the direction of karyogamy and heterokaryotization, respectively. The *blocked arrows* indicate that karyogamy or heterokaryotization is impossible. Interactions of the different systems are described by *thin arrows*. (From Ref.[1])

account of the fact that parasexual events, at least in the large group of fungi imperfecti, are the only means by which exchange of genetic material is achieved. This makes the discussion of the breeding systems more complicated, since one has also to consider genetic systems which control heterokaryosis, the prerequisite for the parasexual cycle.

From these reflections it follows that we have to alter the above definition. This can be done in a very simple way if we relate the definition, not to sexual reproduction, but only to the essential event which is controlled by breeding systems, and this is recombination. In stating that *"breeding systems control the recombination of the genetic material"*, we generalize and, at the same time, clarify the significance of these biological phenomena.

In the scheme of Fig. 1 a survey is presented of the action and interaction of the different breeding systems found in fungi. Before discussing this scheme it seems appropriate to define the terms monoecism, dioecism, and incompatibility.

Considering the fact that the essence of sexual reproduction involves alternation of karyogamy and meiosis, we have felt justified to disregard completely in the definition of monoecism and dioecism the presence or absence of sex organs and have defined both terms on the basis of physiological rather than morphological characters. We have chosen as the *criterion for sex the capacity of an organism to contribute one or both nuclei to karyogamy*. This means that a *monoecious individual can function both as donor and recipient of a nucleus. An individual which exhibits only one or the other potential is called dioecious.*

It follows that the *monoecious fungi* include:

1. All species which *produce male and female sex organs on the same mycelium (i.e.* most *Euascales* and *Uredinales*).

2. All species *without sex organs* in which each mycelium can serve as a *nuclear donor as well as a nuclear acceptor* (mostly *Holobasidiomycetes* but also species of all other fungal taxa).

   Dioecious fungi comprise:

1. All species in which *mycelia bear either female or male sex organs.* This *morphological dioecism* is only described in a few species of *Phycomycetes* and *Ascomycetes.*

2. All *species with two mating types,* in which zygote production follows the fusion of *isogametes or isogametangia.*

Many *Protoascomycetes* and *Mucoraceae* belong to this category. The two mating types of such species (*e.g. Saccharomyces cerevisiae, Mucor mucedo*) cannot be distinguished morphologically and in the sexual process show no nuclear exchange, but simply a fusion of uninucleate or multinucleate cells. We are thus unable to designate them either as morphologically dioecious forms or as self-incompatible monoecious types. Since their sexual behaviour is determined purely through the physiological difference of the two mating types, we call these *physiologically dioecious*.

Incompatibility may be defined as a *genetically determined prevention of karyogamy* which is not caused by sterility defects of the nuclei to be mated. There are two systems for which we have, according to their genetic basis, introduced the terms homogenic and heterogenic incompatibility.

In the *homogenic system, karyogamy is prevented when nuclei carry identical incompatibility factors.* To this system belong the different mechanisms known as bipolar or tetrapolar incompatibility, and furthermore the various mechanisms controlling pollen-tube growth in higher plants often quoted as self-sterility.

The genetic conditions for *heterogenic incompatibility* are exactly opposite: *karyogamy is prevented when nuclei carry non-identical incompatibility factors.* This system, which was first described and analysed in the ascomycete *Podospora anserina*, is in operation only in interracial crosses of strains of different origin and, as a recent review of the literature has revealed, is more common than was at first thought. Heterogenic incompatibility consists of an inability of the incompatible genes to coexist in a common nucleus, or in a common cytoplasm when the genes are located in different nuclei of a heterokaryon. From this it follows that *heterogenic incompatibility controls not only meiotic recombination but also mitotic recombination.* Therefore the examples of heterokaryon or cytoplasmic incompatibility described in some fungi have to be incorporated in this system although they concern only the vegetative phase.

By the *action of the diverse breeding systems* given in Fig. 1, the sexual and parasexual cycles, and therewith the possibilities for recombination, are controlled in two different ways:

1. *Monoecism* allowing self-fertilization, enhances inbreeding and restricts the chances for exchange of genetic material. This is displayed in the most drastic way in haploids, like fungi, where the possibility of recombination after self-fertilization is equal to zero, while in diploids, with their possibility of heterozygosity, a slight chance for recombination is still left. *Heterogenic incompatibility* acts in a similar way between different races of one species, as may be seen from Fig. 2. The diminution of the chances for recombination concerns not only meiotic but also mitotic recombination. The latter is most effective within the *Deuteromycetes*.

Heterogenic incompatibility leads to an isolation of the races and creates a basis for further speciation, since the single race instead of the species as a whole is now the largest unit of breeding. Therefore one has to consider heterogenic incompatibility as a genetic expedient for speciation in the course of evolution.

2. *Dioecism*, independent of its morphological or physiological mechanism, cuts off self-fertilization and enhances the chances for recombination and outbreeding. The same goal is obtained in monoecious species by *homogenic incompatibility*. All the mechanisms of this system in fungi (and also in flowering plants) have one

characteristic in common in spite of numerous genetic and physiological differences, namely, the incompatibility of genetically identical nuclei. Due to the fact that dioecism and homogenic incompatibility limit the inbreeding, wild strains which are controlled by one of these systems are to a large extent heterogenic or, in the case of diplonts, heterozygous.

In general, the action of *heterokaryosis* is not as evident as that of the sexual systems quoted. However, it must not be overlooked that in nature the mixing of nuclei in the vegetative hyphae makes a fairly good contribution to the possibilities of *outbreeding*. Genetically different nuclei brought together by hyphal fusion may easily get admitted to the germ line. The importance of heterokaryosis becomes most effective in the *Deuteromycetes* as the basis for recombination *per se*.

In concluding these considerations, we have to keep in mind that the action of each of the different breeding systems may be partially or totally cancelled by *sterility genes*, causing deficiencies in male or female sex organs or in both. Genes of this kind are often found in natural isolates or occur in laboratory strains after incessant vegetative propagation. Such defects may be removed after heterokaryosis *via* non-allelic complementation, leading to fertile heterokaryons.

It should not be overlooked that the control of recombination in a great number of fungi (and also in higher organisms) is not achieved solely by a single breeding system, but by the *interaction of* various *systems*.

The most common interaction is the *nullification of monoecism by homogenic incompatibility*. This attributes to a species consisting of hermaphrodite individuals the same outbreeding value as is present in a dioecious species. *Heterogenic incompatibility* may *interact with all the systems favouring outbreeding* in such a way that the opportunities for recombination are restricted to particular races. This is attained for mitotic recombination in the *Deuteromycetes* by preventing the formation of heterokaryons. The fact that the interactions of heterogenic incompatibility have not been acknowledged as much as those of the homogenic system is understandable, because the heterogenic system requires a substantial genetic study of the mating competence of a number of different races, in addition to knowledge of the mating behaviour of the laboratory strains in use.

In conclusion one may say that *breeding systems, acting through genetic determinants for the realization of the sexual or parasexual cycle, control the possibilities for recombination of the genetic material achieved in meiosis or mitosis. Whereas dioecism and homogenic incompatibility favour the chances for recombination, this effect is counteracted by monoecism and heterogenic incompatibility.* The consequences for outbreeding and inbreeding, respectively, and therewith for evolution are evident and require no further comment.

We are aware that the application of this model to higher organisms or to prokaryotes could lead to some complications. However, it may serve as a basis not only for understanding the general significance and attributes of sexuality, breeding systems and recombination, but also as a prerequisite for understanding the behaviour and versatility of fungi in theoretical and applied research.

REFERENCES

1. ESSER K.: Mol. Gen. Genet., **110**, 86 (1971).
2. RIEGER R., MICHAELIS A., GREEN M. M., p. 38 in *A Glossary of Genetics and Cytogenetics.* Springer-Verlag Berlin—Heidelberg 1968.

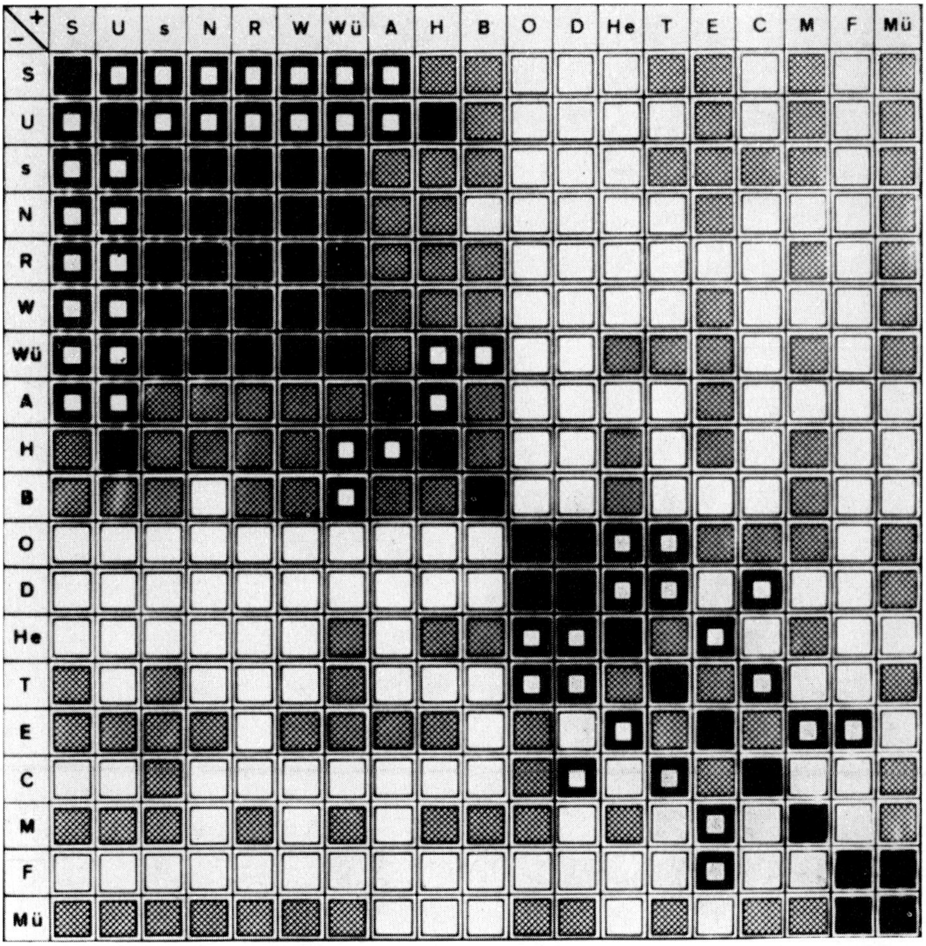

Fig. 2. Scheme for the mating reactions between different races of the ascomycete *Podospora anserina*. The races isolated from different localities in France and Germany are represented by letters.

Black squares = compatibility in both sexual and vegetative phase;

black squares with inserted white square = sexual compatibility, heterokaryon incompatibility;

hatched squares = partial incompatibility (reduced fruit body formation) in sexual phase, heterokaryon incompatibility;

white squares = incompatibility in both sexual and vegetative phase. (From Ref.[1])

Mitotic Recombination and Mitotic Nonconformity in Fungi

J.A. ROPER

Department of Genetics, The University, Sheffield, Great Britain

INTRODUCTION

In the last thirty or so years fungal genetics has vastly extended the under-standing of, among other fundamental phenomena, gene action, genetic fine structure, recombination and extrachromosomal heredity. In addition, the use of genetical ap-proaches has revealed genetic processes of fungi which would have escaped detection by methods less sensitive than those permitted by nutritional mutants and selective techniques. Despite this, fungal genetics has so far contributed relatively little to programmes of industrial strain improvement. The use of mutagens to obtain im-proved strains and of fermentation modification to realise a strain's potential have an obvious basis in genetics but, of necessity, the practical approach to strain im-provement has remained largely empirical.

It may be useful to try to specify the knowledge and biological tools needed for a fundamental and rational approach to strain improvement, though this is to specify an unrealistic ideal which cannot be achieved yet in any organism. There are three, partly interconnected, requirements. First, a knowledge of the relevant biosynthetic pathway and of its regulation. This would probably be almost impossible to determine without the second requirement, genetic recombination. This is needed to analyse the genetic elements which determine the biosynthetic steps and is necessary also for the synthesis of strains likely to give maximum yield. Finally, there is need to understand the factors responsible for adequate strain stability; this could be especially important in certain types of strain bred to give maximum yield. This account is concerned with two of these basic requirements, genetic recombination and aspects of strain stability.

MITOTIC RECOMBINATION

The discovery in *Aspergillus nidulans* of the parasexual cycle, in which re-combination of linked and unlinked genes is achieved during vegetative growth, offered hope of recombination in imperfect filamentous fungi. The individual steps in this "alternative to sex"[1] were elucidated largely because *A. nidulans* has a standard sexual cycle and the consequences of mitotic recombination could be compared and contrasted with those of meiosis. The parasexual cycle has been reviewed in detail[2-4] and it will suffice here to outline its steps briefly. These are: 1) heterokaryosis, 2) diploid formation, 3) mitotic crossing-over, 4) haploidisation.

In the vegetative hyphae of *A. nidulans* there are rare but regular fusions of pairs of haploid nuclei. The resulting diploid nuclei are fairly stable and a small proportion of conidia thus carry diploid nuclei. Diploid selection is possible through heterokaryosis in which the component haploid nuclei differ from wild type and from each other in nutritional requirements and, preferably, in conidial colour. Thus some fusions, between unlike nuclei, produce heterozygous diploids. The conidia of *A. nidulans* are uninucleate and the vast majority produced by a heterokaryon are of parental haploid types. Most mutant alleles are recessive and the rare diploid conidia, each with a heterozygous nucleus, are selected easily by plating on media lacking the nutrients required by the parent haploids. In general, diploid strains lend themselves better to certain studies of gene action than do the corresponding heterokaryons. Tests of diploids are likely to be of particular importance in the study of regulation mechanisms. At some future time, strains with diploid nuclei may prove industrially useful as it would be possible to exercise some control over the balance and dosage of genes responsible for biosynthesis and regulation. It would then be necessary to overcome the vegetative instability of diploids and this might be accomplished partially through the use of balanced lethal systems[5] which effectively prevent breakdown to haploids.

Diploid strains of *A. nidulans* undergo recombination by two rare but regular processes occurring during vegetative growth, mitotic crossing-over and haploidisation. Initially recombinants were recognised as mutant patches (usually mutant conidial colour) in the parental, phenotypically wild-type colony. Now, by a variety of techniques, recombinants can often be selected. Mitotic crossing-over is a rather rare event which occurs between nonsister chromatids at the 4-strand stage of mitosis. At the gross level, and sometimes at a finer, intragenic level, the exchanges are reciprocal. In any one nucleus, crossing-over is almost invariably confined to a single exchange in one arm out of the whole chromosome complement. The subsequent behaviour of centromeres is mitotic and the chromatids of each homologue go to opposite poles. If the chromatid segregation is appropriate the resulting nuclei are homozygous for certain genes for which their parent was heterozygous. Specifically, the nuclei become homozygous for all alleles linked in coupling and distal to the exchange. Markers proximal to the exchange and those on other chromosome arms remain heterozygous. Mitotic crossing-over may be used to order genes on a chromosome arm relative to each other and to their centromere, but linkage cannot be shown between markers on different arms of the same linkage group. Mitotic haploidisation is independent of mitotic crossing-over. Haploidy is usually reached through the successive loss of chromosomes[6]; the intermediate aneuploids, which are very unstable, generally escape detection. Haploidisation is like meiosis in the male *Drosophila*; members of each homologous chromosome pair assort independently without crossing-over so that recombinants are obtained between, but not within, linkage groups. Haploidisation provides a means for the assignment of genes to linkage groups[7] and has been used successfully in *Penicillium chrysogenum* for an initial construction of linkage groups[8]. Haploidisation also offers an effective tool for the substitution, in a strain, of one or more particular linkage groups; as an analytic method, or as a means for strain synthesis, this is sometimes more useful than the "wholesale" recombination which occurs in meiosis. In *A. nidulans*, then, it is possible to undertake detailed genetic analysis *via* the parasexual cycle. The ease of this is such that it is used more extensively than the standard sexual cycle for certain types of analysis.

Parts or all of the parasexual cycle have now been demonstrated in many filamentous fungi, perfect and imperfect. The basic pattern is generally similar to that in *A. nidulans* though there are differences in the frequencies of the various steps and there are occasional technical problems which make analysis difficult.

MITOTIC NONCONFORMITY

Strain instability is of theoretical interest and may even have practical use in a strain improvement laboratory; it is of obvious potential disadvantage in an industrial fermentation. There are several known causes of vegetative instability in filamentous fungi. Heterokaryons are unstable through the reassorting of different nuclear types in the hyphae and spores; aneuploids and polyploids produce variants by mitotic crossing-over and whole chromosome changes. Variants from these systems can show only the genetic variation which existed, perhaps phenotypically unexpressed, in the parent colony. A few years ago a further cause of instability was discovered; in some of its consequences it differs substantially from the above systems. This instability, which might be encountered during strain improvement, is due to imbalance of chromosome segments. It was called "mitotic nonconformity" to convey that parent and daughter nuclei may not conform in genotype[9].

In a study of genetic factors affecting morphology in *A. nidulans*, certain crosses of morphologically normal parents gave progeny of which one third were morphologically abnormal[10]. Pedigree examination showed that in each such cross one parent had an apparently nonreciprocal translocation of a segment of linkage group III to linkage group VIII (Fig. 1). Meiotic segregation, confirmed by ascus analysis, was: two morphologically normal parental types, one with the III—VIII translocation; one inviable product lacking the translocated III segment; one abnormal type with a duplication of the III segment. As might have been expected from their chromosomal imbalance, the duplication progeny had reduced linear growth rate and a modified surface morphology, but they showed one unexpected property, vegetative instability.

Colonies of duplication strains gave frequent sectors which, in varying degree, approached wild type in growth rate and morphology. Most first-order sectors were not fully wild-type; on further growth they gave sectors showing further improvement until, ultimately, types with wild-type morphology and growth rate were obtained. Sectors were tested from duplication progeny likely to be heterozygous for a gene determining thiosulphate requirement; some individual colonies gave both thiosul-

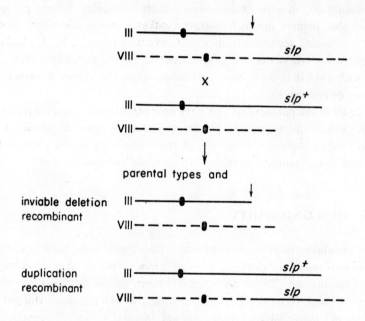

Fig. 1. Results of the cross of a chromosomally-normal strain to one with a segment of linkage group III (*solid line*) translocated and inserted in linkage group VIII (*broken line*). s*lp*, a recessive allele determining thiosulphate requirement.

phate-requiring and thiosulphate-independent sectors. Genetic analysis of the sectors confirmed what was suspected from their phenotypes; the sectors arose by loss of a variable part of *either* duplicate segment and loss of the dominant s*lp*[+] allele gave a thiosulphate-requiring variant from a thiosulphate-independent parent. Different duplication strains have now been obtained by various means; they could arise in haploidisation analysis of imperfect species when one parent has a translocation, though, because of their reduced growth rate, they would usually be missed. In all duplication strains of *A. nidulans* so far tested the same pattern of instability has been observed and it seems likely that the pattern is shown by all such strains.

The most thoroughly studied duplication strain (Fig. 2) has shown a number of features of mitotic nonconformity[11]. This strain gives phenotypically improved green and yellow sectors. The latter have deletions which include the dominant *ylo*[+] locus and occasionally these deletions extend into linkage group II to give hypo-

haploids[12]. The green sectors have deletions elsewhere within either duplicate segment. No sector has so far shown exchange of markers between the homologous
segments; the deletion process is intrachromosomal and not due to, for example,
unequal mitotic crossing-over. Deletions are sometimes, perhaps always, interstitial
as exemplified by types which have lost a segment carrying ylo^+ and ade while retaining
bio. Reduction to a quantitatively haploid state is sometimes one-step but it frequently

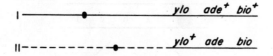

Fig. 2. A strain carrying a duplicate segment of linkage group I (*solid line*) terminally attached to
linkage group II (*broken line*). Mutant alleles determined: *ylo*, yellow (as opposed to wild type
green) conidia, *ade* and *bio*, requirement, respectively, for adenine and biotin.

involves two or more successive, smaller deletions. Deletions from linkage group I
can be mapped as recessive lethals.

The above findings did not distinguish between two possibilities. On the one
hand, deletions might occur throughout any genome, standard or with a duplication;
those from nonduplicated regions would give inviable hypohaploids while nuclei
with deletions from either duplicate segment of a duplication strain would survive
and, in fact, have selective advantage. On the other hand, it was possible that chromosome imbalance provoked deletions. This latter possibility was shown to hold by an
examination of diploid strains with and without a duplicate segment[9]. Standard
diploids gave only rare sectors arising from mitotic crossing-over or haploidisation.
Duplication diploids were even more unstable than duplication haploids; their sectors
arose from deletions which, in this case, were probably all from the translocated,
duplicate segment. Imbalance of chromosome segments provoked instability and
it seemed probable that the primary events were confined largely to the unbalanced
segments. A tentative explanation of this was advanced in terms of attachment sites,
limited in number and specific for chromosome segments, at which replication is
initiated. Specific initiation substances could equally well be substituted for attachment sites. In unbalanced, duplication strains competition for sites, or for limiting
initiation substances, might lead to replication errors.

In formal genetic terms deletions can be explained either by unequal sister
chromatid exchange or by crossing-over within an intrachromosomal loop. Either
process might be expected to yield new tandem duplications as well as deletions
(Fig. 3) and these have probably been detected. In addition to improved sectors,
duplication strains give, though less frequently, variant sectors with deteriorated
morphology. A substantial proportion of these are far more unstable than the
duplication parent and, in a number of them, the locus of increased instability and

morphological deterioration has been located on one or other duplicate segment[11,13]. There is evidence to suggest that part or all of such new, tandem duplications may transpose elsewhere to the nonduplicated parts of the genome. This generally gives reduced instability but the chromosomally-located changes, whatever their nature, continue to provoke yet further mutations which arise in all linkage groups and affect morphology and stability[13]. Some lineages have now been analysed to

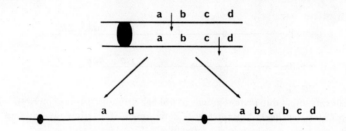

Fig. 3. Unequal sister chromatid exchange resulting from breaks (*arrows*) at different points and giving a deletion and duplication. Crossing-over within an intrachromosomal loop yields similar results and could give, in addition, circular fragments.

fourth-order variants, still unstable, which have accumulated many mutations. In these sequences of events mitotic nonconformity involves novel genetic changes and does not merely expose the mutations present in the parent.

The refined genetic details of mitotic nonconformity remain to be determined and its molecular basis is entirely unexplored. But even at this stage the phenomenon has particular points of interest. There are striking similarities between features of this instability and other cases of genetic instability in microorganisms and higher organisms. In these latter instances duplications are sometimes either known to exist or can be suspected from the origin of the unstable lines. The study of mitotic nonconformity should at least reveal aspects of the control of mitotic fidelity and might eventually also show some underlying unity of mechanism in cases of instability for which diverse explanations are now proposed. The practical significance of the phenomenon may not be negligible. Already a high-titre strain of *P. chrysogenum* has been shown to have instability of a type best interpreted as mitotic nonconformity[8]. Future programmes of strain improvement could indicate the desirability of duplicating, in an otherwise haploid genome, segments of chromosomes carrying, for example, regulatory genes or genes determining a rate-limiting biosynthetic step. It would be technically feasible to synthesise such strains, even in imperfect species, when the basic genetics of linkage group construction is sufficiently advanced. However, it would then be necessary to find means of stabilizing duplication strains since the duplication of even a small fraction of the genome seems likely to lead to instability[14,15].

DISCUSSION AND CONCLUSIONS

When the parasexual cycle was discovered, it was plausible to argue its probable immediate value as a tool in the improvement of industrial, imperfect species of fungi. That view was naive for several reasons. Although vegetative recombination has been shown in a number of asexual species, there have been technical barriers to its effective use and biological barriers such as probable chromosome aberrations in highly-mutated industrial strains. Mutation and fermentation modifications have been too successful to provide the spur to surmount such difficulties. Although useful advances have been made through recombination[8,16-18], they have been insufficient to divert major efforts along these lines. The earlier naivete was based also on a simpler view than that now held of the genetic determination and regulation of biosyntheses. High titres and other desirable characteristics will usually arise from a delicate balance of many genetic elements with fewer, simple, additive effects than might have been supposed originally.

Despite this, recombination still has actual and potential value. In a combined approach with biochemistry it is essential for analysis of the genetic determination of a biosynthesis and its regulation; when this stage is accomplished an approach can be made to genetical engineering of strains. Mitotic nonconformity may become a relevant factor at any stage in strain development. It is unlikely, but not impossible, that the mutations generated by mitotic nonconformity could be exploited usefully. But an understanding of the details of the phenomenon, and eventually of its control, may be essential in future work.

REFERENCES

1. HALDANE J. B. S.: New Biol. **19**, 7 (1955).
2. PONTECORVO G.: Caryologia (Suppl.) **6**, 192 (1954).
3. PONTECORVO G., KÄFER E.: Adv. Genet. **9**, 71 (1958).
4. ROPER J. A., p. 589 in G. C. Ainsworth, A. S. Sussman (Eds.): *The Fungi*, Vol. II Academic Press 1966.
5. AZEVEDO J. L., ROPER J. A.: J. Gen. Microbiol. **49**, 149 (1967).
6. KÄFER E.: Genetics **46**, 1581 (1961).
7. FORBES E.: Heredity **13**, 67 (1959).
8. BALL C.: Paper at 1st Internat. Symp. Genetics of Industrial Microorganisms, Prague 1970. Abstract Book, p. 178; BALL C., *personal communication*.
9. NGA B. H., ROPER J. A.: Genet. Res. **14**, 63 (1969).
10. BAINBRIDGE B. W., ROPER J. A.: J. Gen. Microbiol. **42**, 417 (1966).
11. NGA B. H., ROPER J. A.: Genetics **58**, 193 (1968).
12. ROPER J. A., NGA B. H.: Genet. Res. **14**, 127 (1969).
13. AZEVEDO J. L., ROPER J. A.: Genet. Res. **16**, 79 (1970).
14. MORSE M. L.: Genetics **56**, 331 (1967).

15. HILL C. W., FOULDS J., SOLL L., BERG P.: J. Mol. Biol. 39, 563 (1969).
16. SERMONTI G.: Sci. Reports Ist. Super. Sanità 1, 449 (1961).
17. MACDONALD K. D., HUTCHINSON J. M., GILLETT W. A.: J. Gen. Microbiol. 33, 385 (1963).
18. MACDONALD K. D., HUTCHINSON J. M., GILLETT W. A.: Antonie van Leeuwenhoek J. Microbiol. Serol. 30, 209 (1964).

Repair Models of Recombination

H.L.K. WHITEHOUSE

Botany School, University of Cambridge, Cambridge, Great Britain

In discussing repair models of recombination I propose to consider two questions: (1) How far recombination may be involved in the repair of DNA damaged by UV-light, X-rays or other means. (2) How far the normal mechanism of recombination, whether meiotic or mitotic, involves repair processes.

Several pieces of evidence suggest that recombination may be involved in the repair of damaged DNA in fungi in much the same way as is thought to occur in bacteria. With *Escherichia coli*, HOWARD-FLANDERS and associates[1] found that excision-defective mutants (*uvrA*) can survive considerable doses of UV unless they are also recombination-deficient (*recA*). It was concluded that in normal *rec*+ strains recombination allows survival in cells containing unexcised dimers. HOWARD-FLANDERS *et al.*[1,2] obtained evidence from sedimentation and other studies that when DNA containing dimers replicates a gap is formed opposite each dimer in the newly-synthesized chain. Such a gap would preclude repair by excision of the dimer. It was suggested that the internal free ends somehow trigger off a recombination process, though the details of this process are obscure. It is of interest, however, that in *E. coli*, contrary to what was thought a few years ago, the chemical steps involved in recombination seem to be distinct from those involved in DNA repair. This is suggested by the observation that *recA* mutants survive UV-irradiation to a much greater extent if DNA replication is delayed by starvation[3]. The delay would allow dimer excision and repair synthesis while the complementary strand was still intact. This conclusion has the implication, however, that the excessive DNA breakdown associated with *recA* mutants is controlled in a different way from the DNA breakdown associated with dimer excision.

In fungi the position is much less certain but several observations suggest a similarity to *E. coli*, for example:

(1) HOLLIDAY[4] obtained a UV-sensitive mutant (*uvs-3*) of *Ustilago maydis* which gave about twice the normal frequency of UV-induced mitotic gene conversion at the nitrate reductase (*nar*) locus. A possible explanation for the increased frequency of UV-induced conversion is that pyrimidine dimers are not excised and that, perhaps through single-strand gaps opposite the dimers when the DNA replicates, recombination is stimulated. Conversion would result when such recombination occurred with the homologous

chromosome derived from the other parent, rather than with the sister chromatid following replication.

(2) PARRY and COX[5] found that UV-induced mitotic recombination within the *ade2* locus in *Saccharomyces cerevisiae* was increased by dark holding and reduced by photo-reactivation. This suggests that UV-induced conversion is, at least in part, associated with the repair of lesions in the DNA.

The hypothesis that UV-induced conversion in fungi is associated with repair of the DNA could be tested by studying the phenomenon of polarity in recombination. Polarity is shown by normal meiotic or mitotic recombination and is attributed to fixed opening points in the DNA where recombination is initiated. If UV-induced recombination is initiated at the positions of pyrimidine dimers it would be expected not to show polarity or at least to show a different pattern of polarity from normal recombination. The gene in *S. cerevisiae* which has been most fully studied from the point of view of polarity is probably the histidine-1 (*his1*) locus investigated by HURST and FOGEL[6,7]. It would be instructive to compare flanking marker behaviour in spontaneous and in UV-induced mitotic recombinants when histidine-independent cells were selected from trans heterozygotes for pairs of *his1* mutants.

Turning to the second question, namely, how far the normal mechanism of meiotic or mitotic recombination involves repair processes, there are good reasons for believing that conversion arises from the correction of mispairing in hybrid DNA, as proposed in the models of HOLLIDAY[8] and of HASTINGS and myself[9]. Several other hypotheses to explain conversion have been put forward in the last few years. That of STAHL[10] attributes conversion to the occurrence of two nonreciprocal exchanges in proximity to one another and occurring between segments of the homologues that have undergone an extra duplication. The break-and-copy models of TAYLOR[11], BOON and ZINDER[12] and PASZEWSKI[13] explain conversion in a similar way except that DNA synthesis occurs after the first exchange instead of before. There are several features of conversion, however, which are not explained by these double-exchange models, but can be accounted for if conversion is caused by correction of mismatched bases:

(1) There is plenty of evidence that for certain mutants conversion to wild type and to mutant occur with unequal frequency. One of the first examples was the grey (*g1*) spore colour mutant in *Sordaria fimicola* studied by KITANI, OLIVE and EL-ANI[14]. Such inequality is unaccounted for on the double-exchange models, but is expected with correction of mispairing because it is likely that the nature of the mispairing influences the relative frequency with which the strand with the wild-type nucleotide sequence or that with the mutant sequence is excised.

(2) There is also evidence that the relative frequencies of conversion to wild type and to mutant differ from one mutant to another, even within a gene. Examples of this can be found, for instance, in the work of ROSSIGNOL[15] with *Ascobolus immersus*. Again this is expected with correction of mispairing since mutants are likely to differ in their molecular nature. On the other hand, STAHL's model and the break-and-copy models

do not predict such a result, since with them conversion is not dependent on the nature of the mutation.

(3) LEBLON and ROSSIGNOL[16] have discovered that the conversion pattern of a mutant depends on the mutagen which was used to obtain it. This provides strong support for the hypothesis of correction of mispairing as the cause of conversion. They have obtained mutants at two spore colour loci, *b1* and *b2*, in *A. immersus* using the mutagens nitrosoguanidine (NTG) and the acridine ICR 170, and have found in crosses to wild type that the NTG mutants convert about equally often in either direction, while the ICR 170 mutants convert preponderantly to mutant and also show a much lower frequency of postmeiotic segregation. It is evident that conversion depends on the molecular nature of the mutation, a result which the double-exchange models do not predict.

(4) HOLLIDAY[8,17] has pointed out that, in fungi, well spaced alleles commonly show a recombination frequency greater than the sum of those for intervening intervals, and that this map expansion can be accounted for if (a) there is correction of mispairing and (b) it is associated with extensive excision. Linked conversion, which would be caused by extensive excision, has been observed whenever it has been looked for[15,18−22], and there is good evidence that its frequency decreases with increasing distance between the mutant sites[15,21]. Mutants closer together than the mean excision length will recombine much less often than those separated by more than this length. For example, if 3 mutants *a*, *b* and *c* occur in that sequence and if the intervals *ab* and *bc* are less than the mean excision length while *ac* is greater than it, then in terms of recombination frequency *ac* will be greater than the sum of *ab* and *bc*. Map expansion is unaccounted for on the double-exchange models for conversion[23], and so its occurrence supports the hypothesis of correction of mispairing and also the hypothesis of extensive excision in the correction process. Such excision provides an interesting parallel with the dark repair of DNA in *E. coli*, following UV-irradiation, where there is believed to be extensive excision in the removal of pyrimidine dimers[24].

Until recently the idea of correction of mismatched bases in heteroduplex DNA has not been supported by those working with prokaryotes. HOGNESS *et al.*[25] claimed that such correction took place in phage λ of *E. coli* but their data did not establish that the correction was triggered by the mispairing itself, and some people believed that the leakiness of the *N* mutants used invalidated the experiments. Such criticisms do not apply to the work of SPATZ and TRAUTNER[26] with phage SPP1 of *Bacillus subtilis*. It is of particular interest that they have found that the relative frequency of correction to wild type or to mutant is not dependent solely on the molecular nature of the mispairing, but is much influenced by other factors. They presume that the neighbouring base sequence is important.

There is evidence from fungi, also, that factors additional to the mispairing itself influence the correction process. EMERSON[27] found mutants of *A. immersus* which, when crossed to wild type, gave 6 : 2 ratios in excess of 2 : 6, but 3 : 5 in excess of 5 : 3. Such results can be explained either by postulating that the kind of mispairing in one chromatid differs from that in the other, as HOLLIDAY's model[8] predicts, or by postulating that factors other than the mispairing itself influence the correction process. Support for such additional factors has been obtained by BOND[28] from study of recombination at the *buf* spore colour locus in *Sordaria brevicollis*.

He has studied the behaviour of flanking markers in wild-type recombinant spores when UV-induced *buf* mutants are crossed with one another. The recombinant asci arise with quite low frequency — about 1 ascus in 1250 at the most. About 28% of them had only one wild-type spore and the remainder had a pair of them. BOND found that the wild-type spores which occurred singly showed distal polarity, those with a parental combination of flanking markers nearly always having recombination on the distal side of the alleles in addition to that between them. On the other hand, the wild type spores which occurred in pairs showed the additional recombination equally often on either side of the alleles. In crosses between the individual mutants and wild-type, all the mutants showed postmeiotic segregation about equally often. From the interallelic crosses, it seems that postmeiotic segregation is frequent when hybrid DNA enters the gene from the distal end but is absent, or almost so, when it enters from the proximal end. In other words, the correcting enzyme activity is different in these two categories of events. One can only speculate at present as to the cause of this difference, but it might be related to a difference in their time of occurrence, or to a difference in the distance which the hybrid DNA has extended from the initial opening points.

Although there seems to be good evidence for correction of mispairing as the cause of conversion, there is no evidence to discriminate between the two hybrid DNA models[8,9], both of which involve such correction. One of the differences between these two hypotheses is in their explanation for a parental arrangement of flanking markers associated with recombination of alleles. HOLLIDAY[8] suggests that the same two nucleotide chains, one in each recombining molecule, are cut twice, while on my hypothesis[9] an additional complete crossover is postulated alongside the initial one. In attempting to distinguish these alternatives, use is being made of an inversion in linkage group III of *S. brevicollis*[29]. It has been found that heterozygosity for this structural change greatly reduces the frequency of parental flanking markers in asci showing recombination at the *buf* spore colour locus in linkage group II. It is thought that the recombination events that have parental outside markers may result from a process of cancelling some of the crossovers, as a means of controlling their frequency and distribution in the chromosomes, and that the inversion somehow interferes with this cancelling process. It is hoped that further study of the effects of the inversion on recombination within the *buf* locus may provide information about the cancelling mechanism and so discriminate between the two models.

ESSER in his *Introduction*[30] has emphasized the importance of understanding and making use of recombination for the successful handling of industrial microorganisms. Knowledge of the recombination process in fungi is now advancing rapidly, and one can foresee a time, perhaps only a few years hence, when it will be understood, at least in outline. Moreover, spontaneous mitotic recombination seems to be essentially similar to meiotic, though differing in detail. There is thus a distinct possibility that in both perfect and imperfect fungi it may soon be feasible to exploit recombination to man's advantage with much greater precision than hitherto. In particular, understanding of the mechanism believed to exist in eukaryotes for cancelling crossovers may be of value in controlling recombination in fungi used in industry.

REFERENCES

1. HOWARD-FLANDERS P., RUPP W. D., WILKINS B. M., p. 142 in W. J. Peacock, R. D. Brock (Eds.): *Replication and Recombination of Genetic Material*. Austral. Acad. Sci., Canberra 1968.
2. RUPP W. D., HOWARD-FLANDERS P.: J. Mol. Biol. **31**, 291 (1968).
3. HOWARD-FLANDERS P.: Adv. Biol. Med. Phys. **12**, 299 (1968).
4. HOLLIDAY R.: Mutation Res. **4**, 275 (1967).
5. PARRY J. M., COX B. S.: Genet. Res. **12**, 187 (1968).
6. HURST D. D., FOGEL S.: Genetics **50**, 435 (1964).
7. FOGEL S., HURST D. D.: Genetics **57**, 455 (1967).
8. HOLLIDAY R.: Genet. Res. **5**, 282 (1964).
9. WHITEHOUSE H. L. K., HASTINGS P. J.: Genet. Res. **6**, 27 (1965).
10. STAHL F. W.: Genetics **61**, Suppl., 1 (1969).
11. TAYLOR J. H.: Mol. Genet. **2**, 95 (1967).
12. BOON T., ZINDER N. D.: Proc. Nat. Acad. Sci. U.S.A. **64**, 573 (1969).
13. PASZEWSKI A.: Genet. Res. **15**, 55 (1970).
14. KITANI Y., OLIVE L. S., EL-ANI A. S.: Am. J. Botany **49**, 697 (1962).
15. ROSSIGNOL J. L.: *Thesis*. University of Paris 1967.
16. LEBLON G., ROSSIGNOL J. L., p. 75 in *La recombinaison bactérienne hémotypologie et cytogénétique*. Soc. Franc. Génét., Toulouse 1970.
17. HOLLIDAY R., p. 157 in W. J. Peacock, R. D. Brock (Eds.): *Replication and Recombination of Genetic Material*. Austral. Acad. Sci. Canberra, 1968.
18. CASE M. E., GILES N. H.: Genetics **49**, 529 (1964).
19. MOUSSEAU J.: Compt. rend. sér. D, **262**, 1254 (1966).
20. PASZEWSKI A.: Genet. Res. **10**, 121 (1967).
21. FOGEL S., MORTIMER R. K.: Proc. Nat. Acad. Sci. U.S.A. **62**, 96 (1969).
22. MURRAY N. E.: Genet. Res. **15**, 109 (1970).
23. HOLLIDAY R., WHITEHOUSE H. L. K.: Mol. Gen. Genet. **107**, 85 (1970).
24. HOWARD-FLANDERS P., BOYCE R. P.: Radiation Res., Suppl. **6**, 156 (1966).
25. HOGNESS D. S., DOERFLER W., EGAN J. B., BLACK L. W.: Cold Spring Harbor Symp. Quant. Biol. **31**, 129 (1967).
26. SPATZ H.C., TRAUTNER T.A.: Mol. Gen. Genet. **109**, 84 (1970).
27. EMERSON S.: Genetics **53**, 475 (1966).
28. BOND D. J.: Thesis, University of Cambridge 1969.
29. AHMAD A. F., BOND D. J., WHITEHOUSE II. L. K.: Genet. Res., *in press* (1972).
30. ESSER K.: Paper at 1st Internat. Symp. Genetics of Industrial Microorganisms, Prague 1970. Abstract Book, p. 19; *cf. this book*, p. 73.

4/ Genetic Control of Metabolism

Genetic Control of Enzyme-Protein Structure and Synthesis in Fungi

J.R.S. FINCHAM

Department of Genetics, University of Leeds, Great Britain

Studies of gene—protein relationships in the fungi have an important place in the history of modern genetics. The discovery of auxotrophic mutants in *Neurospora* by BEADLE and TATUM in the mid-1940's led directly to search for similar mutants in bacteria and to the birth of bacterial genetics, while the one gene—one enzyme hypothesis, which came into prominence with the *Neurospora* work, provided one of the main conceptual foundations for the modern era of biochemical genetics. Since these early days it must be admitted that the achievements of fungal geneticists in the field of gene—protein relations have been rather overshadowed by those of their colleagues working with bacteria and bacteriophages. The principle of colinearity of the gene and polypeptide chain has been firmly established in *Escherichia coli*, T4 and λ, the genetic code has been fully worked out and, through the elegant method of analysis of double frameshifts in *E. coli* and T4, a catalogue of the codons actually used *in vivo* is being compiled. Turning from the genetic determination of protein structure to the genetic control of gene activity, we see again the bacterial work setting the pace. The operon model of JACOB and MONOD, with certain embellishments, has been entirely confirmed for the *lac* system of *E. coli* in both genetic and biochemical experiments, some of these being of quite breathtaking sophistication and convincingness. While not all operons in bacteria work in quite the same way in detail, it seems that the main features of the operon model may account very well for regulation of gene action in bacteria, especially taken in conjunction with the newer discovery in bacteriophage-infected cells of special gene products controlling the initiation and termination of messenger production. When one remembers also the great successes which have been achieved in the actual isolation of specific genes or groups of genes in bacteria and bacteriophages, feats which are not at present technically possible in the fungi, one is tempted to ask what advantage there is in working with the fungi at all — why should not every aspiring molecular geneticist go to the bacteria and viruses where the fundamental results are to be had?

There are, I think, two reasons why fungal geneticists should not all be diverted into bacterial genetics. The first is that fungi are important in their own right, both as an important part of nature and because of their industrial importance. We would like to know more about how these important organisms work. The second is that fungi are the most amenable to experiment of eukaryotic organisms, and their study may well reveal general principles of eukaryotic organisation not all of which are shared by prokaryotes.

In this paper I shall review the present state of affairs in four more or less distinct fields of fungal biochemical genetics: the control of protein primary structure, the alteration of enzyme properties by mutation, the mechanism of allelic complementation and the quantitative regulation of gene activity.

GENETIC CONTROL OF PRIMARY STRUCTURE

The main obstacle to progress in this field is always the difficulty of finding an amenable gene—protein system. One needs a gene, within which mutations can be readily selected and mapped, coding for a polypeptide chain which is produced in good yield and which does not have too long an amino-acid sequence. Several promising systems in *Neurospora* have foundered because of failure in one or other of these requirements. The highly interesting and historically important *trp-3* (tryptophan synthase)[1] and *ade-4* (adenylosuccinase)[2] loci are, so far as I know, no longer being seriously investigated from the point of view of gene—protein relationships, the trouble being a combination of the difficulty of purifying the proteins in adequate quantities and the complexity of the protein products. The object of my own studies, the *Neurospora* glutamate dehydrogenase controlled by the *amm* gene[3], has the advantage of being obtainable in pure form in moderately good quantities after a reasonably easy purification, but it is still hard work making enough protein for serious structural studies and the polypeptide chain, with over 400 amino-acid residues is longer than one would like[4]. Progress is slow, and though we have a considerable number of mutant proteins, with a variety of modifications in enzymic properties, we have as yet only preliminary information on the amino-acid replacements in these mutants.

The serious work so far on primary protein structure and coding in fungi has been done, not on enzymes at all, but on yeast cytochrome *c*, by SHERMAN's group[5]. This respiratory pigment seems to be God's gift to the student of gene—protein relations, being easily purified and with a polypeptide chain only 108 residues long. The facultatively anaerobic habit of *Saccharomyces* makes it possible to isolate many mutants totally defective in cytochrome *c*. These primary mutants, selected for their gross deficiency in haem content, are of no use in themselves since they produce no gene product which can be isolated, but they serve as a source of revertants, many of which turn out to have regained not the wild-type pigment but variant cytochromes with single amino-acid replacements each related to the corresponding normal amino acid by two successive changes in a codon. The several fundamental points which have been established by the cytochrome *c* work are reviewed briefly in the following paragraphs.

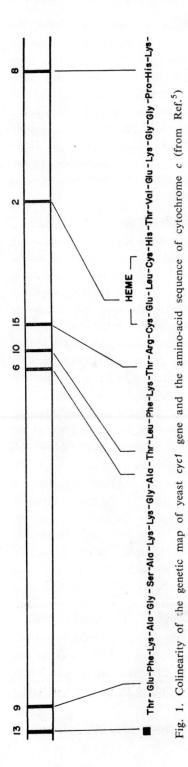

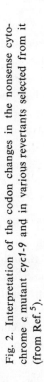

Fig. 1. Colinearity of the genetic map of yeast cyc1 gene and the amino-acid sequence of cytochrome c (from Ref.[5])

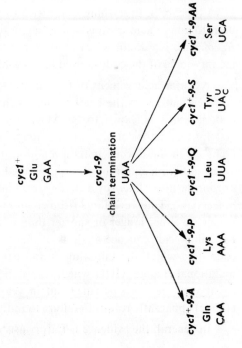

Fig. 2. Interpretation of the codon changes in the nonsense cytochrome c mutant cyc1-9 and in various revertants selected from it (from Ref.[5]).

Colinearity of gene and polypeptide chain. The sites of mutation in *cyc1*, the structural gene for iso-1-cytochrome *c*, have been mapped with fair accuracy by MANNEY and MORTIMER's[6] X-ray technique. Comparison of the resulting fine-structure map with the sequence of the corresponding amino-acid replacements in the polypeptide chain (determined in revertants) shows a perfect agreement (Fig. 1). This result, following the several other examples from the bacteria and viruses, leaves one without a strong motive for undertaking further laborious work simply to obtain more demonstrations of the correctness of the colinearity principle. However, taking colinearity for granted, it is still of value to have the amino-acid-sequence analysis as a check on the validity of the genetic map, since the theoretical basis of fine-structure mapping in fungi is still rather uncertain.

The genetic code — initiation and termination codons. Most of the identified mutants in the yeast *cyc1* gene have lost virtually all cytochrome *c* pigment and many of them have mutant codons which terminate polypeptide chain synthesis before the chain has acquired the cysteine residues which bind the haem group. By selecting respiration-sufficient revertants starting with such a chain termination mutant, SHERMAN and his colleagues were able to isolate a spectrum of cytochrome varieties, each one with a chain of normal length but with a different amino acid inserted in the position corresponding to the previous chain termination (Fig. 2). Assuming that the standard *E. coli* codons apply to yeast, and that all the reversions were due to single nucleotide replacements in the chain-terminating codon, an identification of the latter becomes possible. In two cases, the chain-terminating mutant codon has been identified as UAA ("ochre"). Everything so far is consistent with the yeast dictionary, with respect both to amino-acid codons and to chain termination codons, being the same as the *E. coli* one. No doubt we can look forward to double-frameshift analysis which will identify some of the codons used in the wild type.

A major achievement of the yeast cytochrome *c* studies, and one which has not yet been made with the bacteria, has been the identification of the chain initiation codon, which is shown to be AUG (for methionine) as in the bacterial cell-free systems. A number of mutants are unable to start cytochrome *c* synthesis because of a change in the normal initiation codon; revertants have restored synthesis because a new initiation codon has been generated either one position before the normal one or four positions later (by a change from internal lysine, AAG, to N-terminal methionine, AUG). Thus revertants from chain-initiation mutants have cytochrome *c* chains which are either longer or shorter than usual. The identification of AUG is clinched by amino-acid-sequence analysis of the "long" revertants, which reveals the various codons to which the initiation codon can mutate; all are related to AUG by one mutational change. UUG which has been suspected of being an initiation codon in eukaryotes seems to be ruled out in yeast since the change from AUG to UUG (valine) apparently results in failure to initiate.

In general, the evidence is fully consistent with the genetic code being identical in prokaryotes and eukaryotes as regards both the identity of the codons for individual amino acids and starting and stopping codons. Whether the relative frequencies of utilization of alternative codons for the same amino acid are similar in different groups of organisms is another question — the divergences between groups in overall DNA base composition and between related species (for example species of *Saccharomyces*[7]) in base sequence homology strongly suggest that there may be a good deal of genetic drift within the limits allowed by the degeneracy of the code. One type of investigation which would be technically possible and interesting (though rather expensive) would be the determination, by amino-acid replacement or double-frameshift analysis, of the codons used for identical polypeptide chains, for example of cytochrome *c*, in related species.

ALTERED ENZYME PROPERTIES BY MUTATION

Auxotrophic mutants are usually defective in single specific enzymes. Simply because they are auxotrophic, and thus have enzymes which are almost or quite inactive in the cell, such mutants generally represent rather drastic protein defects. Many of them have prematurely terminated polypeptide chains, because of the direct formation by mutation of chain-terminating codons, or scrambled and probably shortened amino-acid sequences as a result of frameshift mutations. Others have missense mutations which cause the replacement of amino acids which are essential for any enzyme activity.

From some points of view mutants with zero enzyme activity are less interesting than ones with enzyme activity of abnormal kinds. There is available a general method for the isolation of large numbers of different kinds of mutants with altered enzyme activity. This consists in the selection of revertants from enzyme-negative mutants. Some of these revertants are, of course, due to mutation back to the standard wild type but many are of other kinds. The studies in bacteria by GAREN's group[8] on alkaline phosphatase explain this diversity of revertants in molecular terms. There are for example a number of amino-acid codons to which a chain-terminating codon can mutate in one step and each one of these may be compatible with enzyme activity. The detailed properties of the resulting missense enzymes may differ from those of the wild type in subtle or obvious ways. Similarly, unacceptable missense can mutate to more or less acceptable missense, and there is also the possibility that an unacceptable amino acid replacement in one position can be made more acceptable by a compensating amino-acid replacement elsewhere in the chain — both of these possibilities are well exemplified by YANOFSKY's work on *E. coli* tryptophan synthase[9].

In fungi the same principles are demonstrated by the range of missense revertant forms of yeast cytochrome *c* but there are no comparable examples yet of similarly thorough analyses on mutant enzymes. However, there are several studies at the level of enzyme *properties* to show how great a diversity of enzyme types can be generated by the reversion technique. The study of D. WOODWARD *et al.*[2] of revertants at the *ade-4* (adenylosuccinase) gene in *Neurospora* was a pioneering one in this respect. The investigations by several workers over the years on the NADP-linked glutamate dehydrogenase of *Neurospora* also made the point well. Detailed analysis of the revertants obtained from several missense mutants — *amm-2, amm-3, amm-14, amm-19* (Refs[3,10,11]) — revealed a broad spectrum of enzyme types. The variations include decreased heat stability, increased heat stability, activation by heat, extreme sensitivity to mutations, decreased affinity for substrates (in one case accompanied by apparently increased activity at saturating substrate concentrations) and a variety of changes in allosteric properties. (Some of the data are summarized in Table I.) It is likely that these changes represent only a small sample of the ones that could be obtained. The very large number of obtainable variants is obvious when one considers that 19, or nearly a third, of the 61 possible "sense" codons are capable of mutating to "nonsense" codons nearly all of which will eliminate enzyme activity, while each of the nonsense codons can mutate in turn to 7 or 8 different "sense" codons for 6 or 7 different amino acids, the majority of which will probably be compatible with enzyme activity. Thus an enormous number of amino-acid replacements can, in principle, be obtained at most positions in the polypeptide chain of a protein provided that loss of the function of the protein can be made to result in an auxotrophic phenotype. With patience, and using several rounds of mutant-selection and reversion, one should be able to obtain almost any modification of enzyme structure compatible with enzyme activity.

The identification of the nature and position of the amino-acid replacements in mutants with enzyme properties altered in a variety of ways could yield information of great interest from the point of view of enzymology. It must be admitted, however, that mutants in fungi, or, for that matter in bacteria, have so far yielded few results in this field. The protein chemistry is technically demanding and, above all, time-consuming. No doubt the potential usefulness of mutants with altered enzymes in the study of mechanisms of enzyme action will yet be realized.

The mutational modification of enzyme properties may have practical as well as theoretical significance and may even be relevant to the ostensible subject of this book — the industrial uses of microorganisms. The large-scale production of enzymes for experimental or industrial use plays an increasingly important part in modern pharmaceutical industry. Microorganisms are the most common source of such useful enzymes. The normal form of an enzyme in the wild-type organism, while no doubt finely adjusted for optimum *in vivo* function, is by no means necessarily ideal for the special purposes of man. One does not, however, necessarily have to rest content with the wild-type enzyme. If one wants some new property — perhaps increased maximum activity at substrate saturation or increased heat stability, it is very likely that such a property will be represented among the revertants from enzyme-negative mutants. The limiting factor is likely to be not the number and kinds of enzyme varie-

TABLE I

Altered Properties of Various Mutant Varieties of NADP-Linked Glutamate Dehydrogenase in Neuropora[a]

Mutants	Electrophoretic mobility at pH 8.5	Heat stability	Michaelis constant	Requirements for activation	Maximum activity
amm-1	normal	approx. normal	—	—	0
amm-4	slightly fast	very low	—	—	0
amm-2	normal	somewhat low	?	incubation with glutamate or succinate at pH 8.6	ca. 5%
amm-21[b]	normal	low	approx. normal	warming to 30° or exposure to pH 8.0	ca. 30—50%
amm-3	normal[d]	approx. normal	?	incubation with glutamate or succinate at pH 8.5	ca. 50%
amm-19	fast[e]	increased	?	prolonged incubation with succinate at pH 8.7	50% or more
amm-3a[c]	normal	low	greatly increased for NH_4^+, NADP and glutamate	sometimes needs brief warming	about 200% at saturating substrates
amm-3b[c]	normal	approx. normal	somewhat increased for NH_4^+	incubation with 2-oxoglutarate plus NADPH 0.01M-EDTA	approx. 100% in crude extract — loses activity on purification
amm-3-18[c]	normal	approx. normal[f]	?		approx. 100% in crude extract loses activity on purification

[a] Properties are compared with those of wild-type enzyme. Data mainly from Refs[3,10,12].
[b] Revertant from amm-2.
[c] Revertant from amm-3.
[d] Separable from wild type on DEAE-cellulose.
[e] Becomes normal on activation.
[f] Strongly inhibited by Zn^{2+}.

ties potentially obtainable but rather the availability of an efficient screening system for picking out the variety you are interested in.

ALLELIC COMPLEMENTATION

The general mechanism through which allelic mutants are able sometimes to complement each other has been thought to be the formation of hybrid oligomeric protein with the two mutants contributing *complete* polypeptide chains with different conformational defects[13]. In this view, nonsense mutants are not expected to be capable of complementation except in the special case where the chain is terminated so close to the normal C-terminal end that the small deficiency at the C-terminus makes no effective difference to the three-dimensional structure. The time now seems to have come for a reconsideration of this model and perhaps for withdrawing recognition from it as *the* general mechanism while continuing to support it in the more modest role of one *type* of mechanism. It now seems increasingly probable that the hybrid proteins, the formation of which is undoubtedly a common feature of allelic complementation, may quite often involve *incomplete* polypeptide chains. There are two lines of evidence which lead to this view.

On the one hand, *Neurospora* workers who make complementation maps, for example MALLING and DE SERRES[14], working with *ade3B* mutants and, more recently, RADFORD with *pyr-3*[15], have placed emphasis on the apparently polarized patterns shown by the complementation maps. There is a strong tendency for many mutants to show functional overlaps in the complementation map ending at different points at their right-hand ends but all running to the map terminus at the left end. Fig. 3 shows an example of a map of this sort. The interpretation given to these polarized maps was that they were due to nonsense or frameshift mutants with a polarized effect on translation, producing incomplete polypeptide chains which were able to form active hybrids with complete missense chains. Such an interpretation, while consistent with some data on mutagen specificity (mutants induced by "frameshift" reagents, such as nitrogen half-mustards, tend to be polar[15]) did not fit with one's preconception that only chains of normal length were likely to be able to fit properly in an oligomer. It seems, however, that this prejudice must be given up in the light of several recent examples of the ways in which the virtually complete protein structure can be reconstituted from fragments of what is normally a single polypeptide chain.

The possibility of cleaving the polypeptide chain of a monomeric enzyme into two pieces and then reconstituting an active enzyme molecule through non-covalent interaction between the pieces has been known ever since RICHARD's work on the subtilisin fragment of ribonuclease[17]. It is evident that noncovalent bonds can be strong enough to stabilize the three-dimensional structure of a protein even when there are discontinuities of the polypeptide chain. The recent work on staphylococcal nuclease by ANFINSEN's group[18] has shown that stable and fully active enzyme can be formed by interaction between two *overlapping* fragments: the extra material represented by the region of overlap can be accommodated on the outside of the molecule without getting in the way of the specific folding of the rest. The possibility of complementation of defective mutants depending on a mechanism of this kind has been directly demonstrated by JACKSON and YANOFSKY[19] in the case of *E. coli* tryptophan synthase. It was shown that incomplete α-chains, deficient in various lengths of the C-terminal end of the sequence, would complement *in vitro* with several different missense mutant chains with their defects at the N-terminal end of the sequence. The

Fig. 3. A polarized complementation map — *ade3B* of *Neurospora*. After DE SERRES *et al.*[16]. "Nonpolarized" mutants are shown *above* and polarized mutants *below*.

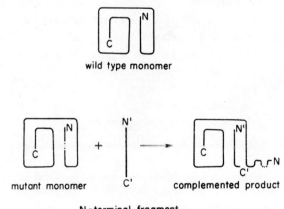

wild type monomer

mutant monomer   complemented product

N-terminal fragment

Fig. 4. Possible mode of association of a mutant polypeptide and a polypeptide fragment to form a functionally normal protein (after JACKSON and YANOFSKY[19]).

interaction presumably involves the formation of a complete monomer with a duplicated region (containing the missense amino-acid replacement) accommodated some-
where on the outside of the structure where it neither disrupts the folding of the monomer nor interferes with the formation of the normal $\alpha_2\beta_2$ tetramer (Fig. 4). A very
similar explanation accounts for complementation between E. coli β-galactosidase
chains deficient respectively in N-terminal and C-terminal parts of the sequence[20].
The mechanism begins to look like a very general one although it is too early to say
how widespread such cases are compared with those involving interactions between
complete missense chains. We may note that, where complementation depends on the
non-overlap of deletions in the polypeptide chain, a much more definite physical
meaning can be attached to the complementation maps — the map should be a linear
representation of the polypeptide chain. Many complementation maps however, are
probably the result of different mechanisms of complementation — some interactions
involving complete chains and others fragments, so the complexities usually found
are not unexpected.

OPERON ORGANIZATION AND REGULATION

There is, of course, an alternative interpretation of complementation maps with
polarized patterns of overlaps — namely that the map represents a polycistronic unit
of transcription, or an operon. Regulation of enzyme synthesis and activity is more
fully discussed elsewhere but I would like to discuss the operon question briefly because it bears on the interpretation of complementation maps.

Before assessing the evidence for the existence of operons in fungi it will be as
well to recall the main features of operons as they have been defined in bacteria.
Characteristically one finds (i) two or more contiguous genes specifying distinct polypeptide chains of related function, (ii) coordinate regulation of the activity of
these genes exercised through interaction of a regulator gene product (usually or
perhaps always a protein) with an operator segment at one end of the operon and
(iii) pleiotropic effects of mutations with nonsense or frameshift mutations in one
gene of the operon depressing, in a coordinated way, the activity of all other genes of
the operon further away from the operator. We may note in parentheses that the
exact mechanism of polarity was until recently in doubt even in the best studied bacterial systems. The simple view that polarity is due simply to unidirectional translation from an obligatory starting point at the operator end of the messenger proved
inadequate; the recent hypothesis of MORSE and YANOFSKY[21] which depends on
a combination of both transcriptional and translational polarity, seems more satisfactory for bacteria but can hardly apply to a eukaryote in which cytoplasmic translation and intranuclear transcription cannot, presumably, be closely coupled.

There are now a number of reports of series of mutants in fungi which look as if
they may represent operons, but the extent to which they really parallel the bacterial

systems is doubtful. The main source of doubt is the fact that in hardly any of the fungal examples is there real proof that the putative distinct genes of the operon specify distinct polypeptide chains or separable enzyme activities. In a number of examples (*e.g.* aro[22], pyr-3[23], his-3[24,25] in *Neurospora* and his-4 of *Saccharomyces*[26]) mutants mapping in different segments of a short region do affect two or more different enzyme activities, but in all the cases mentioned these different activities are associated with a single complex protein or protein aggregate from which components with single activities cannot easily be resolved. So we cannot be absolutely sure in these cases that the multiple enzyme activities really reflect multiple polypeptide chains and genes though, to be sure, to ascribe all six enzyme activities of the *Neurospora* aro aggregate to a single chain would be stretching hypotheses rather far. The difficulty in interpretation is exemplified by the comparison between *pyr-3* in *Neurospora* and *ura-1* in *Coprinus radiatus*[27]. In both these cases two enzyme activities are involved — carbamoylphosphate synthase and aspartate carbamoyltransferase. In the *Neurospora* case the polarized pattern in the complementation map is interpreted on the basis of allelic complementation, presumably involving incomplete polypeptide chains. The *Coprinus* workers, on the other hand, believe there are two genes and that the polarized pleiotropy shown in the complementation map is due to interruptions in the translation of a dicistronic messenger. Their case is greatly strengthened by the report that the two enzymes in *Coprinus* can be physically separated[28]; this has not proved possible in *Neurospora*[23]. It is evident that the choice between the alternative interpretations depends on enzymology and protein chemistry as well as on isolation of a large number of mutants and their accurate mapping. This is no longer an easy field to work in but perseverance in it is amply justified by the prospect of gaining information on coordination of gene activity that may be of general application in eukaryotes.

One respect in which the fungal examples of possible operons have not so far measured up to the bacterial models is in their failure to show any evidence for operator regions. No operator-constitutive mutants are known in fungi so far as I know.* Even constitutive mutants of the regulator-negative (*lacI*) type are not easy to come by. In what is probably the best example of a *lacI* type mutant in fungi (controlling enzymes of methionine synthesis in yeast) the two or more genes whose activity becomes derepressed are scattered and not organized in an operon[29]. An intriguing feature which has cropped up in more than one instance is the implication of an enzymic protein in regulatory functions. Apparently an enzyme can be involved either in regulation of its own synthesis, as in nitrate reductase of *Aspergillus*[30], or in that of other enzymes of the same pathway, as appears to be the case in the purine synthetic pathway in *Saccharomyces*[31].

The field is wide and the variety and complexity of the effects encountered may

---

* *Note added in proof:* J. M. WIAME now has operator mutants affecting arginine katabolism in yeast[32].

be baffling. The need, perhaps, is for concentration on a few fungal regulatory systems. Such investigations may seem esoteric and even trivial in detail, but, as the example of the *lac* system of *E. coli* has shown, the returns from a limited area tend to become the greater the more intensely it is cultivated. It hardly seems possible at this time to specify the systems on which effort should be concentrated. Perhaps they have not been discovered yet.

REFERENCES

1. SUSKIND S. R., LIGON D. S., CARSIOTIS M., in *Molecular Basis of Neoplasia*, 15th Ann. Symp. Fundamental Cancer Research. University of Texas, 1962.
2. WOODWARD D. O., PARTRIDGE C. W. H., GILES N. H.: Genetics **45**, 555 (1960).
3. FINCHAM J. R. S.: J. Mol. Biol. **4**, 257 (1962).
4. WOOTTON J. C., TAYLOR J. G., FINCHAM J. R. S.: Biochem. J., *in press* (1972).
5. SHERMAN F., STEWART J. W., PARKER J. H., PUTTERMAN G. J., AGRAWAL B. B. L., MARGO-LIASH E., in Symp. Soc. Exp. Biol., 24 (*Development and inter-relationships of cell organelles*), 1969.
6. MANNEY T. R., MORTIMER R. K.: Science **143**, 581 (1964).
7. BICKNELL J. N., DOUGLAS H. C.: J. Bacteriol. **101**, 505 (1970).
8. WEIGERT M. G., GALLUCCI E., LANKA E., GAREN A.: Cold Spring Harbor Symp. Quant. Biol. **31**, 145 (1966).
9. YANOFSKY C., ITO J., HORN V.: Cold Spring Harbor Symp. Quant. Biol. **31**, 151 (1966).
10. PATEMAN J. A., FINCHAM J. R. S.: Genet. Res. **6**, 419 (1964).
11. STADLER D. R.: Genet. Res. **7**, 18 (1966).
12. FINCHAM J. R. S., *unpublished results*.
13. FINCHAM J. R. S., p. 143 in Genetic Complementation. W. A. Benjamin, New York 1966.
14. MALLING H. V., DE SERRES F. J.: Mutation Res. **4**, 425 (1967).
15. RADFORD A.: Mol. Gen. Genet. **104**, 288 (1969).
16. DE SERRES F. J., BROCKMAN H. E., BARNETT W. E., KØLMARK H. G.: Mutation Res. **4**, 415 (1967).
17. RICHARDS F. M., VITHAYATHIL P. J.: J. Biol. Chem. **234**, 1459 (1959).
18. OUTJES D. A., ANFINSEN C. B.: J. Biol. Chem. **244**, 6316 (1969).
19. JACKSON D. A., YANOFSKY C.: J. Biol. Chem. **244**, 4539 (1969).
20. GOLDBERG M. E.: J. Mol. Biol. **46**, 441 (1969).
21. MORSE D. E., YANOFSKY C.: Nature **224**, 329 (1969).
22. RINES H. W., CASE M. E., GILES N. H.: Genetics **61**, 789 (1969).
23. HILL J. M., WOODWARD V. W.: Arch. Biochem. Biophys. **125**, 1 (1968).
24. WEBBER B. B.: Genetics **45**, 1617 (1960).
25. MINSON A. C., CREASER E. H.: Biochem. J. **114**, 49 (1969).
26. SHAFFER B., RYTKA J., FINK G. R.: Proc. Nat. Acad. Sci. U.S.A. **63**, 1198 (1969).
27. GANS M., MASSON M.: Mol. Gen. Genet. **105**, 164 (1969).
28. HIRSCH M.-L.: Compt. rend. **267**, 1473 (1968).
29. CHEREST H., EICHLER F., DE ROBICHON-SZULMAJSTER H.: J. Bacteriol. **97**, 328 (1969).
30. COVE D. J., PATEMAN J. A.: J. Bacteriol. **97**, 1374 (1969).
31. DORFMAN B.-Z.: Genetics **61**, 377 (1969).
32. WIAME J. M.: Heredity, *in press*.

Regulation of the Tryptophan Biosynthetic Enzymes in Fungi

RALF HÜTTER

Mikrobiologisches Institut, Eidgenössische Technische Hochschule, Zürich, Switzerland

Abbreviations used:

ABA   4-aminobenzoate
DAH-*P* 3-deoxy-D-*arabino*heptulosonate 7-phosphate
DHQ   5-dehydroquinate
DHS   5-dehydroshikimate
Ery-4-*P* D-erythrose 4-phosphate
ING-*P* 1-(indol-3′-yl)glycerol 3-phosphate
P-DRA N-(5′-phospho-1′-deoxyribulosyl)anthranilate
P-EPS 3-phospho-5-enolpyruvylshikimate
P-EPY phosphoenolpyruvate
P-RAA N-(5′-phosphoribosyl)anthranilate
P-SHA 3-phosphoshikimate

The biosynthesis of aromatic amino acids is well understood in its principal steps. The biosynthetic sequence and the main enzymatic reactions involved are the same in all organisms studied (*see* rewievs[1-4] and Figs 1, 2). However, considerable variations can be demonstrated for regulation, occurrence of isozymes and enzyme complexes[5-7].

Aromatic-amino-acid biosynthesis, especially with respect to tryptophan, will be discussed here. The fungi are of specific importance due to their usefulness for microbial production of tryptophan and for alkaloid production. Furthermore, they are of scientific interest because the biosynthetic enzyme levels are much more rigorously controlled than in bacteria and the mechanisms of maintenance of the balance are not yet understood.

Emphasis will be put on the occurrence of isozymes and enzyme complexes, on transport and on cross-pathway regulation. The discussion will be restricted to *Neurospora crassa* and *Saccharomyces cerevisiae* but results with other organisms will be included to illustrate special points. No attempt will be made to give a complete and exhaustive picture of aromatic-amino-acid biosynthesis.

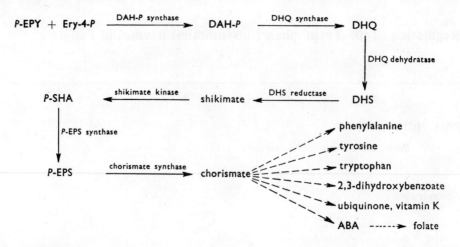

Fig. 1. Enzymes (trivial names) and intermediates in the common pathway of aromatic-amino
-acid biosynthesis.

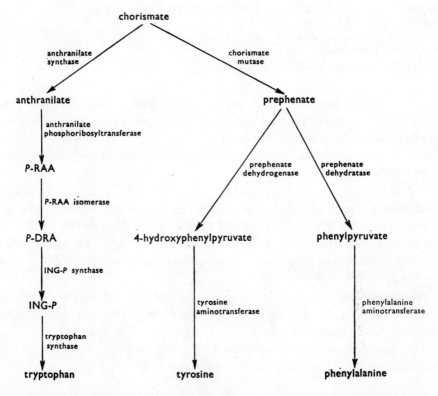

Fig. 2. Enzymes (trivial names) and intermediates in the biosynthesis of aromatic amino acids
from chorismic acid.

ISOZYMES OF AROMATIC-AMINO-ACID BIOSYNTHETIC ENZYMES AND ENZYME COMPLEXES

The occurrence of isozymes for crucial steps in the biosynthesis of amino acids has been demonstrated in several instances. A certain pattern of isozymes is typical of a given organism. Within the aromatic-amino-acid biosynthetic pathways isozymes

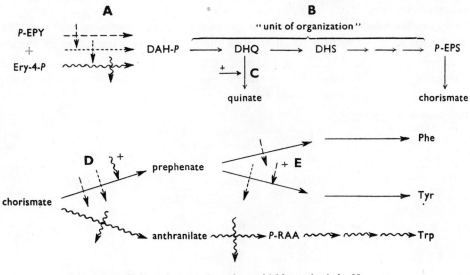

Fig. 3. Regulation of aromatic-amino-acid biosynthesis in *N. crassa*.

————	unknown	———→	repression
– – – –	Phe	—┼→	feedback inhibition
·········	Tyr	↓+→	feedback activation
∿∿∿∿	Trp	—+→	induction

have been observed for several steps, in the common aromatic-amino-acid branch (prior to chorismic acid), as well as for chorismate mutase.

For *N. crassa* three structural genes and three isozymes of DAH-*P* synthase have been demostrated[1,8-10] (Fig. 3A). Each of the three isozymes is inhibited by one of the end-products, L-phenylalanine, L-tyrosine or L-tryptophan. Repression of these enzymes by exogenous aromatic amino acids could not be achieved but, on the other hand, a two- to seven-fold derepression occurs in aromatic-amino-acid auxotrophs (*aro* mutants) under suitable conditions[11]. This indicates a regulative control of DAH-*P* synthase levels by endogenous pools of aromatic amino acids.

The five enzymes following DAH-*P* synthase are organized in an enzyme complex; the structural genes are clustered[5,12,13] (Fig. 3B). As indicated by Doy[8] some mutations affect not only this detected organisational unit but cause also changes in DAH-P synthase. This observation would unify the whole sequence common to aromatic-amino-acid biosynthesis. The organisation in a complex has (at least) two advantages for the organism: first, it increases substantially the catalytic activity[14] (Table I) and, second, it allows for channeling of intermediates. GILES and co-workers[15] have demonstrated the occurrence of two dehydroquinases, a biosynthetic constitutive enzyme and a catabolic inducible enzyme. The biosynthetic enzyme is included in the complex and thus guarantees the successful flow of biosynthetic intermediates (Fig. 3C).

For chorismate mutase (Fig. 3D) no isozymes have been demonstrated; the single enzyme is inhibited by phenylalanine and tyrosine but is stimulated by trypto-

TABLE I

Catalytic Facilitation by Multienzyme Complexes[14]

A. Reactions

Ery-4-P ⟶ anthranilate (8 steps) vs.
shikimate ⟶ anthranilate (4 steps)

Substrate	μM	Anthranilate nmol/h/ml
Ery-4-P	0.1	16
	0.25	49
	1.0	89
Shikimate	0.1	4
	0.2	7.5
	2.0	10

B. Reactions

P-RAA ⟶ ⟶ ING-P vs.
P-DRA ⟶ ING-P

Substrate	μM	ING-P nmol/h/ml
P-RAA	0.125	65
	0.245	77
	0.49	88
P-DRA	0.115	44
	0.23	55
	0.46	55

phan[16]. Unlike observed for the bacteria *Escherichia coli* and *Aerobacter aerogenes* where two isozymes of chorismate mutase are aggregated with prephenate dehydrogenase and prephenate dehydratase respectively, no such aggregates are found in *N. crassa*[17-19]. The following enzyme, prephenate dehydrogenase, is inhibited by tyrosine, but feedback-activated by phenylalanine (Fig. 3E); prephenate dehydratase, on the other hand, is weakly inhibited by phenylalanine, tyrosine having no influence[20]. This general stimulation of tyrosine biosynthesis in two steps by tryptophan (chorismate mutase) and by phenylalanine (prephenate dehydrogenase) has been interpreted by CATCHESIDE[20] as a response to the rapid consumption of tyrosine to melanin during the sexual phase of the fungus.

For *S. cerevisiae* only two DAH-*P* synthases could be demonstrated[21,22], but LINGENS and coworkers[23,24] consider it possible that a third DAH-*P* synthase exists. The two isozymes are phenylalanine- and tyrosine-sensitive (Fig. 4A). No repression has been found by exogenous aromatic amino acids[21,24] in wild-type strains but experiments with *aro* auxotrophs are lacking. The five steps following DAH-*P* synthase are also organized in a unit, genetically as well as enzymatically[5,25] (Fig. 4B). In contrast to *N. crassa* no catabolic dehydroquinase is known

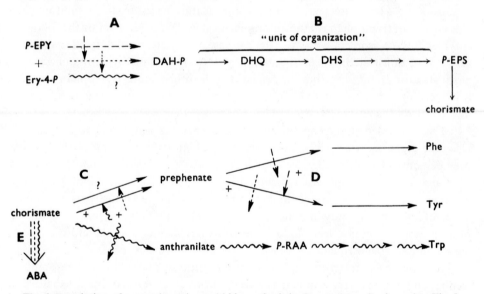

Fig. 4. Regulation of aromatic-amino-acid biosynthesis in *S. cerevisiae*. *See* legend to Fig. 3.

For chorismate mutase two isozymes are postulated[23,24], one feedback-inhibited by tyrosine, the other seemingly insensitive to feedback inhibition. This point, however, needs further clarification. Chorismate mutase is not repressed by phenylalanine or tyrosine but is induced and stimulated by tryptophan (Fig. 4C). Prephenate dehydrogenase is inhibited by tyrosine, but induced and activated by phenylalanine (Fig. 4D); prephenate dehydratase is inhibited and weakly repressed by phenylalanine[23,24]. All three aromatic amino acids together cause an induction of p-amino-benzoate synthase, leading to the excretion of the end product[3] (Fig. 4E).

As can be recognized also for *S. cerevisiae* aromatic-amino-acid biosynthesis is directed to tyrosine by stimulation of chorismate mutase by tryptophan and of prephenate dehydrogenase by phenylalanine.

THE TRYPTOPHAN BRANCH

The tryptophan branch leading from chorismic acid to tryptophan has been previously discussed for *N. crassa*[26,27] and for *S. cerevisiae*[23,24,28,29]. A number of interesting regulatory properties have been found (*see* Figs 3, 4).

Anthranilate synthase, the first enzyme of the tryptophan branch, is known to be feedback-inhibited by the end product in *N. crassa*[30] and in *S. cerevisiae*[23,28]. It may be interesting to see whether anthranilate synthase may also be stimulated by L-histidine as observed for *Bacillus subtilis*[31] (*cf. below*). In addition, the second enzyme anthranilate phosphoribosyltransferase is also inhibited by tryptophan in *N. crassa*[32,107]. This latter may be a regulatory feature necessary to prohibit undesir-able circling of tryptophan *via* the anthranilic-acid cycle (*see below*). For the following enzymes P-RAA isomerase and ING-P synthase no specific regulatory effects have as yet been demonstrated; *e.g.* no inhibition of ING-P synthase by anthranilic acid as postulated for *E. coli*[33,34] (but *cf. below*). On the other hand, tryptophan synthase is induced by high levels of ING-P; tryptophan itself does not exert a direct regula-tory effect on the enzyme[32,36,37].

The whole set of tryptophan biosynthetic enzymes was found to be reduced, when the organism is grown in media containing high levels of tryptophan. This has been demonstrated for several enzymes in *N. crassa*[26,27], in *S. cerevisiae*[24,28] and for *Coprinus radiatus*[38]. This phenomenon is generally called repression. Doubtless, however, it cannot be solely attributed to repression but is at least partially due to a lack of inducing pathway intermediates: Feedback inhibition and partial repression of anthranilate synthase leads to low levels of intermediates and therefore a lack of induction of the subsequent enzymes. The lack of repression observed in *Claviceps paspali*[39] may be so interpreted: In this strain, anthranilate synthase is resistant to feedback inhibition by tryptophan; as a consequence, high levels of intermediates and induced levels of tryptophan enzymes are produced in the absence or presence of external tryptophan. On the other hand, tryptophan enzymes are derepressible by the application of analogues, *e.g.* 5-methyltryptophan (Table II).

Enzyme aggregates occur abundantly in the tryptophan pathway of fungi[7]. The composition of the aggregates may vary: In *N. crassa*, *e.g.*, the complex catalyzes three catalytic steps, anthranilate synthase, P-RAA isomerase and ING-P synth-ase[30,40,41], while in *S. cerevisiae* only anthranilate synthase and ING-P synthase are involved[28,42]. In the example of *N. crassa* it was found[14] that enzyme aggregation

Table II

Derepression of Tryptophan Biosynthetic Enzymes of *S. cerevisiae* X2180-1A by 5-Methyltryptophan (5MT) and 3-Amino-1,2,4-triazole (3AT)

Growth medium[a]	Anthranilate synthase[b]	ING-*P* synthase[c]
MV[d]	63	80
MV + 20 µg Trp/ml	47	45
MV + 10^{-3} M-5MT	149	63
MV + 2×10^{-3} M-5MT	174	110
SC[e]	57	62
SC + 10^{-3} M-3AT	90	113
SC + 10^{-2} M-3AT	176	201

[a] Media used are described by Manney[106].
[b] nmol anthranilate formed/mg protein/h; tested as described by DeMoss[42].
[c] nmol ING-*P* formed/mg protein/h; tested as described by Wegman and DeMoss[109].
[d] Minimal medium.
[e] Synthetic complete medium without L-histidine.

permits higher catalytic activity of the reaction sequence than would separate enzymes, as observed for the "*aro*"-aggregate (Table I). A physiological advantage of the mechanism of the tryptophan synthase reaction in one step ING-*P* → tryptophan instead of in two steps ING-*P* → indole → tryptophan was demonstrated for *S. cere-*

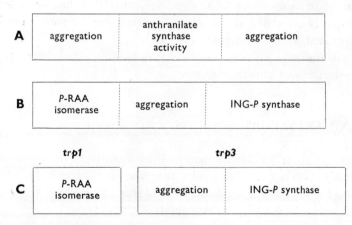

Fig. 5. Functional differentiation of tryptophan loci. **A** *trp2* locus of *N. crassa*, **B** *trp1* locus of *N. crassa*, **C** *trp1* and *trp3* loci in *S. cerevisiae*.

visiae by MANNEY[43]. To what extent enzyme aggregates regulate enzyme synthesis cannot yet be decided.

The genetic loci involved in aggregate formation show an interesting functional diversity (Fig. 5). This was nicely demonstrated for the *trp1* locus of *N. crassa*[44,45], for the *trp2* locus of *N. crassa* by CHALMERS[46] and for the *trp3* locus of *S. cerevisiae*[47]. The direction of polarity within these loci has as yet to be demonstrated.

CONSUMPTION AND DEGRADATION OF TRYPTOPHAN

Tryptophan pool levels in cells are also influenced by consumption of the amino acid for protein synthesis, for the biosynthesis of other metabolites (*e.g.* indolylacetic acid or nicotinic acid), by degradation and by efflux from the cells. Efflux seems to be rather limited under normal conditions (*see below*). In order to maintain desired levels of the pools under variable growth conditions tryptophan may be degraded *via* different pathways. The fate of tryptophan has been intensively studied, mainly because of its connection with indolylacetic acid, nicotinic acid and alkaloids. Most results are summarized in MATCHETT and coworkers[27] and LINGENS[3]. I may, therefore, restrict myself to a short and simplified summary.

Nicotinic acid is mainly derived from tryptophan *via* kynurenine and 3-hydroxyanthranilic acid in fungi, but not in most bacteria[48]. For *S. cerevisiae* this seems only to be true under aerobic conditions; a second pathway is operative especially under anaerobic conditions[49,50]. The inducible kynurenine pathway has been demonstrated for many filamentous fungi[48,51−53]. In addition, tryptophan may be recycled by the inducible "anthranilic-acid cycle"[53,54].

Furthermore, L-tryptophan can be degraded *via* a transamination reaction, leading to indolyllactate and indolylethanol (tryptophol)[55,56].

D-Tryptophan seems either to be directly N-acetylated, first degraded to kynurenine and then N-acetylated, or, possibly after oxidative deamination as in *Pseudomonas*[57], degraded by the same pathways as L-tryptophan[52,56,58,59]. The relative ratios of the degradative pathways must certainly be different in different strains and must vary with unequal growth conditions.

POOLS AND TRANSPORT SYSTEMS

Enzyme regulation is a direct response of the organism to changes in the internal levels of amino acids (amino-acid pools) as well as to changes in the environment; a prerequisite for the latter is the existence of a transport system. We will therefore focus our attention for a while on these points.

Data on the pools of amino acids in fungi under variable growth conditions are scarce. SARACHEK[60] indicates that in wild-type strains of *S. cerevisiae* relatively stable pools of amino acids are maintained. Vitamin nutrition is of main importance: biotin-deficient cells maintain only low levels of amino acids, biotin-sufficient cells are capable of maintaining higher levels[61]. The problem of pools and possible response of the enzymes to changes in the pools is complicated by the fact that different pools may exist for the same amino acid as found for tryptophan in *N. crassa*, namely an expandable and a metabolic pool[62]. Externally added tryptophan is included in the expandable pool first and from there it is either metabolized *via* the anthranilic-acid cycle or is excreted again or is included into the metabolic pool. Biosynthetic tryptophan, on the other hand, is directly included in the metabolic pool and is used primarily for protein synthesis.

It is impossible at the moment to draw more than a generalized conclusion: The levels of internal pools influence enzyme regulation also in fungi. We have to realize that external addition of amino-acids need not necessarily lead to enzyme repression if the internally maintained pools are sufficient for complete repression. If, however, the internal levels are reduced in an auxotroph, derepression is often observed.

How does the organism maintain or change its internal pools in response to externally added amino acids? What are the mechanisms of influx and efflux? Little is known about the mechanism of efflux, more data are available on the influx transport system.

In *N. crassa* four transport systems for amino acids can be recognized[63,64], one specific for neutral amino acids, one specific for basic amino acids, one specific for acid amino-acids, and an unspecific amino-acid transport system. In a general way, it can be stated that every amino acid can be transported by (at least) two systems, a group-specific and an unspecific system, and that different amino acids can be transported by the same system. As transport capacity and transport specificity are elucidated by experiments measuring the uptake of radioactively marked amino acids, single or in combinations (competition experiments), the nonexistence of a unique and separate transport system for an amino acid may severely handicap the interpretation of the results: Uptake ratios obviously depend on the existence of more than one amino-acid transport system (at least a group-specific and the unspecific one) and competition experiments depend on the coexistence and relative ratios of at least two systems. Furthermore, nutrition will influence the levels of the different transport systems. Changes in the internal pools of amino acids may change transport and synthesis of the transport systems. As transport is group-specific this regulation will also be group-specific. In addition, the unspecific transport system is strongly induced in nitrogen-deficient cultures as observed in *Penicillium chrysogenum*[65].

A number of amino acids are transported by the neutral amino-acid transport system in *N. crassa*: tryptophan, phenylalanine, tyrosine, but also leucine, cysteine and methionine, and to a lesser extent histidine and serine, but not indole[66,67]. The rate of transport is regulated by intracellular pool size and the half-life of the transport system was found to be approximately 15 min[68]. The rate of tryptophan transport is modulated through the maintenance of a delicate balance between the synthesis and the breakdown of some component of the transport system. Not only tryptophan but also high leucine pools cause feedback of synthesis of the tryptophan-transporting capacity, illustrating the group-specific regulation of transport. It was further found that leucine competes with tryptophan for transport, not utilization or retention.

Some mutants with altered transport capacity have been isolated. LESTER[69] and STADLER[70,71] have found that in a 4-methyltryptophan-resistant mutant the uptake of aromatic amino acids is damaged. The entry of aromatic amino acids can again be facilitated by a second mutation in the general amino acid transport system[63]; tryptophan has now to compete with lysine, arginine and other amino acids for transport.

ST. LAWRENCE and coworkers[72] reported the isolation of a mutant, *mod-5*, with altered permeability for tryptophan. In *mod-5* mutants the initial rate of entry of aromatic amino acids is increased as compared to the wild type. Since also other similar mutations have been found and no specificity in the increase of permeability was apparent, the authors concluded that their mutation was not due to a change of a specific transport system but rather to a structural alteration of the cell membrane in general.

In *S. cerevisiae*, similar group-specific amino-acid transport systems exist[22,73-77]. The specificities seem to be somewhat different from *N. crassa*; transport of aromatic amino acids is inhibited by some aliphatic amino acids, such as isoleucine, threonine, methionine and, to a lesser extent, by others[78]. Interestingly enough, leucine reverses the inhibiton of tryptophan uptake by isoleucine or threonine[78], but this reversion is seemingly not caused by a reversion of uptake inhibition. It was concluded that isoleucine and threonine not only inhibit uptake of tryptophan but also its subsequent utilization and that leucine reversion works at the metabolic and not at the transport level[79]. Undoubtedly, this point needs further clarification. The aromatic-amino-acid transport system is inducible by external additions but also the internal pool size governs transport velocity[73].

The efflux of amino acids seems rather weak under normal conditions and independent of uptake. Entry mechanism and intracellular pool maintenance are two biochemically distinct processes[67,77].

As we have seen few detailed conclusions can be drawn as yet from pool, influx (transport) and efflux studies as to their participation in biosynthetic enzyme regulation. However, a generalized picture can be drawn: biosynthetic enzyme activity, pool size and transport system are interconnected; through the group specificity of transport systems an interconnection of very different amino-acid systems is achieved.

CROSS-PATHWAY REGULATION TRYPTOPHAN—HISTIDINE

The biosynthesis of tryptophan is not only dependent on the aromatic amino acids but also depends on cross-pathway regulation with other amino acids. Apart from the common permeation systems discussed *above* cross-pathway regulation has especially been demonstrated with histidine. "Metabolic interlock" between tryptophan and histidine biosynthesis has been investigated primarily in *B. subtilis* and *N. crassa*.

In *N. crassa* a reduction of histidine synthesis leads to derepression of tryptophan biosynthetic enzymes[80]. Conversely, histidine enzymes are derepressed under conditions of tryptophan deprivation[81]. A further analysis of this phenomenon revealed that tryptophan synthase is derepressed by a factor of 2—3 under conditions which allow internal accumulation of imidazolylglycerol phosphate (IMG-*P*). The action is specific for tryptophan synthase; ING-*P* synthase was not influenced under these conditions either by using a heterokaryon with a defective IMG-*P* dehydratase or by inhibiting this enzyme with 3-amino-1,2,4-triazole[35,36,82].

A more detailed study is available for the bacterium *B. subtilis*. We will treat these data because they bear on the possible interpretation of cross-pathway regulation tryptophan—histidine in fungi. The effects of histidine on aromatic-amino-acid synthesis in this bacterium are manifold. 1. Histidine feedback inhibitis and represses a number of enzymes in the aromatic branch, especially prephenate dehydrogenase and prephenate dehydratase[83]. 2. In the same organism, single mutants can be isolated which are at the same time maximally derepressed for histidine and aromatic -amino-acid enzymes and which are no more repressible by either histidine or tyrosine[84]. In these mutants, tryptophan synthase is still repressed. The mutation is supposed to influence a common aporepressor, probably a tRNA synthetase. 3. Anthranilate synthase is stimulated by histidine; this stimulation does not counteract the inhibition by tryptophan. As a consequence to histidine stimulation more tryptophan is synthesized and the tryptophan enzymes become slightly repressed[31]. Thus histidine plays a dual role for tryptophan synthesis in *B. subtilis*, activation of anthranilate synthase activity (increasing its *V*) and, indirectly, repression of the enzymes.

For *Saccharomyces* a connection between histidine and tryptophan was also postulated by ZAMBONELLI and GUERZONI[85]. The derepression of tryptophan enzymes by 3-amino-1,2,4-triazole illustrates a situation analogous to *N. crassa* (Table II)[108].

MICROBIAL PRODUCTION OF TRYPTOPHAN

Industrial production of tryptophan by microorganisms has been successfully achieved in the last years. An excellent presentation of several problems is given by TERUI and NIIZU[86] and EBIHARA and coworkers[87].

The production of tryptophan by bacteria, although possible, has not been equally promising as the procedures using fungi[88-90]. The processes using yeasts, especially *Candida utilis* and *Hansenula anomala*, have proved to be most successful. Usually anthranilic acid or indole are fed as precursors. An inhibitory effect of high levels of anthranilic acid was observed[91]. This may be due to the inhibition of tryptophan biosynthesis by anthranilic acid, as described for *E. coli* (*see above*). In order to allow for effective feeding by anthranilic acid it is therefore necessary to use a mutant resistant to high levels of anthranilic acid, as has been done by TERUI and NIIZU[86]. A combined feeding of anthranilic acid and indole is more beneficial than precursor alone[87]. The exact nature of the mutation has not yet been elucidated. An effect of cysteine on the consumption of anthranilic acid has been described by LOBYREVA and RUBAN[92]. The authors interpret their observation in that anthranilic acid may be used either for tryptophan synthesis or may be diverted by another pathway; L-cysteine inhibits utilization by this second pathway and permits higher tryptophan yields from the precursor.

The excretion process of the synthesized tryptophan should be of special interest. Amino-acid excretion is an active process and seems to depend on an exchange with external amino acids[67,77]. To what extent the diversion of polysaccharide biosynthesis and the change of cell wall composition to augmented mannan content in the presence of high tryptophan levels, as observed in *Hansenula schneggii* by SUNDHAGUL and HEDRICK[93], contributes to the accumulation while the excretion process remains open.

The production of substituted tryptophans, potential sweetening agents, by *Claviceps purpurea*, may open a new use for the tryptophan-synthesizing capacity of fungi[94].

MICROBIAL ALKALOID PRODUCTION AND TRYPTOPHAN BIOSYNTHESIS

Alkaloid production in microorganisms has long been recognized to be connected to tryptophan biosynthesis[95-97]. A detailed discussion of this point and of alkaloid production in general will not be attempted; these aspects are discussed extensively elsewhere. Two aspects will be discussed shortly: tryptophan biosynthetic and degradative capacities in alkaloid-producing *Claviceps* strains.

An alkaloid-producing strain of *Claviceps paspali* studied by LINGENS and coworkers[39] has an exceptionally high tryptophan biosynthetic capacity: (1) 60% of the DAH-*P* synthase are tryptophan-specific as compared to only 10% in *N. crassa*, thus guaranteeing ample synthesis of chorismic acid even in the presence of phenylalanine and tyrosine, (2) this chorismic acid can be effectively used by a tryptophan-feedback-resistant anthranilate synthase, (3) diversion of chorismic acid to tyrosine seems to be restricted by a reduced level of prephenate dehydrogenase.

On the other hand, all usual degradative pathways (to indolylacetic acid, kynurenine or formylanthranilic acid) are present in the alkaloid-producing *Claviceps*[97,98]; but they do not play an important role under production conditions. Therefore, alkaloid production is made possible by a high tryptophan-synthesizing capacity *and* a high rate of specific tryptophan consumption. This high degree of specific reservation of tryptophan (and its first degradation products) for alkaloid production is due to specific transamination[99] and specific methylation[95,100]. Further studies will be necessary to establish the mechanism of excretion of the synthesized product.

GENERAL REMARKS

In aromatic amino acid and tryptophan biosynthesis a number of controlling elements can be recognized:

1. Isozymes and enzyme complexes permit and guarantee a channeling of precursors to end products. They also facilitate catalytic activity of the reaction sequence. Furthermore they separate biosynthetic from degradative pathways.

2. Usually the activity of the first enzyme of a common pathway or a branch is regulated by feedback inhibition. A repression of these enzymes may be achieved, if the internal pools can be sufficiently augmented; derepression is possible by reducing the pool size.

3. Subsequent enzymes are not at all or not only controlled by repression and derepression, but by induction by their substrates. Such sequential induction has already been observed in other pathways of fungi (*see* review[101].)

4. Biosynthesis of tryptophan is connected in a number of ways to other amino acids, not only to the other aromatics phenylalanine and tyrosine, but especially also to histidine and, due to a common transport system, to leucine. This "metabolic interlock"[31] must lead to a multitude of regulatory signals working on the system.

5. Normally aromatic-amino-acid biosynthesis of fungi seems to be directed towards tyrosine. It may be tuned to tryptophan, however, and allow considerable excretion of this amino acid. Such a change has to be accompanied by an effective and selective consumption of the amino acid (transamination and methylation) in order to permit high alkaloid yields. A number of points, however, remain to be clarified *e.g.*:

6. Do enzyme aggregates influence regulation of enzyme synthesis? And do complexes between biosynthetic and degradative enzymes occur as observed in arginine biosynthesis of yeast[102,103]?

7. What are the roles of tRNA synthetase and the relative levels of charged and uncharged tRNA? They seem to be involved, as observed for isoleucine biosynthesis in yeast[104], but the response of different tRNA's to changes in nutrition may vary[105].

8. In what degree does cross pathway regulation contribute to the maintenance of the levels and activities of the tryptophan biosynthetic enzymes?

REFERENCES

1. Doy C. H.: Rev. Pure Appl. Chem. 41 (1968).
2. Gibson F., Pittard J.: Bacteriol. Rev. 32, 465 (1968).
3. Lingens F.: Angew. Chem. Internat. Ed. 7, 350 (1968).
4. Truffa-Bachi P., Cohen G. N.: Ann. Rev. Biochem. 37, 79 (1968).
5. Ahmed S. I., Giles N. H.: J. Bacteriol. 99, 231 (1969).

6. BERLYN M. B., GILES N. H.: J. Bacteriol. **99**, 222 (1969).
7. HÜTTER R., DeMoss J. A.: J. Bacteriol. **94**, 1896 (1967).
8. DOY C. H.: Biochim. Biophys. Acta **198**, 364 (1970).
9. HALSALL D. M., DOY C. H.: Biochim. Biophys. Acta **185**, 432 (1969).
10. JENSEN R. A., NASSER D. S.: J. Bacteriol. **95**, 188 (1968).
11. DOY C. H., HALSALL D. M.: Biochim. Biophys. Acta **167**, 422 (1968).
12. BURGOYNE L., CASE M. E., GILES N. H.: Biochim. Biophys. Acta **191**, 452 (1969).
13. CASE M. E., BURGOYNE L., GILES N. H.: Genetics **63**, 581 (1969).
14. GAERTNER F. H., ERICSON M. C., DeMoss J. A.: J. Biol. Chem. **245**, 595 (1970).
15. GILES N. H., PARTRIDGE C. W. H., AHMED S. I., CASE M. E.: Proc. Nat. Acad. Sci. U.S.A.
    58, 1930 (1967).
16. BAKER T. I.: Biochemistry **5**, 2654 (1966).
17. BAKER T. I.: Genetics **58**, 351 (1968).
18. COTTON R. G. H., GIBSON F.: Biochim. Biophys. Acta **100**, 76 (1965).
19. EL-ERYANI A. A.: Genetics **62**, 711 (1969).
20. CATCHESIDE D. E. A.: Biochem. Biophys. Res. Comm. **36**, 651 (1969).
21. DOY C. H.: Biochim. Biophys. Acta **151**, 293 (1968).
22. MEURIS P : Bull. Soc. Chim. biol. **49**, 1573 (1967).
23. LINGENS F., GOEBEL W., UESSELER H.: Biochem. Z. **346**, 357 (1966).
24. LINGENS F., GOEBEL W., UESSELER H.: European J. Biochem. **1**, 363 (1967).
25. DeLEEUW A.: Genetics **56**, 554 (1967).
26. AHMAD M., CATCHESIDE D. G.: Heredity **15**, 55 (1960).
27. MATCHETT W. H., TURNER J. R., WILEY W. R.: Yale J. Biol. Med. **40**, 257 (1968).
28. DOY C. H., COOPER J. M.: Biochim. Biophys. Acta **127**, 302 (1966).
29. LINGENS F.: Angew. Chem. Internat. Ed. **6**, 811 (1967).
30. DeMoss J. A., WEGMAN J.: Proc. Nat. Acad. Sci. U.S.A. **54**, 241 (1965).
31. KANE J. F., JENSEN R. A.: J. Biol. Chem. **245**, 2384 (1970).
32. LESTER G.: J. Bacteriol. **96**, 1768 (1968).
33. HELD W. A., SMITH O. H.: J. Bacteriol. **101**, 209 (1970).
34. HELD W. A., SMITH O. H.: J. Bacteriol. **101**, 202 (1970).
35. CLEARY M. J., CARSIOTIS M.: Bacteriol. Proc. 136 (1968).
36. TURNER J. R., MATCHETT W. H.: J. Bacteriol. **95**, 1608 (1968).
37. TURNER J. R., MATCHETT W. H.: Bacteriol. Proc. 144 (1970).
38. GUERDOUX J. L.: Compt. rend. **264**, 1323 (1967).
39. LINGENS F., GOEBEL W., UESSELER H.: European J. Biochem. **2**, 442 (1967).
40. ARROYO-BEGOVICH A., DeMoss J. A.: Proc. Nat. Acad. Sci. U.S.A. **64**, 1072 (1969).
41. GAERTNER F. H., DeMoss J. A.: J. Biol. Chem. **244**, 2716 (1969).
42. DeMoss J. A.: Biochem. Biophys. Res. Comm. **18**, 850 (1965).
43. MANNEY T. R.: J. Bacteriol. **102**, 483 (1970).
44. AHMAD M., KHALIL M., KHAN N. A., MOZMADAR A.: Genetics **49**, 925 (1964).
45. DeMoss J. A., JACKSON R. W., CHALMERS J. H.: Genetics **56**, 413 (1967).
46. CHALMERS J. H.: Dissertation. University of California, San Diego 1968.
47. SCHÜRCH-RATHGEB Y.: Diss. ETH, Zürich, N° 4828; Arch. Genetik, *in press* (1972).
48. YANOVSKY C., in W. D. McElroy, Bentley Glass (Eds.): *Amino Acid Metabolism.* The
    Johns Hopkins Press, Baltimore 1955.
49. AHMAD F., MOAT A. G.: J. Biol. Chem. **241**, 775 (1966).
50. HEILMANN H. D., LINGENS F.: Z. physiol. Chem. **349**, 231 (1968).
51. GUERDOUX J. L.: Compt. rend. **265**, 681 (1967).
52. LUCKNER M.: Z. allg. Mikrobiol. **3**, 93 (1963).
53. MATCHETT W. H., DeMoss J. A.: Biochim. Biophys. Acta **71**, 632 (1963).

54. Wainwright S. D., Bonner D. M.: Canad. J. Biochem. Physiol. 37, 741 (1959).
55. Glombitza K. W., Hartmann T.: Planta 69, 135 (1966).
56. Hagemann F.: Naturwiss. 52, 626 (1965).
57. Okazaki H.: Agric. Biol. Chem. 32, 254 (1968).
58. Hagemann F.: Arch. Mikrobiol. 49, 150 (1964).
59. Shiehr D. J., Chang C., Cheng H. L.: Phytochemistry 8, 397 (1969).
60. Sarachek A.: Experientia 14, 360 (1958).
61. Moat A. G., Ahmad F., Alexander J. K., Barnes I. J.: J. Bacteriol. 98, 573 (1969).
62. Matchett W. H., DeMoss J. A.: Biochim. Biophys. Acta 86, 91 (1964).
63. Pall M. L.: Genetics 60, 209 (1968).
64. Pall M. L.: Biochim. Biophys. Acta 173, 113 (1969).
65. Benko P. V., Wood T. C., Segel I. H.: Arch. Biochem. Biophys. 129, 498 (1969).
66. Brockman H. E.: J. Gen. Microbiol. 34, 31 (1964).
67. Wiley W. R., Matchett W. H.: J. Bacteriol. 92, 1698 (1966).
68. Wiley W. R., Matchett W. H.: J. Bacteriol. 95, 959 (1968).
69. Lester G.: J. Bacteriol. 91, 677 (1966).
70. Stadler D. R.: Genetics 54, 677 (1966).
71. Stadler D. R.: Genetics 57, 935 (1967).
72. St. Lawrence P., Maling B. D., Altwerger L., Rachmeler M.: Genetics 50, 1383 (1964).
73. Greasham R. L., Kurlans A., Moat A. G.: Bacteriol. Proc. 126 (1970).
74. Grenson M.: Biochim. Biophys. Acta 127, 339 (1966).
75. Grenson M., Mousset M., Wiame J. M., Bechet J.: Biochim. Biophys. Acta 127, 325 (1966).
76. Meuris P.: Genetics 63, 569 (1969).
77. Surdin Y., Sly W., Sire J., Bordes A. M., de Robichon-Szulmajster H.: Biochim. Biophys. Acta 107, 546 (1965).
78. Moat A. G., Kateiva A.: Bacteriol. Proc. 119 (1968).
79. Greasham R. L., Moat A. G.: Bacteriol. Proc. 146 (1969).
80. Carsiotis M., Lester G.: Bacteriol. Proc. 136 (1968).
81. Jones R. F., Carsiotis M.: Bacteriol. Proc. 136 (1968).
82. Jones R., Carsiotis M.: Bacteriol. Proc. 114 (1967).
83. Nester E. W.: J. Bacteriol. 96, 1649 (1968).
84. Chapman L. F., Nester E. W.: J. Bacteriol. 96, 1658 (1968).
85. Zambonelli C., Guerzoni M. E.: Arch. Mikrobiol. 70, 288 (1970).
86. Terui G., Niizu H.: Biotechnol. Bioeng. Symp. No. 1, 33 (1969).
87. Ebihara Y., Niitsu H., Terui G.: J. Ferment. Technol. 47, 733 (1969).
88. Bandiera M., Morpurgo G., Ricci R.: Experientia 23, 724 (1967).
89. Lim P. G., Mateles R. I.: J. Bacteriol. 87, 1051 (1964).
90. Verkhovtseva M. I., Ruban E. L., Suvorov N. N.: Izv. Akad. Nauk S.S.S.R., Ser. Biol. 374 (1967).
91. Ruban E. L., Lobyreva L. B.: Prikl. Biokhim. Mikrobiol. 1, 74 (1965).
92. Lobyreva L. B., Ruban E. L.: Izv. Akad. Nauk S.S.S.R., Ser. Biol. 512 (1969).
93. Sundhagul M., Hedrick L. R.: J. Bacteriol. 92, 241 (1966).
94. Fukuda D., Mabe J. A., Brannon D. R.: Bacteriol. Proc. 4 (1970).
95. Bu'Lock J. D., Barr J. G.: Lloydia 31, 342 (1968).
96. Floss H. G., Mothes U.: Arch. Mikrobiol. 48, 213 (1964).
97. Kaplan H., Hornemann U., Kelley K. M., Floss H. G.: Lloydia 32, 489 (1969).
98. Teuscher E.: Pharmazie 21, 320 (1966).
99. Teuscher E.: Z. allg. Mikrobiol. 10, 137 (1970).

100. HORNEMANN U., SPEEDIE M. K., KELLEY K. M., HURLEY L. H., FLOSS H. G.: Arch. Biochem. Biophys. **131**, 430 (1969).
101. GROSS S. R.: Ann. Rev. Genet. **3**, 395 (1969).
102. BECHET J., GRENSON M., WIAME J. M.: European J. Biochem. **12**, 31 (1970).
103. MESSENGUY F., WIAME J. M.: F.E.B.S. Letters **3**, 47 (1969).
104. MCLAUGHLIN C. S., MAGEE P. T., HARTWELL L. H.: J. Bacteriol. **100**, 579 (1969).
105. EHRESMANN B., WEIL J. H.: Compt. rend. Soc. Biol. **162**, 2290 (1968).
106. MANNEY T. R.: Genetics **50**, 109 (1964).
107. DEMOSS J. A., *personal communication*.
108. SCHÜRCH A.: Diss. ETH, Zürich, N° 4862; Juris-Verlag, Zürich, 1972.
109: WEGMAN J., DEMOSS J. A.: J. Biol. Chem. **240**, 3781 (1965).

Genetic Control of Tryptophan Biosynthesis in *Streptomyces coelicolor*

PAULINUS P. ENGEL

Department of Biology, St. Benedict's College*, Atchison, Kansas, U.S.A.

INTRODUCTION

Mutant sites of tryptophan auxotrophs occur in two clusters on the chromosome of *Streptomyces coelicolor*. It has not been determined whether other loci, not involved in tryptophan metabolism, occur among the closely linked *trp* mutant sites within the clusters. Nor is it known whether there are other, yet undiscovered, clusters of mutant sites affecting the structure and regulation of tryptophan biosynthetic enzymes. Evidence for the chromosomal location of two clusters of *trp* mutants and the order of mutant sites within each cluster is presented in this paper. Progress made in the study of complementation reactions between mutants within the same cluster is also reported.

Tryptophan auxotrophs and regulatory mutants affecting tryptophan metabolism have been investigated in *Bacillus subtilis*[1], *Escherichia coli*[2-5], *Pseudomonas aeruginosa*[6], *Pseudomonas putida*[7-10] and *Salmonella typhimurium*[11-13] to elucidate the pathway of tryptophan biosynthesis, the arrangement of structural and regulatory genes on the chromosome and the regulatory mechanisms involved in controlling the synthesis of tryptophan biosynthetic enzymes. Investigation of mutants affecting structural genes and regulation of the tryptophan biosynthetic pathway in *Streptomyces coelicolor* may provide insights into: how the regulation of gene expression is effected and coordinated in an organism in which genes with related functions are located in more than one cluster; what adaptive advantages such an arrangement of genes might confer on an organism; and the differences between *S. coelicolor* and other prokaryotes in the regulation of tryptophan biosynthesis. Being primarily water and soil bacteria, perhaps quite different regulatory mechanisms for tryptophan metabolism evolved in the pseudomonads and streptomycetes from those found in enteric bacteria. Knowledge gained from comparative studies of genetic regulatory mechanisms must eventually provide a more adequate approach for the solution of health-related problems and other improvements in the quality of human life.

---

* Present address: Department of Biology, Virginia Polytechnic Institute, Blacksburg, Virginia 24061, U.S.A.

MATERIALS AND METHODS

BACTERIAL STRAINS AND CULTURE MEDIA

All strains employed in this investigation were derived either by mutation or recombination from wild-type *S. coelicolor* A3(2)[14]. Mutant strains *trp-1, 2, 5, 7* and *8* (Table I) were obtained by D. A. HOPWOOD and *trp-v1* by A. VIVIAN. These strains were formerly designated *try-1, 2, 5, 7, 8* and *v1*[15]. Following the recommendations of DEMEREC, ADELBERG, CLARK and HARTMAN[16] the abbreviation *trp* is used to represent the genotype of tryptophan auxotrophs in this paper. These strains were generously provided by D. A. HOPWOOD. The other mutants in Table I whose isolation numbers are prefixed with an e were obtained by this investigator.

The culture media employed are those described by HOPWOOD[15].

INDUCTION OF MUTANTS

Ethyl methanesulfonate was used to induce mutants following the procedure described by FREESE[17]. N-Methyl-N'-nitro-N-nitrosoguanidine-induced mutants were obtained employing the method described by DELIĆ, HOPWOOD and FRIEND[18].

GENETIC ANALYSIS

The genetic procedures employed in this investigation were described in detail by HOPWOOD[15]. The approximate location of mutants on the linkage map was determined by crossing a newly isolated *trp* mutant which was streptomycin-sensitive, with multiply marked strain 876 which bears *proA1, his C9, argA1, cys C3, pheA1* and *strA1*. The chromosomal locations of these markers may be seen in Fig. 1. The selected markers in these crosses were *strA1* and *his⁺*. A new *trp* mutant is located in that region of the map which requires the fewest crossovers to account for the frequency and linkage relations of *trp* and *trp⁺* alleles with other nonselected markers.

Two types of marker selection were employed in four-point reciprocal crosses to order mutant sites within each cluster: one method involved the selection of *pro⁺*, allele of *proA1*, and *arg⁺*, allele of *argA1*; the second method employed selection of haploid recombinants between *trp⁺* markers within a cluster. Outside markers *pro⁺* and *arg⁺* are separated by approximately 33 map units; the entire linkage map illustrated in Fig. 1 contains 260 map units[15].

The order of mutant sites within each cluster was determined from the frequency of *trp⁺* colonies among progeny of reciprocal crosses in which outside markers were selected. The order of mutant sites indicated by this selective method was compared with the order of sites determined by the frequency of nonselected outside markers *pro⁺* and *arg⁺* among selected *trp⁺* markers. In most cases these two methods permit an unambiguous assignment of order of sites within a cluster.

The selection of *trp⁺* markers also provides a complementation test of the *trp* mutants involved in the cross.

COMPLEMENTATION ANALYSIS

HOPWOOD[15] described the methods of complementation analysis employed in this investigation. These methods involve either (i) the identification of heteroclones or (ii) the observation of a large value for the ratio of colonies on media selecting distant markers to colonies on media selecting trp^+ markers indicating that complementation did not occur when heteroclones are absent; thus few colonies appear on media selecting pro^+ and arg^+ markers precluding the analysis of a significant number of colonies to determine whether they are of heteroclonal origin.

Complementation analysis based on the identification of heteroclones. A heteroclone is a colony containing haploid cells with different genotypes which are generated by recombination in heterozygous regions of partial diploids. The complementation test which depends on the identification of heteroclones was done in conjunction with four-point reciprocal crosses, described in the previous section, which are also designed to detect the order of mutant sites since the selection of trp^+ alleles of trp mutants whose complementation reaction is being tested also permits the growth of heteroclones.

Complementation tests involve plating spores from a cross of appropriately marked parents (*proA1 trp-x* × *argA1 trp-y*) on two media: the first medium, containing tryptophan but lacking proline and arginine, selects distant markers pro^+ and arg^+; the second medium, lacking tryptophan but containing proline and arginine, selects haploid recombinants between trp^+ markers and permits growth of heteroclones. Since heteroclones can only be formed by complementing mutants, the detection of complementation relies on the identification of heteroclones.

Haploid trp^+ colonies on the second medium, lacking tryptophan, are distinguished from heteroclones by streaking spores and mycelial fragments from a colony onto an appropriate culture medium over an area about one centimeter square. Such master plates, after two to three days incubation, are then replicated to media diagnostic for the outside markers. A heteroclone gives rise to non-uniform replicates on different diagnostic media indicating a mixture of genotypes among the cells of a single colony.

Failure to detect heteroclones is interpreted to indicate that two mutants either affect the same reaction or the mutational lesion has a polarized effect which prevents complementation. However, the failure to detect heteroclones may also be the result of some incompatibility of parents not directly related to the trp mutants being tested. The recovery of haploid trp^+ progeny provides a control to determine whether the failure to complement is a consequence of incompatibility of parents in the formation of merozygotes disomic for the region including the trp mutant sites being tested for complementation. The presence of haploid trp^+ progeny, generated by recombination between selected trp^+ markers, indicates that such merozygotes were formed. Thus the failure to detect heteroclones and the recovery of haploid progeny from a cross selecting trp^+ markers constitute satisfactory evidence indicating that two mutants do not complement.

Failure to detect heteroclones might also be expected if two mutants are not closely linked since recombination early in the development of the colony might give rise to haploid cells of a phenotype which would out-grow cells with other phenotypes. Such colonies might escape detection by the method of streaking employed in this investigation. However, when complementation tests are done between cluster 1 and cluster 2 mutants which are not so closely linked as mutants within the same cluster about 10% of the colonies are heteroclones. It seems improbable that the failure to detect heteroclones is a consequence of fast-growing haploid progeny generated by recombination between trp^+ markers within the same cluster.

Complementation analysis based on the ratio of colonies on media selecting distant markers to colonies on media selecting trp^+ markers. Frequently the total number of colonies on media selecting distant markers is at least fifty times greater than the number of colonies on media selecting trp^+ markers. Since few colonies appear on media selecting trp^+ markers this severely limits the number of colonies that may be tested to determine whether they are of heteroclonal origin. Nevertheless, the few haploid colonies that appear on media selecting trp^+ markers must result from recombination between the selected markers. This suggests that, although the trp mutants fail to complement since heteroclones are not detected, this failure is not the result of some incompatibility of the parents for the formation of merozygotes disomic in the region of the trp mutant sites. In such cases one may conclude that the trp mutants in question do not complement and presumably affect the same reaction or are polarity mutants.

Distinguishing allelic from genic complementation. This investigation has not progressed to the stage where a distinction can be made between these two types of complementation. Perhaps such a distinction will be possible after some of the trp mutants are identified with the enzymic reactions and/or regulatory functions they affect.

ANALYSIS OF ACCUMULATED INTERMEDIATES

Preparation of cells for study of accumulated intermediates in culture filtrates. A blue pigment which interferes with some of the accumulation tests is formed rapidly in liquid cultures of germinating spores. The culture filtrate of a sufficiently large inoculum of rapidly growing cells may be analyzed for accumulated intermediates before an appreciable quantity of the pigment is present. Such inocula are prepared by incubating approximately 2.5×10^7 spores in 25 ml of complete medium containing 50 µg/ml L-tryptophan in a 250 ml Erlenmeyer flask left standing for 48 h at 30 °C. Standing cultures are used rather than shake cultures because it is difficult to fragment the mycelial balls formed in shaking cultures to get a uniform suspension of cells for the next step. At the end of 48 h the mycelia are filtered and washed with minimal medium. The mycelial pad and the small cotton plug used as a filter are then put into a milk dilution bottle containing 50 ml of minimal medium and 0.5 g of glass beads (0.5 mm diameter) and shaken gently to obtain a uniform suspension of cells. The 50 ml of minimal medium containing the suspension of cells is then decanted into a 250 ml Erlenmeyer flask and incubated at 30 °C with shaking in a New Brunswick Model G-25 gyrotory shaker. The accumulation tests are done at the end of 16—18 h incubation.

Detection of indolylglycerol and indole in culture filtrates. The ferric chloride reagent[19] was used to detect indolylglycerol in culture filtrates of *trp-8*. None of the other mutants appear to accumulate this intermediate. Ehrlich's reagent[20] was employed in several unsuccessful attempts to detect indole accumulation in mutants which do not utilize indole.

RESULTS

DESCRIPTION OF *trp* MUTANTS

The mutagens used to induce *trp* mutants, the location of these mutants in either cluster 1 or 2 and their growth response to anthranilic acid and indole are described in Table I. Mutants that utilize anthranilic acid have not been found.

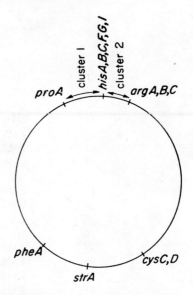

Fig. 1. Linkage map of *S. coelicolor* showing location of clustered *trp* mutant sites relative to markers routinely employed in the genetic analysis. Mutant sites in cluster 1 and cluster 2, represented by lines terminating with arrows, occur between *proA* and *hisC* for cluster 1 and between *hisC* and *argA* for cluster 2. This map is redrawn to scale from the linkage map published by HOPWOOD[15]. Mutants auxotrophic for arginine, cystine, histidine, phenylalanine and proline are represented by *arg*, *cys*, *his*, *phe* and *pro*, respectively. Streptomycin resistance is represented by *str*.

No mutant hunt yielded more than one mutant with the same growth response; thus the clonal origin of mutants in Table I is excluded.

CROSSES TO DETERMINE THE LOCATION OF *trp* MUTANTS

The location on the linkage map of all the mutants described in Table I, except *v1*, was determined from the frequency and linkage relations of nonselected markers after selecting histidine prototrophs and streptomycin-resistant colonies among progeny from crosses of *trp* with strain 876 which bears markers *proA1*, *hisC9*, *argA1*, *cysC3*, *pheA1* and *strA1* (Fig. 1).

As a consequence of the circular linkage map[15,21] an even number of crossovers and a minimum of one crossover in each arc separating the selected markers are required to generate progeny bearing the selected markers. More than one crossover in each arc on both sides of the selected markers is referred to as a multiple exchange in Table II.

The results of crosses to determine the location of *trp* mutants are summarized in Table II. The cross of 876 × *trp-e1* gave progeny whose genotypes are explained by invoking a relatively large number of crossovers. Attempts will be made, if these initial observations are corroborated, to study this increased frequency of genetic exchange which seems to be associated with *trp-e1*.

TABLE I

Tryptophan-Requiring Mutants of *S. coelicolor*

Isolation number	Mutagen[a]	Located in cluster	Growth response[b]	
			anthranilic acid	indole
1	UV	2	−	−
2	UV	2	−	−
v1	UV	?	−	−
e1	NTG	1	−	+
e2	EMS	1	−	+
e3	EMS	2	−	−
e4	EMS	1	−	+
5	UV	1	−	+
e5	EMS	1	−	+
e6	EMS	1	−	+
7	UV	1	−	+
8	UV	2	−	+
e7	EMS	1	−	+
e9	EMS	1	−	+
e10	EMS	1	−	+
e11	EMS	2	−	−
e12	EMS	1	−	+
e13	NTG	2	−	−
e14	NTG	2	−	−
e16	NTG	1	−	+

[a] UV, ultraviolet light; NTG, N-methyl-N′-nitro-N-nitrosoguanidine; EMS, ethyl methanesulfonate.
[b] + growth on media containing the supplement;
− no growth on media containing the supplement.

Generally the difference in number of multiple crossovers is not very great when *trp* mutants are hypothetically assigned to the regions bounded by *proA* and *hisC* or *hisC* and *argA*, as may be seen in the results for crosses **2, 3, 5, 7, 12, 13, 15, 17, 18** and **19** (cross reference numbers in Table II) in Table II. It was desirable, therefore, to do four-point reciprocal crosses to determine from the frequency of nonselected *his*[+] and *trp*[+] markers among selected *pro*[+] and *arg*[+] markers whether a mutant site is in cluster 1 or cluster 2.

TABLE II

Crosses to Determine the Map Position of *trp* Mutants

Reference N° of cross	Cross[b]	Number of colonies scored	Number of multiple exchanges[a] required if order is as indicated						Map position
			pro	trp	his	his	trp	arg	
1	876 × trp-e1	156		64			71		pro trp-e1 his
2	876 × trp-1	48		3			3		not determined
3	876 × trp-2	298		12			14		pro trp-2 his
4	876 × trp-e2	250		8			19		pro trp-e2 his
5	876 × trp-e3	299		20			22		pro trp-e3 his
6	876 × trp-e4	448		17			34		pro trp-e4 his
7	876 × trp-5	346		30			34		pro trp-5 his
8	876 × trp-e5	298		12			18		pro trp-e5 his
9	876 × trp-e6	80		5			10		pro trp-e6 his
10	876 × trp-7	392		51			62		pro trp-7 his
11	876 × trp-e7	100		5			8		pro trp-e7 his
12	876 × trp-8	300		3			3		not determined
13	876 × trp-e9	243		7			9		pro trp-e9 his
14	876 × trp-e10	297		8			18		pro trp-e10 his
15	876 × trp-e11	360		13			13		not determined
16	876 × trp-e12	100		2			5		pro trp-e12 his
17	876 × trp-e13	650		4			4		not determined
18	876 × trp-e14	139		8			7		his trp-e14 arg
19	876 × trp-e16	250		13			15		pro trp-e16 his

[a] Multiple exchange is defined in the text.

[b] 876 bears *proA1*, *hisC9*, *argA1*, *cysC3*, *strA1* and *pheA1*; the selected markers in all these crosses are *his*[+] and *strA1*.

TABLE III

Four-Point Reciprocal Crosses to Determine the Order of Nonselected Markers $hisC9^+$ and trp^+ with Respect to Selected Outside Markers $proA1^+$ and $argA1^+$

	Cross: *pro trp* × *his arg*		Cross: *pro his* × *trp arg*		Order of markers
Parents	Number of colonies scored	Frequency of his^+ trp^+ colonies %	Parents	Number of colonies scored	Frequency of his^+ trp^+ colonies %	
trp-1 × *his*	450	0.22	*his* × *trp-1*	450	2.9	*his trp-1 arg*
trp-2 × *his*	111	0	*his* × *trp-2*	296	11.5	*his trp-2 arg*
trp-e3 × *his*	376	0	*his* × *trp-e3*	366	1.4	*his trp-e3 arg*
trp-8 × *his*	350	0	*his* × *trp-8*	350	1.7	*his trp-8 arg*
trp-e11 × *his*	244	0.82	*his* × *trp-e11*	246	2.45	*his trp-e11 arg*
trp-e13 × *his*	225	0.44	*his* × *trp-e13*	150	5.3	*his trp-e13 arg*
trp-e14 × *his*	358	0	*his* × *trp-e14*	250	1.2	*his trp-e14 arg*
trp-e4 × *his*	200	12.5	*his* × *trp-e4*	200	0	*pro trp-e4 his*
trp-7 × *his*	200	12	*his* × *trp-7*	200	1	*pro trp-7 his*
trp-e10 × *his*	300	9.7	*his* × *trp-e10*	300	0.67	*pro trp-e10 his*
trp-e12 × *his*	104	14.4	*his* × *trp-e12*	190	1.1	*pro trp-e12 his*

FOUR-POINT RECIPROCAL CROSSES TO DETERMINE THE POSITION OF *trp* MUTANTS WITH RESPECT TO REGIONS BOUNDED BY *proA* AND *hisC* OR *hisC* AND *argA*

It was established that mutants *1, 2, e3, 8, e11, e13* and *e14*, all cluster 2 mutants, are located between *hisC* and *argA* from the frequency of *his*⁺ and *trp*⁺ colonies among progeny from four-point reciprocal crosses in which *pro*⁺ and *arg*⁺ were the selected markers. The data from these reciprocal crosses are summarized in Table III and constitute the evidence for assigning these seven mutants to cluster 2.

The data for the last four crosses in Table III are included to show that mutants *e4, 7, e10* and *e12* of cluster 1 are located between *proA* and *hisC*. Evidence will be presented later to show that *e4, 7, e10* and *e12* are among those mutants in cluster 1 which are nearest *hisC*, corroborating the evidence presented in Table II for assigning mutants *e1, e2, 5, e5, e6, e7, e9* and *e16* to cluster 1, in the region bounded by *proA* and *hisC*.

FOUR-POINT RECIPROCAL CROSSES TO DETERMINE THE ORDER OF MUTANT SITES IN CLUSTER 1

The order of mutant sites in cluster 1 was determined from the frequency of *trp*⁺ colonies among progeny of reciprocal crosses in which outside markers *pro*⁺ and *arg*⁺ were selected. The results of these crosses are summarized in Table IV.

The order of mutant sites in cluster 1 was also determined from the frequency of nonselected *pro*⁺ and *arg*⁺ markers among progeny of reciprocal crosses in which haploid recombinants between *trp*⁺ markers were selected. The results of these crosses are summarized in Table V.

A comparison of the order of mutant sites revealed by the two types of marker selection is presented in Fig. 2. In general, there is satisfactory agreement regarding the order of mutant sites as determined by these selective methods, although *trp-e12* appears to the left of *trp-e16* when *pro*⁺ and *arg*⁺ are the selected markers and to the right of *trp-e16* when *trp*⁺ markers are selected. Perhaps additional crosses will resolve this discrepancy.

Nearly all the necessary evidence to determine the sequence of mutant sites has been obtained and should be of considerable value in correlating the genetic map of cluster 1 with the complementation reactions of mutants within this cluster. Evidence is presented in Table III indicating that mutants *e4, 7, e10* and *e12* are located between *proA* and *hisC*. Reference to Fig. 2 suggests that all other mutants assigned to cluster 1 are also located between *proA* and *hisC*.

TABLE IV

Order of *trp* Mutant Sites with Respect to Selected Outside Markers *pro*$^+$ and *arg*$^+$

	Cross *pro trp-x* × *trp-y arg*			Cross *pro trp-y* × *trp-x arg*			Order of mutants
trp parents	*trp* parents	Number of colonies scored	Frequency of *trp*$^+$ colonies among selected *arg*$^+$ and *pro*$^+$ colonies %	*trp* parents	Number of colonies scored	Frequency of *trp*$^+$ colonies among selected *arg*$^+$ and *pro*$^+$ colonies %	
e6 × e2		604	0.33	e2 × e6	617	0.65	*pro trp-e2 trp-e6 his*
5 × 7		925	0.32	7 × 5	962	2.61	*pro trp-7 trp-5 his*
e9 × 7		986	0.41	7 × e9	600	0.17	*pro trp-e9 trp-7 his*
e5 × e16		400	0.75	e16 × e6	600	0.16	*pro trp-e6 trp-e16 his*
7 × e16		274	0	e16 × 7	600	1.70	*pro trp-e16 trp-7 his*
e10 × 7		200	0	7 × e10	200	3.50	*pro trp-7 trp-e10 his*
e4 × e10		449	0.22	e10 × e4	743	2.30	*pro trp-e10 trp-e4 his*
e16 × e9		1248	0	e9 × e16	690	3.65	*pro trp-e9 trp-e16 his*
e12 × e10		985	19.40	e10 × e12	300	0.33	*pro trp-e12 trp-e10 his*
e12 × e4		772	0.78	e4 × e12	593	0.68	*pro trp-e12 trp-e4 his*
e2 × e10		120	8.40	e10 × e2	297	0.34	*pro trp-e2 trp-e10 his*
7 × e1		349	0	e1 × 7	179	1.10	*pro trp-e1 trp-7 his*
e12 × e5		89	1.10	e5 × e12	289	0	*pro trp-e12 trp-e5 his*
e7 × e12		600	0.17	e12 × e7	182	0	*pro trp-e7 trp-e12 his*
e12 × e16		350	2	e16 × e12	521	0.77	*pro trp-e12 trp-e16 his*
e1 × e2		298	0.33	e2 × e1	460	0	*pro trp-e1 trp-e2 his*
7 × e6		268	0.37	e6 × 7	443	0.68	*pro trp-e6 trp-7 his*
e9 × e6		392	0.25	e6 × e9	177	0	*pro trp-e9 trp-e6 his*
e4 × e1		500	0	e1 × e4	265	0	not established
e7 × e2		622	0.32	e2 × e7	34	0	not established

Cross			Cross			
e10 × e9	400	0	e9 × e10	484	0	not established
e12 × 7	166	0	7 × e12	186	0	not established
e9 × e2	297	0.34	e2 × e9	1200	0.25	not established
5 × e10	400	0.75	e10 × 5	261	0.76	not established
7 × e5	442	1.10	e5 × 7	548	1.10	not established
e1 × e10	199	3	e10 × e1	249	2.8	not established
7 × e2	203	0.49	e2 × 7	196	0	not established
e2 × e12	23	0	e12 × e2	31	0	not established
e6 × e12	26	0	e12 × e6	5	0	not established
e4 × e7	400	0	e7 × e4	400	0	not established
e1 × e6	89	0	e6 × e1	400	0	not established

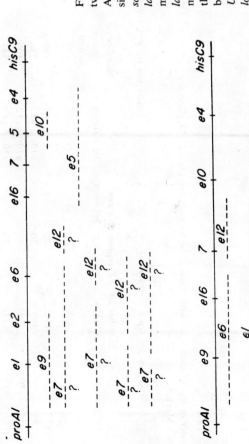

Fig. 2. The order of cluster 1 mutant sites based on the two types of marker selection described in the text. A *dashed line* indicates that the position of the mutant site has not been established within the region on the solid line immediately above the dashed line. *Non-overlapping* dashed lines indicate that the order of two mutants, represented by dashed lines, is known; *overlapping* dashed lines indicate that the order of two mutants is not known. The space between mutants on the solid line does not represent relative map distance between mutant sites.

Upper: selected markers: pro^+/arg^+, based on Table IV; *lower*: selected markers: trp^+/trp^+, based on Table V.

TABLE V

Order of trp Mutant Sites with Respect to Nonselected Outside Markers pro+ and arg+ Determined by Selecting Haploid Recombinants between trp+ Markers

Cross pro trp-x × trp-y arg			Cross pro trp-y × trp-x arg			Order of mutants
trp parents	Number of colonies scored	Frequency of pro+ trp+ arg+ colonies %	trp parents	Number of colonies scored	Frequency of pro+ trp+ arg+ colonies %	
e10 × 7	152	42	7 × e10	160	56	pro trp-7 trp-e10 his
e4 × e10	360	30.5	e10 × e4	256	64	pro trp-e10 trp-e4 his
e12 × e10	499	39	e10 × e12	109	26.5	pro trp-e12 trp-e10 his
e1 × e10	132	22	e10 × e1	343	0.58	pro trp-e1 trp-e10 his
e12 × e16	121	8.3	e16 × e12	176	47	pro trp-e16 trp-e12 his
e16 × e9	469	14	e9 × e16	120	37.5	pro trp-e9 trp-e16 his
7 × e6	159	0	e6 × 7	176	25	pro trp-e6 trp-7 his
e1 × e6	208	0	e6 × e1	8	25	(pro trp-e6 trp-e1 his)[a]
e7 × e12	250	36.5	e12 × e7	13	0	(pro trp-e7 trp-e12 his)[a]
e12 × e4	355	5.4	e4 × e12	33	12.2	(pro trp-e4 trp-e12 his)[a]
e9 × e2	10	20	e2 × e9	261	12.8	(pro trp-e9 trp-e2 his)[a]
e2 × e10	278	27.5	e10 × e2	339	29.3	(pro trp-e10 trp-e2 his)[a]
5 × 7	144	58	7 × 5	19	21	(pro trp-5 trp-7 his)[a]
e6 × e16	173	12.7	e16 × e6	6	33	(pro trp-e16 trp-e6 his)[a]

[a] Suggested.

TABLE VI

Order of *trp* Mutant Sites with Respect to Selected Outside Markers *pro*$^+$ and *arg*$^+$

	Cross *pro trp-x* × *trp-y arg*		Cross *pro trp-y* × *trp-x arg*			
trp parents	Number of colonies scored	Frequency of *trp*$^+$ colonies among selected *arg*$^+$ and *pro*$^+$ colonies %	*trp* parents	Number of colonies scored	Frequency of *trp*$^+$ colonies among selected *arg*$^+$ and *pro*$^+$ colonies %	Order of mutants
1 × e3	350	0	e3 × 1	150	1.3	*his trp-e3 trp-1 arg*
1 × e11	350	2.3	e11 × 1	349	0.28	*his trp-1 trp-e11 arg*
e11 × e3	200	1	e3 × e11	250	9.3	*his trp-e3 trp-e11 arg*
8 × 1	613	13	1 × 8	618	0.32	*his trp-8 trp-1 arg*
8 × e3	743	2.15	e3 × 8	732	0.68	*his trp-8 trp-e3 arg*
e13 × 8	409	0.74	8 × e13	417	0	*his trp-e13 trp-8 arg*
e14 × e11	311	9.3	e11 × e14	400	6.5	*his trp-e14 trp-e11 arg*
e11 × 2	549	2	2 × e11	341	0.29	*his trp-e11 trp-2 arg*
e13 × e11	300	0	e11 × e13	400	2	*his trp-e11 trp-e13 arg*
8 × e14	74	0	e14 × 8	199	1	*pro trp-e14 trp-8 arg*
8 × e11	350	0.57	e11 × 8	397	0.75	*pro trp-e11 trp-8 arg*
2 × e14	5	0	e14 × 2	8	0	*not established*

TABLE VII

Order of *trp* Mutant Sites with Respect to Nonselected Outside Markers *pro*+ and *arg*+ Determined by Selecting Haploid Recombinants between *trp*+ Markers

Cross *pro trp-x* × *trp-y arg*			Cross *pro trp-y* × *trp-x arg*			Order of mutants
trp parents	Number of colonies scored	Frequency of *pro*+ *trp*+ *arg*+ colonies %	*trp* parents	Number of colonies scored	Frequency of *pro*+ *trp*+ and *arg*+ colonies %	
e3 × 8	203	5.9	8 × e3	275	26.3	*his trp-8 trp-e3 arg*
e13 × 8	81	48	8 × e13	51	37.5	*his trp-e13 trp-8 arg*
e14 × e11	43	54	e11 × e14	299	27	*his trp-e14 trp-e11 arg*
e11 × 2	206	28.7	2 × e11	509	0.79	*his trp-e11 trp-2 arg*
8 × 1	184	9.2	1 × 8	118	34.8	*his trp-1 trp-8 arg*
e13 × e11	396	1	e11 × e13	169	25	*his trp-e11 trp-e13 arg*
e13 × 1	259	23.2	1 × e13	230	62	*his trp-1 trp-e13 arg*
8 × e14	177	0	e14 × 8	214	35	*his trp-e14 trp-8 arg*
8 × e11	287	15.4	e11 × 8	265	21.5	*his trp-e11 trp-8 arg*
1 × e3	45	11.1	e3 × 1	0	—	not established
1 × e11	108	21	e11 × 1	16	18.7	(*his trp-1 trp-e11 arg*)[a]
e11 × e3	11	9.1	e3 × e11	100	30.7	(*his trp-e3 trp-e11 arg*)[a]

[a] Suggested.

TABLE VIII

Complementation of *trp* Mutants in Cluster 1

Cross *pro trp-x* × *trp-y arg*		Cross *pro trp-y* × *trp-x arg*		Total number of colonies scored for presence of heteroclones	Hetero-clones %	Complementation[b]
trp parents	Ratio of colonies on selective media[a]	*trp* parents	Ratio of colonies on selective media[a]			
e9 × e2	99	e2 × e9	29	271	3.7	+
e7 × e2	128	—	—	519	2.7	+
e4 × e10	156	e10 × e4	14	616	9.4	+
e12 × e10	8	e10 × e12	4	608	4.9	+
e10 × 7	46	—	—	312	2.9	+
e2 × e10	16	e10 × e2	97	617	4.4	+
e1 × e10	1400	e10 × e1	0.32[c]	475	1.5	+
e12 × e4	44	e4 × e12	47	388	1.3	+
e16 × e9	194	e9 × e16	36	589	5.1	+
7 × e1	157	e1 × 7	0.0004[d]	110	1.8	+
7 × e6	0.76[e]	e6 × 7	118	335	1.2	+
e6 × e16	14	e16 × e6	800	179	1.7	+
e12 × e16	13	e16 × e12	44	297	0.67	—
e16 × e1	131	—	—	267	0.37	—
e7 × e12	153	e12 × e7	15	263	0	—
e4 × e6	0.14[f]	—	—	299	0	—
7 × e12	6	e12 × 7	6	47	0	?
e12 × e5	2	e5 × e12	22	35	0	?

[a] Ratio of colonies on medium selecting distant markers pro^+ (allele of *proA1*) and arg^+ (allele of *argA1*) to colonies on medium selecting trp^+ markers.

[b] 1% heteroclones among 100 colonies analyzed is arbitrarily considered as evidence for complementation; + complementation, — no complementation, ? insufficient evidence to determine whether or not complementation occurs.

[c] 96% of 343 colonies analyzed had genotype $pro\ trp^+\ arg^+$.

[d] 100% of 106 colonies analyzed had genotype $pro\ trp^+\ arg^+$.

[e] 100% of 159 colonies analyzed had genotype $pro\ trp^+\ arg^+$.

[f] 100% of 299 colonies analyzed had genotype $pro\ trp^+\ arg^+$.

TABLE IX

Complementation of trp Mutants in Cluster 2

Cross pro trp-x × trp-y arg		Cross pro trp-y × trp-x arg		Total number of colonies scored for presence of heteroclones	Hetero-clones %	Comple-mentation[b]
trp parents	Ratio of colonies on selective media[a]	trp parents	Ratio of colonies on selective media[a]			
8 × 1	5	1 × 8	50	302	5.6	+
e3 × 8	13	8 × e3	9	478	7.8	+
e13 × 8	9	8 × e13	11	132	3.8	+
e11 × e13	2	e13 × e11	1	565	1.8	+
e14 × e11	10	e11 × e14	5	342	2.3	+
8 × e11	7	e11 × 8	7	552	17.8	+
2 × e13	0.7[c]	—	—	339	1.48	+
e11 × 2	25	2 × e11	1	715	0.42	—
8 × e14	0.1[d]	e14 × 8	84	391	0.51	—
e11 × e3	73	e3 × e11	6	305	0	—
e13 × 1	19	1 × e13	2	489	0·82	—
e14 × e13	9	—	—	115	0	?

[a] See footnote a in Table VIII.
[b] See footnote b in Table VIII.
[c] 87% of 339 colonies analyzed had genotype pro^+ trp^+ arg.
[d] 62% of 177 colonies analyzed had genotype pro trp^+ arg^+.

TABLE X

Noncomplementing *trp* Mutants in Cluster 1

Cross *pro trp-x* × *trp-y arg*			Cross *pro trp-y* × *trp-x arg*			Number of colonies tested for complementation and recombination[b]
trp parents	Number of colonies on medium selecting markers[a]		*trp* parents	Number of colonies on medium selecting markers[a]		
	pro^+ arg^+	trp^+		pro^+ arg^+	trp^+	
e9 × e6	560	2	e6 × e9	80	0	9
e9 × e4	570	4	—	—	—	8
e10 × e9	143,000	2	e9 × e10	97,000	2	10
e6 × e2	78	1.3	e2 × e6	302	1.3	18
e6 × e1	1,540	2	—	—	—	216
e2 × e4	450	4	—	—	—	13
e1 × e2	200	0	e2 × e1	380	5	18
5 × e10	170	0	e10 × 5	30	1,480	100[c]
7 × e2	113	1	e2 × 7	160	0	82
7 × e5	35	0	e5 × 7	295	3.8	42
e9 × 7	340	2	7 × e9	290	3	24
7 × e16	110	0	e16 × 7	3,440	170	160
5 × 7	114	0.66	7 × 5	147	2.3	163[d]
e1 × e4	670	0	e4 × e1	740	2	120[e]
e4 × e7	690	0	e7 × e4	1,850	0	0[f]
7 × e4	265	0	—	—	—	0[f]

[a] Values based on the average number of colonies on three plates; whenever necessary, dilutions giving 50—150 colonies per plate were used.

[b] Values indicate number of recombinant colonies analyzed; heteroclones were not found except as indicated in footnotes *d* and *e*.

[c] 100% of 100 colonies analyzed had genotype pro trp^+ arg^+.

[d] 1 heteroclone was found among 163 colonies analyzed.

[e] 1 heteroclone was found among 120 colonies analyzed.

[f] Apparently neither recombination between trp^+ markers nor complementation occurred since colonies did not appear on plates selecting trp^+ markers; incompatibility of parents for the formation of recombinants and heteroclones as a consequence of some characteristic unrelated to tryptophan metabolism is therefore not excluded.

TABLE XI

Noncomplementing *trp* Mutants in Cluster 2

Cross *pro trp-x* × *trp-y arg*			Cross *pro trp-y* × *trp-x arg*			Number of colonies tested for complementation and recombination[b]
trp parents	Number of colonies on medium selecting markers[a]		*trp* parents	Number of colonies on medium selecting markers[a]		
	pro^+ arg^+	trp^+		pro^+ arg^+	trp^+	
1 × e3	1,110	16	*e3 × 1*	51	0	45
1 × e11	670	52	*e11 × 1*	250	5	124

[a] *See* footnote *a* in Table X.

[b] Values indicate number of recombinant colonies analyzed; heteroclones were not found.

TABLE XII

Summary of Complementation Analysis of *trp* Mutants in Cluster 1[a]

	e1	e7	e9	e2	e6	e12	e16	7	e5	5	e10
e4	−	○	−	−	−	+	○	○	○	○	+
e10	+	○	−	+	○	+	○	+	○	−	
5	○	○	○	○	○	○	○	−	○		
e5	○	○	○	○	○	○	○	−			
7	+	○	−	−	+	○	−				
e16	−	○	+	○	+	−					
e12	○	−	○	○	○						
e6	−	○	−	−							
e2	−	+	+								
e9	○	○									
e7	○										

[a] + complementation, − no complementation, ○ complementation analysis has not been done.

TABLE XIII

Summary of Complementation Analysis of *trp* Mutants in Cluster 2[a]

	e14	e11	e13	8	e3	1
2	○	−	+	○	○	○
1	○	−	−	+	−	
e3	○	−	○	+		
8	−	+	+			
e13	○	+				
e11	+					

[a] + complementation, − no complementation, ○ complementation analysis has not been done.

FOUR-POINT RECIPROCAL CROSSES TO DETERMINE THE ORDER OF MUTANT SITES IN CLUSTER 2

The order of mutant sites in cluster 2 was determined from the frequency of trp^+ colonies among progeny of reciprocal crosses in which outside markers pro^+ and arg^+ were selected. The results of these crosses are summarized in Table VI.

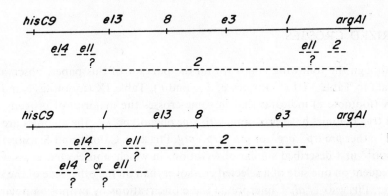

Fig. 3. The order of cluster 2 mutant sites based on the two types of marker selection described in the text. A *dashed line* indicates that the position of the mutant site has not been established within the region on the *solid line* immediately above the dashed line. *Non-overlapping* dashed lines indicate that the order of two mutants, represented by dashed lines, is known. The space between mutants on the solid line does not represent relative map distance between mutant sites.

Upper: selected markers: pro^+/arg^+, based on Table VI;

lower: selected markers: trp^+/trp^+, based on Table VII.

The order of mutant sites in cluster 2 was also determined from the frequency of nonselected pro^+ and arg^+ markers among progeny of reciprocal crosses in which haploid recombinants between trp^+ markers were selected. The results of these crosses are summarized in Table VII.

A comparison of the order of mutant sites revealed by the two types of marker selection is presented in Fig. 3. The ambiguities in the location of trp-1 and trp-$e11$ will probably be resolved by additional crosses.

COMPLEMENTATION REACTIONS OF trp MUTANTS

Evidence for complementation based on the identification of heteroclones among progeny of crosses involving mutants in cluster 1 is presented in Table VIII. The results of complementation tests based on the same type of analysis for mutants in cluster 2 are summarized in Table IX.

Evidence indicating that mutants do not complement since large values were observed for the ratios of colonies on media selecting pro^+ and arg^+ markers to colonies on media selecting trp^+ markers and heteroclones were not formed is summarized in Table X for mutants in cluster 1 and in Table XI for mutants in cluster 2.

The results of complementation tests of mutants in cluster 1 and cluster 2 are summarized in Tables XII and XIII. The analysis of complementation reactions is not sufficiently complete to reveal groups of noncomplementing mutants.

POLARIZED CROSSES

Although the following is not within the purview of this paper, observation of some data in Table VIII (footnotes c, d, e and f), Table IX (footnotes c and d) and Table X (footnote c) indicates that in some crosses the majority of progeny among selected trp^+ colonies have predominantly one phenotype, $i.e.$, the colonies are predominantly either $pro\ trp^+\ arg^+$ or $pro^+\ trp^+\ arg$. PIPERNO, CARERE and SERMONTI[22] and HOPWOOD[15] first described similar observations in which nonselected crossovers are more frequent on one side of a selected crossover than on the other side of the selected crossover. PIPERNO et al.[22] interpreted these observations by proposing polarization of the recombination process. HOPWOOD's[15] interpretation of these observations is based on the evidence that a merozygote genome consists of either a complete circular genome or a circularly permuted genome[21] contributed by one parent and a partial linear genome contributed by the other parent[15]. Thus, by HOPWOOD's interpretation, polarized crosses result whenever the same parent contributes the complete genome in a majority of merozygotes; and the higher frequency of nonselected crossovers on one side of the selected crossover is then a function of the order of the selected markers.

Polarized crosses also seem to be associated with an unexpected large number of colonies on media selecting *trp*⁺ markers, *i.e.* with a low ratio of colonies on media selecting *pro*⁺ and *arg*⁺ markers to colonies on media selecting *trp*⁺ markers. This may be seen by referring to footnotes *c, d, e* and *f* in Table VIII; footnotes *c* and *d* in Table IX and *c* in Table X. This investigator is confident that reversion and/or suppression as possible explanations for the low ratio of colonies on media selecting *pro*⁺ and *arg*⁺ markers to colonies on media selecting *trp*⁺ markers have been satisfactorily excluded.

Perhaps investigators of polarized crosses in other laboratories will suggest a plausible hypothesis for the mechanism by which one parent is favored in contributing a complete genome to merozygotes. A unitary hypothesis to explain the low ratios of colonies on different selective media observed in this investigation and polarized crosses is not immediately apparent.

DISCUSSION

The pathway of tryptophan biosynthesis based primarily on studies carried out with *Aerobacter aerogenes*[23], *Escherichia coli*[19,20,24,25] and *Neurospora crassa*[26-28] is illustrated in Fig. 4. Enzymes catalyzing reactions in the pathway have been described in several bacteria and fungi.

This investigator[29] reported that cell-free extracts of two mutants in cluster 1 (*trp-7* and *trp-e10*) fail to carry out the conversion of anthranilate to phosphoribosylanthranilate. The evidence on which this report is based may be spurious since APR--Tase activity in cell-free extracts of other mutants in cluster 1 and wild type is also

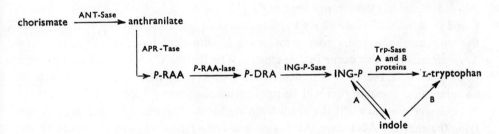

Fig. 4. A summary of the enzymes and intermediates in the tryptophan biosynthetic pathway.

P-RAA	N-(5'-phosphoribosyl)anthranilate
P-DRA	N-(5'-phospho-1'-deoxyribulosyl)anthranilate
ING-*P*	1-(indol-3'-yl)glycerol 3-phosphate
ANT-Sase	anthranilate synthase
APR-Tase	anthranilate phosphoribosyltransferase
P-RAA-Iase	N-(5'-phosphoribosyl)anthranilate isomerase
ING-*P*-Sase	1-(indol-3'-yl)glycerol 3-phosphate synthase
Trp-Sase	tryptophan synthase

very low. KANE[30] has isolated a *dispersed* mutant in *S. coelicolor* whose cell-free extracts possess satisfactory activity to catalyze reactions in the aromatic pathway. Perhaps utilization of this mutant will help circumvent some of the problems encountered in the identification of *trp* mutants with enzymatic reactions in the tryptophan pathway.

Although *trp-8* grows poorly on indole-supplemented minimal medium, it is the only mutant in cluster 2 which gives any indication of growth in the presence of indole. It is also the only mutant that accumulates a substance in culture filtrates which gives a positive test with the ferric chloride reagent indicating the accumulation of indolylglycerol. This evidence suggests that *trp-8* affects the reaction by which ING-*P* is converted to indole, the reaction catalyzed by the A subunit of tryptophan synthase[31].

The other mutants in cluster 2, which do not utilize indole, apparently crossfeed indole-utilizing mutants indicating the accumulation of indole by these mutants. However, culture filtrates of these mutants fail to give a positive test with Ehrlich's reagent; this failure is probably the result of some technical problem which needs investigation. To suggest, on the basis of the crossfeeding evidence, that the mutants in cluster 2 which do not utilize indole affect the reaction in which indole is converted to tryptophan, the reaction catalyzed by the B subunit of tryptophan synthase[31], is premature.

Of particular interest is that some of the mutants in cluster 2 which do not utilize indole map to the right of *trp-8* while others map to the left of *trp-8* (Fig. 3). If mutants in cluster 2 which do not utilize indole represent structural gene mutations of the B subunit of tryptophan synthase one might expect these mutants to map on one side of *trp-8*. Identification of enzymological and/or regulatory functions affected by mutants in cluster 2 is required to interpret the significance of the arrangement of these mutants on the chromosome with respect to *trp-8*.

It is clear that the arrangement of *trp* genes in *S. coelicolor* is different from that found in *B. subtilis*[1] since all six *trp* genes occur in one cluster in this bacterium. All five *trp* genes, some genes coding for enzymes with dual functions, in *E. coli*[2] and *S. typhimurium*[13] also occur in one cluster. Whether the chromosomal arrangement of *trp* genes in *S. coelicolor* is similar to that found in *P. putida*[9] and *P. aeruginosa*[6] is an interesting question that will be resolved as the investigation of *trp* mutants in *S. coelicolor* progresses. It is known from studies of *P. putida* that: the genes specifying the enzymes ANT-Sase, APR-Tase and ING-*P*-Sase are closely linked[8,9]; the gene controlling *P*-RAA isomerase is not linked to other *trp* genes[9]; the two genes controlling the A and B subunits of tryptophan synthase are closely linked but are not linked to any other *trp* structural genes, and regulatory mutants closely linked to these two structural genes affect constitutive synthesis of the A and B subunits of tryptophan synthase (tryptophan synthase is induced by ING-*P* in wild-type cells[7]).

Investigation of *trp* mutants in *S. coliecolor* may reveal yet other *trp* gene arrangements and modes of regulation.

I gratefully acknowledge the help of Dr. David A. Hopwood in whose laboratory this investigation was initiated and the excellent technical assistance of Joyce Clark and Cecilia Donham.

This research was supported by National Science Foundation Grant GB 6765, a grant from the Brown-Hazen Fund of Research Corporation and two Title III grants from the U.S. Office of Education.

REFERENCES

1. CARLTON B. C., WHITT D. D.: Genetics **62**, 445 (1969).
2. YANOFSKY C., LENNOX E. S.: Virology **8**, 425 (1959).
3. YANOFSKY C.: Bacteriol. Rev. **24**, 221 (1960).
4. ITO J., CRAWFORD I. P.: Genetics **52**, 1303 (1965).
5. MATSUSHIRO A., SATO K., ITO J., KIDA S., IMAMOTO F.: J. Mol. Biol. **11**, 54 (1965).
6. FARGIE B., HOLLOWAY B. W.: Genet. Res. **6**, 284 (1965).
7. CRAWFORD I. P., GUNSALUS I. C.: Proc. Nat. Acad. Sci. U.S.A. **56**, 717 (1966).
8. CHAKRABARTY A. M., GUNSALUS C. F., GUNSALUS I. C.: Proc. Nat. Acad. Sci. U.S.A. **60**, 168 (1968).
9. GUNSALUS I. C., GUNSALUS C. F., CHAKRABARTY A. M., SIKES S., CRAWFORD I. P.: Genetics **60**, 419 (1968).
10. ENATSU T., CRAWFORD I. P.: J. Bacteriol. **95**, 107 (1968).
11. DEMEREC M., HARTMAN Z.: Carnegie Inst. Wash. Publ. **612**, 5 (1956).
12. BAUERLE R. H., MARGOLIN P.: Bacteriol. Proc. **65**, 94 (1965).
13. BLUME A. J., BALBINDER E.: Genetics **53**, 577 (1966).
14. HOPWOOD D. A.: Ann. N. Y. Acad. Sci. **81**, 887 (1959).
15. HOPWOOD D. A.: Bacteriol. Rev. **31**, 373 (1967).
16. DEMEREC M., ADELBERG E. A., CLARK A. J., HARTMAN P. E.: Genetics **54**, 61 (1966).
17. FREESE E., p. 3 in W. J. Burdette (Ed.): *Methodology in Basic Genetics.* Holden Day, San Francisco 1963.
18. DELIĆ V., HOPWOOD D. A., FRIEND E. J.: Mutation Res. **9**, 167 (1970).
19. YANOFSKY C.: J. Biol. Chem. **223**, 171 (1956).
20. YANOFSKY C., p. 233 in S. P. Colowick, N. O. Kaplan (Eds.): *Methods in Enzymology*, Vol. II. Academic Press, New York 1955.
21. HOPWOOD D. A.: J. Mol. Biol. **12**, 514 (1965).
22. PIPERNO R., CARERE A., SERMONTI G.: Ann. Ist. Super. Sanità **2**, 393 (1966).
23. GIBSON F.: Biochem. J. **90**, 256 (1964).
24. SMITH G. H., YANOFSKY C.: J. Biol. Chem. **235**, 2051 (1960).
25. DOY C. H., RIVERA A. Jr., SRINIVASAN P. R.: Biochem. Biophys. Res. Comm. **4**, 83 (1961).
26. TATUM E. L., BONNER D., BEADLE G. W.: Arch. Biochem. **3**, 477 (1943).
27. DEMOSS J. A.: Biochim. Biophys. Acta **62**, 279 (1962).
28. WEGMAN J., DEMOSS J. A.: J. Biol. Chem. **240**, 3781 (1965).
29. ENGEL P. P.: Genetics **61**, s16 (1969).
30. JENSEN R. A., *personal communication.*
31. CRAWFORD I. P., YANOFSKY C.: Proc. Nat. Acad. Sci. U.S.A. **44**, 1161 (1958).

Regulatory Aspects of Threonine and Methionine Biosynthesis in *Saccharomyces cerevisiae*

H. de Robichon-Szulmajster, Y. Surdin-Kerjan, H. Cherest

Laboratoire d'Enzymologie du C.N.R.S., Gif-sur-Yvette, France

Abbreviations used:

ASA	aspartate semialdehyde	OAS	O-acetylserine
CST	cystathionine	SAH	S-adenosylhomocysteine
HCS	homocysteine	SAM	S-adenosylmethionine
HSR	homoserine	SEC	S-ethylcysteine
OAH	O-acetylhomoserine	SMC	S-methylcysteine

ASA-Dase	aspartate semialdehyde dehydrogenase
ASP-Aase	aspartate aminotransferase
ASP-Kase	aspartokinase
ATP-Sase	ATP-sulfurylase
GLU-Dase	glutamate dehydrogenase
HCS-Sase	homocysteine synthase
HSR-Aase	homoserine acetyltransferase
HSR-Dase	homoserine dehydrogenase
HSR-Kase	homoserine kinase

In yeasts, as in other eukaryotic organisms such as *Neurospora crassa* or *Aspergillus nidulans*, genes coding for enzymes which catalyze the successive steps in biosynthetic pathways are usually unlinked[1]. In the case of the branched pathway leading to threonine and methionine, which involves homoserine as a branching point (Fig. 1), absence of linkage has been observed for genes leading to auxotrophy towards threonine alone, methionine alone or homoserine (which can be replaced by threonine + methionine) in *Saccharomyces cerevisiae*[3,4].

The synthesis of threonine from aspartate can be divided in two parts: first, there are three steps involved in the formation of HSR, which are shared by threonine and methionine; secondly, two steps lead to threonine biosynthesis from HSR. Each of the genes formerly named *thr3*, *thr2*, *thr6*, *thr1* and *thr4* have been assigned to one of the enzymes corresponding to the five steps concerned. In addition, gene *thr5* has been shown to deal with the synthesis of ASP-Aase[3]. In fact, it was somehow incidental, due to peculiar media conditions, that such mutants could be selected on the basis of threonine + methionine auxotrophy. Truly, the ASP-Aase deficiency leads to auxotrophy for aspartate, which is the precursor of metabolites other than threonine and

TABLE I   Gene—Enzyme Relationships in Threonine Biosynthesis in *S. cerevisiae*[a]

Gene nomenclature		Requirement		Step	Enzyme	E. C. N°
Former	Present	Compound	Concentration[b] M			
thr5	asp5	aspartate	10^{-3}	1	aspartate aminotransferase	2.6.1.1
thr3	hom3	homoserine or threonine + methionine	10^{-3}	2	aspartokinase	2.7.2.4
thr2	hom2	homoserine or threonine + methionine	$10^{-3} + 10^{-4}$ 10^{-3}	3	aspartate semialdehyde dehydrogenase	1.2.1.11
thr6	hom6	homoserine or threonine + methionine	$10^{-3} + 10^{-4}$ 10^{-3}	4	homoserine dehydrogenase	1.1.1.3
thr1	thr1	threonine	$10^{-3} + 10^{-4}$ 10^{-3}	5	homoserine kinase	2.7.1.39
thr4	thr4	threonine	10^{-3}	6	threonine synthase	4.2.99.2

[a] For details in enzyme nomenclature, reactions, *etc.*, *see* Ref.[6].   [b] L-Form.

TABLE II   Phenotypic Expression of Mutations in Methionine Biosynthesis Corresponding to Known Enzyme Deficiencies

Step involved	Addition to minimal medium						Mutated allele	Strain number	Enzyme studied Ref.
	OAH	CST	HCS	Cys	Met	SAM			
7	+	−	+	−	+	+	met2	D6 (Grenson)	7
8	−	−	+	+	+	+	met8	X963-18C (Mortimer)	9
10	−	−	+	+	+	+	?	EY 9 (Mortimer)	15
11	−	−	+	+	+	+	?	3-27 (Naiki)	15
12	−	−	+	+	+	+	?	15 (Naiki)	15
14	−	−	−	−	+	−	met5-Parks	12-22 (Naiki)	16
17	−	−	−	−	+	−	met1-Parks		16

methionine[3]. So far, we have not found any regulatory effect on ASP-Aase synthesis or activity either by methionine, threonine or both. Consequently, and in view of the potent regulation found on the next step, involving ASP-Kase (*see further*), ASP-Aase will not be included in the study of regulatory aspects of the threonine and methionine pathway.

In order to obey new terminology rules[5], gene – enzyme relationship involved in threonine biosynthesis should be arranged as in Table I.

As far as methionine *per se* is concerned, the situation is more complex than for the threonine branch since at least 10 independent loci have been found to be involved in its biosynthesis. On the biochemical basis, methionine biosynthesis can be divided into different parts:

1) Synthesis of homocysteine from homoserine. Two reactions are implicated. First, an acetylation of homoserine takes place, which is catalyzed by HSR-Aase and is lacking in *met a* mutants[7] recently identified with mutants mapping at the locus *met2* (Ref. [4]). Then direct homocysteine synthesis occurs from OAH and sulfide; this reaction, catalyzed by HCS-Sase[8] was found to be present in all wild type and methionine auxotrophs so far studied, but *met8* mutants[9]. The existence of such a mutant and the fact that all methionine prototrophic revertants from this strain have recovered HCS-Sase activity made us conclude that this enzyme is indispensable for homocysteine biosynthesis in *S. cerevisiae*. This conclusion is reinforced by the findings (described *below*) that synthesis and activity of HCS-Sase are regulated by methionine*.

2) Synthesis of sulfide from sulfate. The implication of sulfide in the synthesis of homocysteine, as described in the upper section, led to the consideration that sulfide synthesis is a part of the methionine pathway in *S. cerevisiae*. There are four steps to be considered, the last one (step 13 in Fig. 1) probably involving more than one protein

---

* Since the transsulfuration pathway which involves CST as a key intermediate has been shown to participate in methionine biosynthesis in bacteria and *N. crassa*[2,10,11], the exact role of this compound in yeast metabolism deserves some comment. All facts presented above rendered unnecessary to implicate cystathionine in the direct pathway leading to methionine biosynthesis in *S. cerevisiae*. This conclusion is also supported by other findings:

1. CST cannot be used by any methionine auxotroph, although it penetrates the cells[12,13];

2. CST γ-synthase which catalyzes the condensation between O-acetylhomoserine and cysteine could hardly be detected in extracts of either wild type or *met8* mutants of *S. cerevisiae*[14].

Still, the existence of a rather inefficient CST γ-synthase might account for the ability of strains blocked in HCS-Sase to use cysteine instead of homocysteine as compared to the incapacity (unique among all other methionine auxotrophs) of strains blocked in HSR-Aase to grow at the expense of cysteine. In this line, CST γ-synthase might be considered as a rescue enzyme enabling cells devoid of HCS-Sase to grow, provided excess cysteine is present. In addition, the existence of CST β-synthase (catalyzing condensation of homocysteine and serine) might explain how methionine can serve as a source of cysteine for mutants blocked in the sulfate assimilation pathway (Table II).

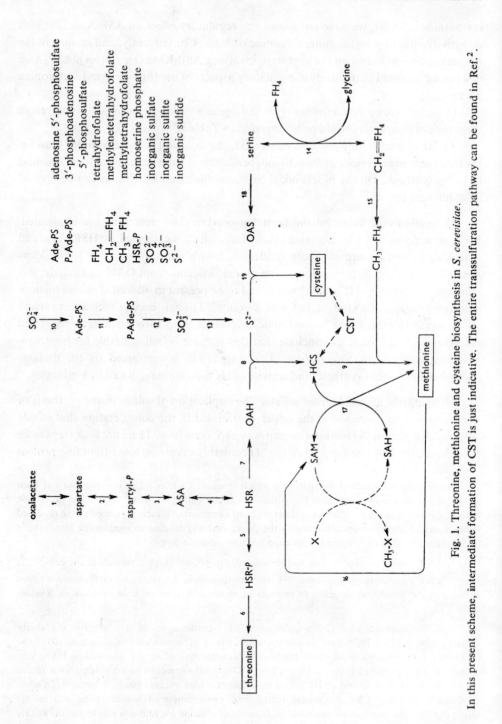

Fig. 1. Threonine, methionine and cysteine biosynthesis in *S. cerevisiae*.

In this present scheme, intermediate formation of CST is just indicative. The entire transsulfuration pathway can be found in Ref.[2]

component[17]. Independent mutants isolated on the basis of methionine auxotrophy have been shown to lack activity corresponding to one of the first enzymes (step 10, 11, 12 in Fig. 1)[15]. Moreover, methionine-mediated regulation of ATP-Sase synthesis (step 10) gives additional evidence that sulfate assimilation is primarily involved in methionine biosynthesis[9,18].

3) Synthesis of methionine from homocysteine. Formally, only one step is concerned involving addition of a methyl group to homocysteine. This methyl group originates from serine through tetrahydrofolate derivatives. The existence of a mutant lacking serine hydroxymethyltransferase (step 14) and exhibiting an absolute requirement for methionine shows that biosynthesis of the proper methyltetrahydrofolate derivative is also implied in the methionine pathway[16].

In addition to the complexity of this terminal part of the pathway, one has to account for the existence of a methionine auxotroph which lacks S-adenosylmethionine—homocysteine methyltransferase (step 17)[16], indicating that S-adenosylmethionine is directly implicated in the last step of methionine biosynthesis. Methionine requirement can also be exhibited by mutants not directly concerned with the biosynthesis of this amino acid, as shown by the existence of a temperature-sensitive mutant in which methionine auxotrophy results from the presence of an impaired methionyl-tRNA synthetase[19].

At this point, it seemed helpful to include a list of the phenotypes which are typical of the different blocks already related to a precise enzyme deficiency (Table II). So far, three phenotypes seem to be unique to only one enzyme deficiency, *viz.* inability of the HSR-Aase-deficient mutants (step 7) to grow on cysteine, ability of S-adenosylmethionine—homocysteine methyltransferase mutants (step 17) to grow on methionine exclusively, and ability of mutants blocked in the synthesis of tetrahydrofolate derivatives to grow as well on S-adenosylmethionine or methionine. On the opposite, the same phenotype (inability to use O-acetylhomoserine) is shared by HCS-Sase-deficient mutants (step 8) and mutants blocked in different steps in the sulfate assimilation pathway.

Our studies on the threonine—methionine pathway have been mostly concerned with five of the six steps involved in threonine biosynthesis and the two steps directly involved in homocysteine biosynthesis. Evidence that the genes listed in Tables I and II are indeed structural genes coding for the corresponding enzymes is given by a study of gene dosage in homozygote and heterozygote diploids (*see* Table III). It can be seen that in each case, the specific activity recovered is directly proportional to the number of wild-type alleles present in the diploid.

For each of the haploid mutants studied, enzyme activity has been found to be almost totally lacking. On this basis, there has been no evidence for isozymes to be present, contrary to the situation with the same pathway in *E. coli*[25,26]. In this regard, two previous findings from this laboratory deserve some comment: a) Although aspartokinase activities with slightly distinct properties were found after

TABLE III

Gene Dosage Effect on Enzyme Level in Homozygous Wild-Type and Heterozygous Diploids[a]

Step	Enzyme	Specific activity[b]		Mutated allele present in heterozygote
		homozygote +/+	heterozygote +/−	
1	ASP-Aase	140	34	asp5
2	ASP-Kase	32	15	hom3
3	ASA-Dase	33	17	hom2
4	HSR-Dase	220	120	hom6
5	HSR-Kase	23	11	thr1
7	HSR-Aase	0.290	0.120	met2
8	HCS-Sase	236	130	met8

[a] All diploids have been grown on minimal medium and cells collected at the end of the exponential phase.

[b] In nmol $\times$ min^{-1} $\times$ mg prot^{-1}. Extracts were prepared and enzymes assayed using standard technics described[7,9,20-24].

growth in the presence of 10^{-2}M methionine as compared with other growth conditions[21], further experiments have shown that no aspartokinase activity could be detected in *hom3* mutants even after growth in the presence of high methionine concentrations; b) HSR-Dase activity was found to accept NADH and NADPH as cofactors[27,23]; nevertheless, activities towards both nucleotides were lost in *hom6* mutants and recovered to the same extent in revertant strains. Moreover, thermal inactivation affects both activities at the same rate and NADH was found to protect both activities as well[24]. In addition to a previous finding that the ratio between activities towards both nucleotides was not modified during partial purification[27], the present results show that there must be only one structural gene for HSR-Dase as well as only one structural gene for aspartokinase in *S. cerevisiae*.

The absence of linkage between structural genes of the same pathway and the absence of isozymes imply that the mechanisms which are responsible for the overall regulation of threonine and methionine biosynthesis can neither operate through the operon organization or through multienzyme systems. However, metabolic regulation of this branched pathway must be as efficient as in any other microorganism or other pathways in yeast, since no overproduction of end products occurs when wild-type cells are grown in minimal medium. A recent study of amino-acid pool composition and size in different conditions shows that spontaneous levels of threonine and methionine in the pool during exponential growth on minimal medium are of the order of 5×10^{-3}M[28]. This concentration represents at most 2% of the threonine and 12% of the methionine present in yeast cell proteins at the same moment. In addition, only traces of these metabolites can be detected in the culture medium at any moment.

EFFECTS ON ENZYME SYNTHESIS

Regulation of synthesis of each of the enzymes concerned was examined in wild type and mutant strains after growth either in excess or under limitation of threonine or methionine. For the sake of convenience, we shall consider in turn the enzymes concerned with threonine biosynthesis and the enzymes concerned with methionine biosynthesis. However, it has to be recalled that three enzymes common to the synthesis of both end products (steps 2, 3, 4) will have to be considered in the two sections.

ENZYMES INVOLVED IN THREONINE BIOSYNTHESIS

Table IV summarizes results obtained for the enzymes catalyzing steps 2, 3, 4 and 5 after growth of a wild-type strain in the presence of an excess of different amino acids. It can be seen that excess of threonine essentially represses aspartokinase synthesis, and has very little or no effect on the synthesis of the other three enzymes. On the opposite, methionine principally affects synthesis of these last three enzymes. However, the effects are not identical. As could have been expected, a repressive effect of methionine is observed which is exerted upon HSR-Dase synthesis. More surprising was the finding that "induction" could be observed upon one of the enzymes common to threonine and methionine, *i.e.* ASA-Dase, and upon the first enzyme specific for threonine biosynthesis, *i.e.* homoserine kinase[24]. It is remarkable that homoserine is able to reproduce the repressive effects of both end products, that is, the effect of threonine upon aspartokinase synthesis, and the effect of methionine upon HSR-Dase synthesis. These effects might be explained if one assumes that an excess of exogenous homoserine increases endogenous production of both amino

TABLE IV

Specific Activities of Four Enzymes Involved in Threonine Biosynthesis in Different Growth Conditions[a]

Enzyme	Growth conditions[b]			
	MM	Thr	Met	HSR
ASP-Kase	27	14	23	11
ASA-Dase	35	28	81	—
HSR-Dase	610	530	350	420
HSR-Kase	18	20	35	—

[a] Strain 4094-B: haploid (*mat-α ade2 ura1*).

[b] MM: minimal medium. Each amino acid was added to the minimal medium at the final concentration of 10 mM (L-form). Cells were harvested during exponential growth. Extracts were prepared and enzyme assayed using standard technics already described[21-24].

TABLE V

Effect of Threonine Limitation upon Specific Activities of the Three Enzymes Involved in Homoserine Biosynthesis[a]

Enzyme	Growth conditions[d]	
	Control[b]	Chemostat[c]
ASP-Kase	14	28
ASA-Dase	23	45
HSR-Dase	655	680
HSR-Kase	—	—

[a] Strain T 1-6: haploid (*mat-α thr1 ade2 ura1*).
[b] Growth on minimal medium containing (in addition to the required amounts of adenine and
   uracil) 4×10^{-4}M L-threonine; cells have been collected during exponential phase of growth.
[c] Same basal medium as in the control but containing 10^{-4}M L-threonine. Cells used to inoculate
   the chemostat were identical to those used for control; they were allowed to one doubling in
   the chemostat.
[d] Other conditions were as already described (*see* Table III).

acids. However, a direct repressive effect exerted by homoserine itself cannot be
excluded.

In another type of experiment, a mutant deficient in HSR-Kase was used to study
the effect of a limitation in threonine concentration upon synthesis of the three
enzymes involved in homoserine biosynthesis.

It can be seen in Table V that under these conditions aspartokinase synthesis is
derepressed together with ASA-Dase whereas HSR-Dase synthesis is not modified.

From these findings, it already appears that both end products participate in the
regulation of synthesis of the three enzymes which are common to their biosynthesis
and that one of the end products, methionine, can apparently act in opposite directions, repressing one of the enzymes and inducing the other. Such conflicting effects will
be further documented when methionine auxotrophic mutants will be considered.

ENZYMES INVOLVED IN METHIONINE BIOSYNTHESIS[29]

The effect of methionine upon the rate of synthesis of various enzymes involved
in its biosynthesis has been studied using two mutants: D6, which is devoid of homoserine acetyltransferase activity, and JE-2 devoid of ASA-Dase activity. With the last
mutant, only 3 methionine enzymes have been considered since this strain was mainly
used to study the variations of HSR-Aase. GLU-Dase (NADPH dependent) was
chosen as a reference enzyme, not metabolically involved in the pathway studied.

In each case, the control experiment consisted in a culture with a precursor
(O-acetylhomoserine or homoserine) present in no excess, in order to prevent as much

as possible excessive endogenous synthesis of both end products. Two other cultures were compared, one with an excess of methionine, the other being a chemostat in methionine limitation. Results of these experiments are summarized in Table VI. Many effects can be observed:

Aspartokinase. Synthesis of this enzyme which appeared in the previous section is repressible by threonine and must be under a double repressive control since its synthesis can also be enhanced when methionine becomes the limiting factor.

ASA dehydrogenase. Previous observation that methionine acts as an "inducer" for this enzyme is confirmed by the finding that a lowering of methionine concentration results in a marked decrease in the amount of enzyme synthesized.

Homoserine dehydrogenase. In agreement with the previous finding that synthesis of this enzyme is placed under repressive control by methionine, it can be seen that derepression occurs in methionine limitation.

TABLE VI

Effect of Exogenous Methionine Concentration upon Synthesis of Enzymes Involved in Methionine Biosynthesis

Strain[a]	Enzyme	Growth conditions			Ratio[d]
		10^{-3}M-Met	Control[b]	Chemostat[c]	
D6	ASP-Kase	16	18	62	3.9
	ASA-Dase	139	58	27	0.19
	HSR-Dase	120	210	645	5.4
	HSR-Aase	—	—	—	—
	HCS-Sase	14	97	433	31
	ATP-Sase	0	100	280	∞
	GLU-Dase	17	24	19	1
JE2	HSR-Tase	0.075	0.460	1.43	19
	HCS-Sase	61	170	340	5.6
	ATP-Sase	0	81	220	∞
	GLU-Dase	9	10	7	1

[a] D6: haploid (*mat-α ura met2*); JE 2: haploid (*mat-α ade2 ade thr2*).

[b] Cells were grown in the presence of 10^{-3}M-DL-OAH in the case of strain D6 and in the presence of 2×10^{-3}M-DL-HSR in the case of strain JE 2.

[c] The minimal medium contained, in addition to uracil, 8×10^{-5}M-DL-methionine in the case of strain D6 and 4×10^{-5}M-DL-methionine plus 2×10^{-3}M-DL-threonine in the case of strain JE 2. Four fractions, each one corresponding to approximately one doubling of cellular mass were collected and analyzed. Maximal effect was observed in fraction III and the enzyme levels attained were retained in fraction IV. Only specific activities measured in this last fraction have been reported here. Enzyme assays have been described previously (*see* Table III and Refs[18,20].

[d] Chemostat/10^{-3}M methionine.

Methionine-specific enzymes. Homocysteine synthase and ATP sulfurylase (first step in the sulfate assimilation pathway) are very efficiently repressed by excess of methionine and derepressed by methionine limitation.

The response of the last enzyme certainly reveals one of the most potent repressive controls so far evidenced for enzymes involved in amino-acid biosynthesis in yeast[1,30]. In the second part of Table VI, it can be observed that HSR-Aase displays the same type of response as the other two methionine-specific enzymes already mentioned.

The main conclusion to be derived from these experiments resides in the pleiotropic repressive effects of methionine on the synthesis of many enzymes: some, like aspartokinase, shared with threonine and methionine biosynthesis, some (enzymes catalyzing steps 7 and 8) being specific for the methionine biosynthetic pathway, and some, like ATP-Sase, being normally expected to be involved in methionine and cysteine biosynthesis.

EFFECTS ON ENZYME ACTIVITY

The results described above have shown that each of the enzymes studied, regardless of its location in the pathway, is subjected to some regulation of its synthesis. Both end products participate in this control, alone or in combination, either in a negative (repression) or in a positive (induction) fashion. However, it must not be forgotten that repression of enzyme synthesis is a slow and cumulative process, the complete expression of which requires at least three generations, *i.e.* an 8-fold increase of the cellular mass. In addition, it can be pointed out that, in yeast cells, there must be at least twice as many units of each enzyme considered since gene-dosage studies (*see* Table II) have shown that a 2-fold decrease in specific activity has no consequence on the growth rate. It follows that the most efficient and immediate type of control that end products can exert on a biosynthetic pathway resides in regulation of enzyme activity.

We have examined all the enzymes already considered in the previous section and found that end products exert their main control at crucial points in this pathway, namely threonine on reactions 2 and 5, and methionine on reaction 8. Besides this very potent inhibitory effect, other additional or unexpected effects were observed, the biological meaning of which remains obscure, such as the strong positive effect of bicarbonate anions upon reaction 3[22].

THREONINE AS A FEEDBACK INHIBITOR

Aspartokinase. It was found a few years ago that yeast aspartokinase can be totally and specifically inhibited by 10 mM L-threonine[21]. More recently[24], it could be shown, using partially purified preparations, that cooperativity between threonine molecules takes place ($n = 2$). An apparent K_i of 7mM was estimated which is of the same order of magnitude as the K_m's for both substrates (6mM for aspartate and 2,5mM for ATP). Threonine inhibition of aspartokinase activity was found to be competitive towards aspartate only (*see* Fig. 2). Methionine and other amino acids tested have little or no effect.

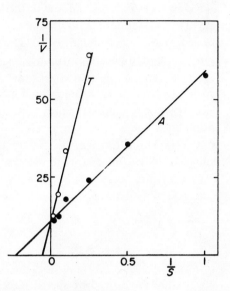

Fig. 2. Kinetic parameters concerned with aspartokinase from *S. cerevisiae*. A activities in the presence of various concentrations of L-aspartate (in 1/mM), T activities in the presence of 7 mM L-threonine. Activities are expressed in $1/V$ (arbitrary units). K_m's are, respectively, 6 mM and 33 mM in the absence and the presence of threonine.

The susceptibility of aspartokinase to allosteric feedback inhibition by one of the end products of this pathway, susceptibility which is not exhibited by either the preceding or the next enzyme, aspartate aminotransferase and ASA dehydrogenase, respectively, certainly designates this enzyme as the "first" enzyme in threonine and methionine biosynthesis.

Homoserine kinase. As could be expected, the first step after the branching point, which is the first one specific for threonine biosynthesis, is also put under feedback inhibition exerted by threonine[6,31]. More attention was recently paid to yeast homoserine kinase in this laboratory, using a more accurate assay procedure than the one used previously, and a purified enzyme. Saturation curves towards both substrates (homoserine and ATP) and the feedback inhibitor threonine were perfectly Michaelian under all the conditions used and threonine inhibition was found strictly competitive towards homoserine[32].

METHIONINE AS A FEEDBACK INHIBITOR

A limited inhibitory effect of methionine* has been found on different steps of the pathway (aspartokinase, after culture in the presence of 10^{-2}M methionine[21], homoserine dehydrogenase[23]).

However, only one enzyme in this pathway has been found to be very sensitive in yeast to methionine inhibition so far[8,9]. This is homocysteine synthase, also studied in *N. crassa* and *E. coli*[8]. The inhibition was found to be competitive toward O-acetylhomoserine with a $K_i = 1.8 \times 10^{-3}$M.

Some special properties of yeast homocysteine synthase recently described[33] are of interest here. Ethionine which had been used as an analogue of methionine to select regulatory mutants mainly concerned with methionine biosynthesis, was found to inhibit homocysteine synthase activity almost as well as methionine ($K_i = 3.3 \times \times 10^{-3}$M). This effect was first attributed to structural analogy between these two amino acids. In the meantime, we discovered that ethionine could also be used as a substrate for yeast homocysteine synthase. The following reactions were shown to be reversibly catalyzed *in vitro* by this enzyme:

$$CH_3SH \ + \ \text{ethionine} \ \rightleftharpoons \ \text{methionine} \ + \ C_2H_5SH$$

$$SMC \ \ + \ \text{ethionine} \ \rightleftharpoons \ \text{methionine} \ + \ SEC$$

It could be proved by specific labeling experiments that direct methionine synthesis could be attained through an exchange of mercaptyl groups and the C_4 moiety of ethionine (or methionine). Such reactions were shown to occur *in vivo* in mutants blocked at, or before, O-acetylhomoserine synthesis, providing that homocysteine synthase activity and a resistance allele, *eth-2r*, which prevents repressibility of homocysteine synthase[34], were present. In addition, it can also be pointed out that these secondary activities catalyzed by homocysteine synthase are repressed by exogenous methionine to the same extent as the main (*i.e.* homocysteine synthesis) activity and that the extent of repression observed corresponds to the regulatory gene *eth-2r/eth-2s* for the three reactions as well.

---

* Some inhibition due to S-adenosylmethionine on HSR-Aase was also described[7]. However, recent examination of this effect has shown that pure SAM was not inhibitory as such. Inhibition could be restored after boiling a pure sample of SAM at pH 7.0 for a few minutes. The amount of homoserine formed by this treatment is in agreement with calculations, expected to have caused isotopic dilution of ^{14}C-homoserine used as substrate in the exchange reaction. Since methionine alone did not cause any inhibition and did not modify the pseudo-inhibition attributed to SAM[7], it seems though, that HSR-Aase is not subjected to any end-product regulation of its activity.

DISCUSSION

Among the 13 steps directly involved (according to Fig. 1) in methionine and threonine biosynthesis in yeast, 8 have been investigated so far. Gene—enzyme relationships in haploid mutant strains and in appropriate heterozygous/homozygous diploids have led to identification of the structural genes involved and have shown that only one enzyme catalyzes each of the steps studied.

The biochemistry of this pathway in yeast has shown a very unique feature, which resides in the exclusion of cystathionine as an obligatory intermediate in methionine biosynthesis. Various consequences result from this finding: Homocysteine synthase activity which has been found also in *N. crassa* or bacteria[8] but shown inoperative *in vivo* in these organisms[35,11], becomes a very important enzyme in yeast. The importance of homocysteine synthase is assessed (1) by its position in the pathway which ensures insertion of inorganic sulfur into the immediate precursor of methionine; (2) by its susceptibility to methionine inhibition, being the main point for feedback inhibition exerted by this end product on enzyme activity in the whole pathway and by its susceptibility to methionine-mediated repression; (3) by its ability to catalyze secondary reactions leading to synthesis of methionine from ethionine. These reactions might provide, under special circumstances, a detoxication process for ethionine.

More important, these reactions might reveal a participation of compounds such as methylmercaptan and S-methylcysteine to yeast amino acid metabolism. In fact, these compounds have already been found in different organisms[36-38] but no precise role could be assigned to them until now. Their recognition as substrates by homocysteine synthase and the observation that they are not required in addition to ethionine to permit its utilization for methionine biosynthesis, renders very promising a study of their possible *de novo* synthesis in yeast.

Another consequence of the exclusion of cystathionine from this pathway resides in the direct connection of the sulfate assimilation path with methionine biosynthesis. The regulations so far observed, *i.e.* complete repression of ATP sulfurylase* exerted by exogenous methionine, together with the general observation that cysteine can always arise from methionine**, renders it likely that the main (if not the only) utilizer of endogenous sulfide is methionine biosynthesis.

The study of regulatory mechanisms in such eukaryotic organisms and in this branched and complex pathway has proved to be very interesting. First, it should be pointed out that all the enzymes studied so far appear to be synthesized spontaneously (*see* control experiments in Tables IV, V and VI) at intermediate levels between

---

* Very recently, preliminary experiments from this laboratory have shown that sulfite reductase is even more susceptible to methionine-mediated repression than ATP sulfurylase.

** Cysteine synthase has been found in extracts of *S. cerevisiae* with activity towards serine almost equivalent to that towards acetylserine. However, no mutant which exhibits an exclusive requirement for cysteine has been described in yeast.

complete repression and complete derepression. It follows that maximal potentialities are not expressed in wild-type cells grown on minimal medium and that suitable adjustment can be expected to occur either by repression or by derepression processes.

It is also worth to note that the absence of linkage between structural genes did not cause the regulations to be less efficient in this system than when they are, as in bacteria, sequentially arranged and put under command of a single operator gene. In fact, it seems that end products might exert a more flexible and diversified control over enzyme synthesis through gene dispersion than through gene assembly. Different types of regulation have been observed:

1) Bivalent repression over synthesis of aspartokinase is exerted by both end products, methionine and threonine. That this type of control, first evidenced in isoleucine—valine—leucine biosynthesis in bacteria[39], applies to aspartokinase, reinforces the metabolic importance of this enzyme, previously defined on feedback basis as the first in the pathway.

2) Antagonistic control: The finding that threonine represses and methionine induces the synthesis of two enzymes, ASA dehydrogenase and homoserine kinase, shows that selection of corrective regulatory mechanisms has been necessary, probably to permit the finest possible adjustment to yeast environment. Nevertheless, it is striking that such an effect can be observed for an enzyme located, in the pathway, after the branching point. Such an observation is not isolated since a parallel antagonistic control was described, also in *S. cerevisiae*, for some enzymes involved in aromatic -amino-acid biosynthesis. It was found that phenylalanine represses the first enzyme in its own branch but, surprisingly enough, induces chorismate mutase responsible for the synthesis of prephenate, the common precursor of phenylalanine and tyrosine and also induces the first enzyme in the tyrosine branch[40].

If one tries to rationalize these findings it can be said that in pathways with multiple end products, regulation seems to have been devised, not only to prevent overproduction but also to keep, under any circumstance, an appropriate balance between end products of a common pathway.

3) Univalent repression is exerted over different enzymes. With homoserine dehydrogenase, being located in the pathway before the branching point, it seems surprising that only methionine participates in its repression. However, the observed repression is never complete. Although no other amino acid tried alone or in combination with methionine has shown any increment in the repressive effect caused by methionine, it cannot be excluded that some other, yet unknown, effector could also modify the rate of synthesis of this enzyme.

The situation looks very different when specific methionine enzymes are considered. In this case, not only complete repression is attained but a much lower concentration of methionine is required to produce this effect than is necessary to affect the synthesis of enzymes located in other parts of the pathway. It will be shown in the following paper[34] that, in fact, these two classes of enzymes respond to different repressor systems.

As a last remark, it is of general interest to point out that the dispersion of structural genes concerned with threonine and methionine biosynthesis, each of them being subject to regulation of its function, there must exist, for each one of them, a genetic sequence which recognizes the appropriate aporepressor—corepressor complex. Since similar conclusions seem to apply to most of the known biosynthetic pathways in yeast it means that the amount of genetic information devoted to regulation of gene function should be greater in unicellular eukaryotes than in prokaryotes. Indeed, calculations made so far lead to the conclusion that only 10% of the total yeast DNA is necessary to ensure synthesis of all the enzymes and structural macromolecules[41]. Thus 90% of the gene products in one of the simplest eukaryotic cells are unknown. It is tempting to assume that a large part of them are involved in regulatory mechanisms. If this surprising conclusion happens to be not too far from the truth, it would signify that the metabolism of eukaryotic cells is a very elaborate society in which as much as 90% of the population is busy planning, directing and correcting the work of 10% slaves.

The authors are thankful to F. de la Torre and D. Henry for their skilful assistance.

This work was supported by grants from the Délégation Générale à la Recherche Scientifique et Technique (66.00.140) and from the C.E.A., France.

REFERENCES

1. DE ROBICHON-SZULMAJSTER H., SURDIN-KERJAN Y., in A. H. Rose, J. S. Harrison (Eds.): *The Yeasts*, Vol. II. Academic Press, London and New York 1970.
2. DELAVIER-KLUTCHKO C., FLAVIN M.: J. Biol. Chem. **240**, 2537 (1965).
3. DE ROBICHON-SZULMAJSTER H., SURDIN Y., MORTIMER R. K.: Genetics **53**, 603 (1966).
4. MORTIMER R. K., HAWTHORNE D. C.: Genetics **53**, 165 (1966).
5. VON BORSTEL R. C. (Ed.): *Microbial Genetics Bulletin* No 3. Oak Ridge 1969.
6. DE ROBICHON-SZULMAJSTER H.: Bull. Soc. Chim. Biol. **49**, 1431 (1967).
7. DE ROBICHON-SZULMAJSTER H., CHEREST H.: Biochem. Biophys. Res. Comm. **28**, 256 (1967).
8. WIEBERS J. L., GARNER H. R.: J. Biol. Chem. **242**, 5644 (1967).
9. CHEREST H., EICHLER F., DE ROBICHON-SZULMAJSTER H.: J. Bacteriol. **97**, 328 (1969).
10. ROWBURY R. J., WOODS D. D.: J. Gen. Microbiol. **36**, 341 (1964).
11. KERR D. S., FLAVIN M.: Biochem. Biophys. Res. Comm. **31**, 124 (1968).
12. SORSOLI W. A., BUETTNER M., PARKS L. W.: J. Bacteriol. **95**, 1024 (1968).
13. KJELLIN-STRABY K., PHILLIPS J. H.: J. Bacteriol. **100**, 679 (1969).
14. KERR D. S., FLAVIN M.: J. Biol. Chem. **245**, 1842 (1970).
15. NAIKI N.: Plant & Cell Physiol. **5**, 71 (1964).
16. PIGG C. J., SPENCE K. D., PARKS L. W.: Arch. Biochem. Biophys. **97**, 491 (1962).
17. WAINWRIGHT T.: J. Gen. Microbiol. **61**, 107 (1970).
18. DE VITO P. C., DREYFUSS J.: J. Bacteriol. **88**, 1314 (1964).
19. MCLAUGHLIN C. S., HARTWELL L. H.: Genetics **61**, 557 (1969).

20. COLOWICK S. P., KAPLAN N. O. (Eds.): *Methods in Enzymology*, Vol. II, p. 220. Academic Press, New York 1962.
21. DE ROBICHON-SZULMAJSTER H., CORRIVAUX D.: Biochim. Biophys. Acta **73**, 248 (1963).
22. SURDIN Y.: European J. Biochem. **2**, 341 (1967).
23. KARASSEVITCH Y., DE ROBICHON-SZULMAJSTER H.: Biochim. Biophys. Acta **73**, 414 (1963).
24. SURDIN-KERJAN Y.: Thèse de doctorat d'Etat. Paris 1969.
25. STADTMANN E. R., COHEN G. N., LE BRAS G., DE ROBICHON-SZULMAJSTER H.: J. Biol. Chem. **236**, 2033 (1961).
26. PATTE J. C., LE BRAS G., COHEN G. N.: Biochim. Biophys. Acta **136**, 245 (1967).
27. BLACK S., WRIGHT N. G.: J. Biol. Chem. **213**, 51 (1955).
28. BOURGEOIS C.: Bull. Soc. Chim. Biol. **51**, 935 (1969).
29. CHEREST H.: Thèse de doctorat d'Etat. Paris 1970.
30. BECHET J., WIAME J. M., GRENSON M.: Arch. Internat. Physiol. Biochem. **70**, 546 (1962).
31. WORMSER E. H., PARDEE A. B.: Arch. Biochem. Biophys. **78**, 416 (1958).
32. TALBOT G., *unpublished results.*
33. CHEREST H., TALBOT G., DE ROBICHON-SZULMAJSTER H.: J. Bacteriol. **102**, 448 (1970).
34. CHEREST H., DE ROBICHON-SZULMAJSTER H.: Paper at 1st Internat. Symp. Genetics of Industrial Microorganisms, Prague 1970. Abstract Book, p. 95; *cf. this book*, p. 165.
35. FLAVIN M., SLAUGHTER C.: Biochim. Biophys. Acta **132**, 400 (1967).
36. BIRKINSHAW J. H., FINDLAY W. P. K., WEBB R. A.: Biochem. J. **36**, 526 (1942).
37. MORRIS C. J., THOMPSON J. F.: Chem. & Ind. 951 (1955).
38. RAGLAND J. B., LIVERMAN J. L.: Arch. Biochem. Biophys. **65**, 574 (1956).
39. FREUNDLICH M., BURNS R. O., UMBARGER H. E.: Proc. Nat. Acad. Sci. U.S.A. **48**, 1804 (1962).
40. LINGENS F., GOEBEL W., UESSELER H.: European J. Biochem. **1**, 363 (1967).
41. HARTWELL L. H.: J. Bacteriol. **93**, 1662 (1967).

The Roles of Two Independent Genes in One of the Regulatory Systems Involved in Methionine Biosynthesis in *Saccharomyces cerevisiae*

H. Cherest, H. de Robichon-Szulmajster

Laboratoire d'Enzymologie du C.N.R.S., Gif-sur-Yvette, France

The previous paper[1] was mostly concerned with a survey of the various regulatory mechanisms involved in threonine and methionine biosynthesis in *Saccharomyces cerevisiae*. The pleiotropic effect of exogenous methionine upon synthesis of at least six enzymes motivated a more detailed study of methionine as a regulatory signal for enzyme synthesis. Since wild-type strains of *S. cerevisiae* are very sensitive to ethionine, we have used this structural analogue of methionine for selecting regulatory mutants specifically concerned with methionine biosynthesis.

Beside changes in permeability[2], resistance to a structural analogue like ethionine, which has shown to be able to replace methionine in all its functions[3], can be due to various mechanisms: increase in specificity towards methionine of key enzymes involved in methionine utilization, desensitization towards ethionine as a feedback inhibitor, modification in susceptibility to methionine-mediated repression. The last two processes are expected to permit methionine biosynthesis in the presence of the analogue. As far as methionine utilization is concerned, two activating enzymes are metabolically important: methionyl-tRNA synthetase, dealing with incorporation of methionine into proteins, and S-adenosylmethionine synthase, dealing with transfer of methyl groups to all methylated compounds. Although changes of specificity of one or the other of these two enzymes are expected to permit a better competition between methionine and ethionine, it seems probable that as long as the regulatory mechanisms remain unimpaired, such a change would not be sufficient to cancel ethionine toxicity. On the opposite, since abolition of ethionine toxicity can already be obtained by a 10-fold lower methionine concentration than that of ethionine[4,5], it seemed conceivable that even a small modification in endogenous methionine production, brought about by changes in regulatory mechanisms, would be accompanied by ethionine resistance and then preferentially selected.

EVIDENCE FOR A REGULATORY ROLE OF THE eth-2 GENE

Among the numerous ethionine-resistant mutants obtained in this laboratory, only one was extensively studied, the resistance to ethionine of which resulted from

TABLE I   Repressibility of Homoserine Acetyltransferase and Homocysteine Synthase in Strains of Different Genotypes[a]

Enzyme	Medium[b]	Strain			
		4094-B eth-1s; eth-2s	CH 82-7A eth-1r; eth-2s	CC30-1D eth-1s; eth-2r	CH 82-7D eth-1r; eth-2r
Homoserine acetyltransferase	MM	0.380	0.405	0.359	0.310
	MM + Met	0.125	0.140	0.330	0.320
	Mean repression		66%		3%
Homocysteine synthase	MM	444	425	416	525
	MM + Met	62	74	236	273
	Mean repression		85%		46%

[a] Activities are expressed in nmol × min^{-1} × mg prot^{-1}. Growth and assay conditions are identical with those previously mentioned[1].
[b] MM: minimal synthetic medium, Met: DL-methionine 2×10^{-3} M.

TABLE II   Repressibility of ATP Sulfurylase and Homoserine Dehydrogenase in Strains of Different Genotypes[a]

Enzyme	Medium	Strain			
		4094-B eth-1s; eth-2s	CH 82-7A eth-1r; eth-2s	CC 30-1D eth-1s; eth-2r	CH 82-7D eth-1r; eth-2r
ATP sulfurylase	MM	148	81	181	246
	MM + Met	15	0	12	37
	Mean repression		93%		88%
Homoserine dehydrogenase	MM	333	430	314	365
	MM + Met	232	262	193	251
	Mean repression		35%		35%

[a] Activities are expressed in nmol × min^{-1} × mg prot^{-1}. See legend of Table I.

three independent mutations. One of these was shown to reside in a lowering of amino
-acid uptake[2], while the two others conferred specific resistance toward ethionine and
were not accompanied by a permeability modification[5,6]. These three genes, *aap*,
eth-1 and *eth-2*, segregate independently. The mutated resistance allele *eth-1r* is dominant and is only expressed in the presence of the mutated resistance allele *eth-2r*
which is recessive. Moreover, the presence of *eth-1r* enhanced the resistance due to
eth-2r.

In vivo isotopic experiments[7] using either ^{35}S-sulfate or differently labelled methionine led to the following conclusions: (1) exogenous methionine completely suppresses endogenous methionine production in a wild-type strain, showing that regulation of methionine biosynthesis is very efficient in *S. cerevisiae*; (2) in strains carrying the *eth-2r* allele, methionine is synthesized even in the presence of exogenous
methionine; (3) the presence of *eth-1r* does not by itself show any effect on methionine endogenous production. However, in the presence of exogenous ethionine, the
amount of methionine formed is higher in strains carrying *eth-1r* and *eth-2r* than in
strains carrying *eth-2r* alone.

These findings seemed to indicate that at least the *eth-2* gene is concerned with
a regulatory function in methionine biosynthesis. Nevertheless, it was thought useful
to verify that no modification in the two activating enzymes had occurred. For this
purpose, the activities of S-adenosylmethionine synthase and methionyl-tRNA synthetase (activity assayed for activation and charge of tRNA) were estimated in sensitive (*eth-1s*; *eth-2s*) and resistant (*eth-1s*; *eth-2r* and *eth-1r*; *eth-2r*) strains. No difference
could be detected in either enzyme[7,8]. These results prompted us to study the rate
of synthesis of the five enzymes found to be repressible by methionine, in strains carrying the *eth-2r* allele in the presence of either *eth-1s* or *eth-1r*.

Using 2×10^{-3}M DL-methionine, it was observed that repressibility of homoserine acetyltransferase and homocysteine synthase would vary greatly whether
eth-2s or *eth-2r* were present. Moreover, the nature of the *eth-1* allele did not affect
the observed repressibility. It can be noticed in Table I that, at the chosen methionine
concentration, the repressibility of homocysteine synthase is only lowered while the
repressibility of homoserine acetyltransferase is completely suppressed[7]. In the same
growth conditions, homoserine dehydrogenase and ATP sulfurylase did not show any
significant difference in repressibility (*see* Table II). As far as aspartokinase is concerned, no significant difference between repressibility of strains carrying either *eth-2s*
or *eth-2r* could be observed either. However, in view of the findings that repression
caused by exogenous methionine is not very important even in wild-type strains and
that the repressive effect of methionine is mainly evidenced by methionine limitation
using methionine auxotrophs in chemostat[1], it seemed difficult to expect conclusive
results for this enzyme in such a study.

The finding that homocysteine synthase was still repressible in strains carrying
the allele *eth-2r* suggested that the presence of this allele does not lead to a complete loss
of repressibility but to a modified response toward exogenous methionine concentra-

tion. Consequently, a study was undertaken on the effect of a wide range of methionine concentrations upon the synthesis of homocysteine synthase. As shown in Fig. 1 *above*, a much greater difference could be observed between the two groups of strains (*i.e.*, *eth-2s* and *eth-2r* strains) at low methionine concentrations (2×10^{-4}M) than at the concentration previously chosen (2×10^{-3}M). These results prompted us to reexamine the case of the three enzymes, the synthesis of which was not previously found to

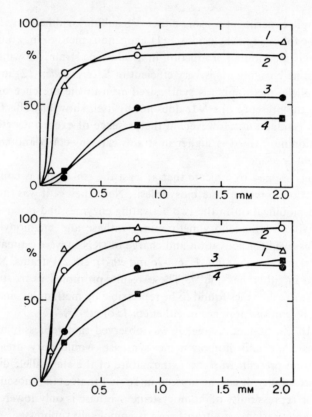

Fig. 1. Effect of various concentrations of DL-methionine (mM) upon repressibility (%) of two methionine-specific enzymes. *Above:* homocysteine synthase, *below:* ATP sulfurylase; *1* strain 4094-B (*eth-1s eth-2s*), *2* strain CH82-7A (*eth-1r eth-2s*), *3* strain CC30-1D (*eth-1s eth-2r*), *4* strain CH82-7D (*eth-1r eth 2r*).

be modified by the presence of *eth-2r* allele. In the case of aspartokinase and homoserine dehydrogenase it can be seen that repressibility decreases promptly and concomitantly in strains carrying various genotypes when the concentration of methionine is lowered, while no enhancement of repressibility can be attained by a 10-fold increase in exogenous methionine concentration. On the other hand[7], ATP sulfurylase exhibited a differential response towards methionine concentrations lower than 10^{-3}M, whether *eth-1s* or *eth-1r* were present (Fig. 1 *below*). Finally, if one takes

2×10^{-4}M DL-methionine as a basis for comparison, it can be seen that, in strains carrying eth-2r, the repressibility of homocysteine synthase and ATP sulfurylase is decreased by a factor of six and a factor of three, respectively. It can also be seen in Fig. 1 that the curves appear sigmoidal. It follows that a very small change in exogenous methionine concentration can lead to a great variation in enzyme levels.

All the results reported here indicate that eth-2 has to be considered as a regulatory gene in methionine biosynthesis. Furthermore, the finding that the mutated allele eth-2r modifies the repressibility of three different enzymes in the pathway shows its pleiotropic function. Appropriate crosses have shown that eth-2 segregates independently from the structural genes met 2 and met 8 (encoding respectively for homoserine acetyltransferase and homocysteine synthase) which have already been mapped on different chromosomes[9]. These findings, together with the recessivity of the eth-2r allele, imply that the product of eth-2r is a soluble compound. The probable nature of this compound will be discussed later.

In any case, the fact that eth-2 has no effect on the repressibility of two other enzymes which have been found susceptible to methionine-mediated repression (aspartokinase and homoserine dehydrogenase) renders it probable that there might exist at least two different repressors using methionine or a methionine derivative as a corepressor.

INFLUENCE OF METHIONYL-tRNA SYNTHETASE ACTIVITY ON REGULATION OF METHIONINE BIOSYNTHESIS

Studies of heat-sensitive strains from various organisms, carrying a modified aminoacyl-tRNA synthetase, pointed to the importance of these enzymes in cellular regulations. This seems to be the case for histidyl-tRNA synthetase[10], valyl-tRNA synthetase[11] in E. coli, arginyl-tRNA synthetase in N. crassa[12] and isoleucyl-tRNA synthetase[13] in S. cerevisiae. These results have provided some evidence that the control exerted by a given amino acid over its own biosynthesis might require intermediate synthesis of the corresponding aminoacyl-tRNA.

The existence of a heat-sensitive mutant of S. cerevisiae devoid of methionyl-tRNA synthetase in vitro[14] renders possible a study of the regulatory consequence of such a mutation on the rate of biosynthesis of enzymes susceptible to methionine-mediated repression[7]. This strain, H.19.3.4, which carried the recessive mutation ts⁻296, was isolated by McLaughlin and Hartwell[14], and was shown to be unable to grow at 37 °C. Moreover, even at the permissive temperature (28 °C), this mutant shows a requirement for methionine, a requirement which always segregates with heat sensitivity. The authors suggested that methionine might be required either to saturate or stabilize the modified methionyl-tRNA synthetase. Since methionine was required by the mutant and not by the parental strain, A 364 A, it seemed easier to start with a comparative study of these two strains from cultures in YPGA* medium.

It was hoped that by using a shift of culture from the permissive to the nonper-missive temperature the best conditions would be provided for showing different rates of enzyme synthesis between wild-type and heat-sensitive mutant. Unfortunately, it was found even for the wild-type strain, that the level of some of the enzymes studied was lower after the shift. It seems likely that this effect might be due to increased proteo-lysis rather than to real instability of the enzymes. In any case, these observations prompted us to switch to permissive (28 °C) or semipermissive (32 °C) temperature, conditions which did not show striking differences in specific activities on the wild-type strain (*see* Table III). Glutamate dehydrogenase which is not metabolically related to the pathway, has been used as a control enzyme for such a study.

It can be seen in Table III that, whatever the temperature of growth, the rate of synthesis of three of the enzymes studied, namely homoserine acetyltransferase, homo-cysteine synthase and ATP sulfurylase, was greatly enhanced in the mutant as compared with the wild-type strain in the same conditions. On the other hand, the two other enzymes dealing with threonine — methionine biosynthesis show a very small variation, which has the same range as the variation observed for the control enzyme. It can be pointed out that even the highest activities obtained are still lower than the maximally derepressed levels attained in chemostat experiments previously reported[1] with other strains.

It can be seen in Table IV that in the parent strain A 364 A, the specific activities observed after growth in YPGA medium are similar to the one found for a reference wild-type strain 4094-B. Comparison with specific activities observed in minimal me-

TABLE III

Influence of Growth Temperature upon Synthesis of Various Enzymes in Wild-Type and Heat-Sensitive Strains[a]

Enzyme	Wild type (A364A)		Mutant (H.19.3.4)	
	28 °C	32 °C	28 °C	32 °C
Aspartokinase	12.3	14.6	35.3	50.9
Homoserine dehydrogenase	197	303	247	404
Homoserine acetyltransferase	0.025	0.060	0.307	0.600
Homocysteine synthase	13.1	11.6	95.0	108
ATP sulfurylase	5.5	15.5	70.7	77.6
Glutamate dehydrogenase	5.86	4.50	7.40	6.70

[a] Activities are expressed in nmol $\times$ min^{-1} $\times$ mg prot^{-1}.
Assay conditions are identical with those mentioned previously[1].

* YPGA: Difco Yeast extract 5 g/l, Bacto-peptone 5 g/l, glucose 30 g/l, adenine 20 mg/l.

TABLE IV

Effect of Different Growth Conditions on Specific Activities of Various Enzymes in Wild-Type Strains[a]

Enzyme	Strain and medium[b]					
	4094-B			A364A		
	YPGA	MM	MM + Met	YPGA	MM	MM + Met
Aspartokinase	13.0	32.0	34.0	12.3	33.0	34.0
Homoserine dehydrogenase	170	290	250	197	586	503
Homoserine acetyl-transferase	0.120	0.420	0.450	0.025	0.618	0.672
Homocysteine synthase	35	405	390	13.0	205	183
ATP sulfurylase	6.5	150	120	5.50	140	110
Glutamate dehydrogenase	3.0	2.34	2.64	5.86	8.0	5.0

[a] Growth was carried out at 28 °C. Specific activities are expressed in $nmol \times min^{-1} \times mg\ prot^{-1}$. Assay conditions are identical with those mentioned previously[1].

[b] YPGA: Difco Yeast extract 5 g/l, Difco Bacto-peptone 5 g/l, glucose 30 g/l, adenine 20 mg/l, MM: minimal medium, Met: DL-methionine 10^{-4}M.

dium shows that YPGA medium is highly repressive for the enzymes of the threonine — methionine pathway. Thus it seemed necessary to prove, by cultivation in a synthetic medium, that methionine itself could reproduce the difference in the repressive effect previously observed between the parent wild-type strain and the mutant, after growth on a YPGA medium. Since the mutant strain requires methionine for growth, we have looked for a concentration of this amino acid which, while supporting growth of the mutant, did not produce any significant repressive effect on the parent and the reference wild-type strains. Previous studies have shown that 0.1 mM DL-methionine is sufficient to permit growth of the mutant strain. Results in Table IV indicate that, for the enzymes studied, this concentration does not significantly modify the specific activities of the two wild-type strains. Consequently, minimal medium + + 0.1 mM DL-methionine was chosen as reference minimal medium for the parent and mutant strains. It can be noticed also, in Table IV, that there are some intrinsic differences between enzyme levels in the two wild-type strains, after growth in minimal medium. In order to have a better comparison between the parent (A 364 A) and the mutant strain (H.19.3.4), results obtained from cultures in a synthetic medium with repressive concentrations of methionine and from cultures in a YPGA medium, have been referred in per cent to the activities obtained after growth in the reference minimal medium.

TABLE V

Activity of Various Enzymes after Growth of Wild-Type (WT) and Mutant (M) Strains in Different Conditions[a]

Temperature °C	Enzyme	Growth conditions					
		YPGA		MM + DL-methionine			
				2×10^{-4}M		4×10^{-3}M	
		WT	M	WT	M	WT	M
28	Glutamate dehydrogenase	122	151	104	90	107	120
	Aspartokinase	36	94	72	89	70	79
	Homoserine dehydrogenase	39	61	77	99	77	82
	Homoserine acetyltransferase	3.7	61	27	91	7.3	40
	Homocysteine synthase	7.1	61	16	89	9.0	18
	ATP sulfurylase	5.0	81	15	89	11	23
32	Glutamate dehydrogenase	150	300	101	125	87	111
	Aspartokinase	29	83	72	85	69	61
	Homoserine dehydrogenase	63	125	59	113	71	87
	Homoserine acetyltransferase	8.0	185	15	66	1.8	21
	Homocysteine synthase	6.4	77.2	13	87	4.0	31
	ATP sulfurylase	12	135	12	87	8.0	24

[a] Results are expressed as per cent of the corresponding specific activity measured for each strain
    in the reference minimal medium (containing DL-methionine 10^{-4}M).
    See legend to Table IV and the text.

Results in Table V show striking differences in the repressibility of three of the enzymes studied, homoserine acetyltransferase, homocysteine synthase and ATP sulfurylase, especially when cultures are made in the presence of 2×10^{-4}M DL-methionine.

On the other hand, the synthesis of two other enzymes, aspartokinase and homoserine dehydrogenase, although they belong to the same pathway as the previous three, is not modified more than the control enzyme, glutamate dehydrogenase. These results are in complete agreement with experiments obtained in the study of the effects of the regulatory gene eth-2. Again, the three methionine-specific enzymes exhibit a pleiotropic response which is not shared by the enzymes involved in the common part of the pathway leading to threonine and methionine. These conclusions are further evidenced by comparison of ratios between enzymatic activities found in mutant and wild type, as shown in Table VI.

If one accounts for a general nonspecific repressive effect of YPGA over unrelated enzymes (of the order of two-fold as evidenced by the results in the upper part of Table VI) it seems that the specific effects obtained by adding 2×10^{-4}M methionine

TABLE VI

Ratios between Enzymatic Activities in Mutant and Wild-Type Strains after Growth in Different Conditions[a]

Enzyme	YPGA		MM + DL-methionine			
			2×10^{-4}M		4×10^{-3}M	
	28 °C	32 °C	28 °C	32 °C	28 °C	32 °C
Glutamate dehydrogenase	1.2	2.0	0.9	1.2	1.1	1.4
Aspartokinase	2.6	2.9	1.2	1.2	1.1	0.9
Homoserine dehydrogenase	1.6	2.0	1.3	1.9	1.1	1.2
Homoserine acetyltransferase	16.0	23.0	3.4	4.3	5.5	11.7
Homocysteine synthase	8.6	11.4	5.5	6.7	2.0	7.7
ATP sulfurylase	16.0	11.2	5.9	7.2	2.1	3.0

[a] *See* explanation in the text.

to a minimal medium are quite similar to those obtained with the YPGA medium: ratios of 3.4, 5.5 and 5.9 as compared with 8.9, 5.0 and 8.9 at 28 °C or 4.3, 6.7 and 7.2 as compared with 10.5 and 4.9 at 32 °C for homoserine acetyltransferase, homocysteine synthase and ATP sulfurylase, respectively.

All these results can be taken as evidence that methionyl-tRNA synthetase activity is somehow connected with the ability of exogenous methionine to exert its repressive effect on the synthesis of methionine-specific enzymes.

The use of a much higher DL-methionine concentration (4×10^{-3}M) shows that the repressive effect of this amino acid is restored at 28 °C but not at 32 °C for homocysteine synthase and is restored at 28 °C as well as at 32 °C for ATP sulfurylase. These results confirm, even with the heat-sensitive mutant, the greater sensitivity of ATP sulfurylase to methionine-mediated repression. If one assumes that heat sensitivity of the mutant H.19.3.4 (mutation ts⁻296) is caused by progressive impairment of the *in vivo* activation of methionine when temperature is raised above the optimum, these results should be interpreted as showing that the conditions which improve the *in vivo* formation of methionyl-tRNA also lead to recovery of the repressive effect of methionine*. Results obtained by using different concentrations of methionine thus lead to the assumption that a product of methionyl-tRNA synthetase activity participate to the same pleiotropic regulatory system discovered by the study of the *eth-2* gene.

Since previous results have shown that methionyl-tRNA synthetase did not seem to be modified in strains carrying the *eth-2r* allele it seemed unlikely that the two

---

* In this light, the fact that derepression was never observed with the mutant H.19.3.4 (ts⁻296) can be explained since any condition which permits growth, necessarily leads to at least a partial acylation of tRNA^Met.

genes, *eth-2* and ts⁻296, could be allelic. However, owing to the similitude in the effects of these two genes, it was very interesting to prove their genetic independence. An appropriate diploid (*eth-2s*; ts⁻296/*eth-2r*; +) was formed, the phenotype of which was heat-resistant, methionine-independent and ethionine-sensitive. Moreover, segregation of this diploid led to equal recovery, among the methionine-independent

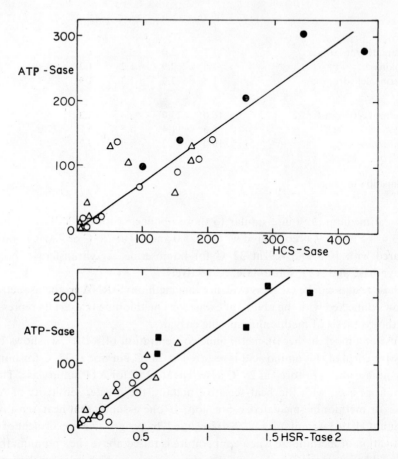

Fig. 2. Coordinate synthesis of enzymes specifically involved in methionine biosynthesis (group I enzymes). *Above:* ATP sulfurylase activity (ATP-Sase) *vs.* homocysteine synthase activity (HCS-Sase), *below:* ATP sulfurylase activity (ATP-Sase) *vs.* homoserine acetyltransferase activity (HSR-Aase). Specific activities are expressed in nmol $\times$ min^{-1} $\times$ mg prot^{-1}.

●   Values obtained from strain D6 (*met2*, homoserine acetyltransferase-deficient) grown in methionine limitation.

■   Values obtained from strain JE2 (*thr2*, aspartate semialdehyde dehydrogenase-deficient) grown in methionine limitation.

○   Values obtained from strains H.19.3.4 and A364A (the methionyl-tRNA synthetase-deficient and the corresponding parent strain respectively) grown at 28 °C, in different media.

△   Values obtained from strains H.19.3.4 and A364A grown at 32 °C in different media.

part of the population, of ethionine-sensitive and ethionine-resistant strains.These results show that the two mutations are unlinked[7].

All the results so far obtained pointed to a classification of the enzymes subjected to methionine-mediated repression, into two distinct regulatory groups. In bacteria it was found that, when synthesis of many enzymes involved in the same biosynthetic

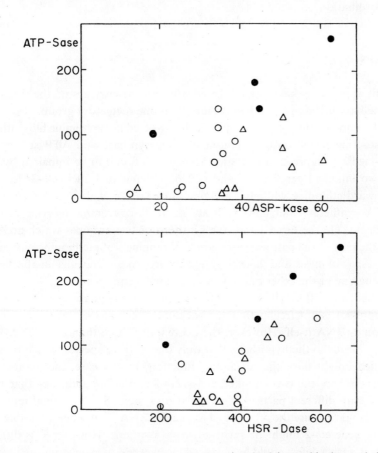

Fig. 3. Absence of coordination between enzymes involved in methionine and threonine bio-synthesis and one methionine-specific enzyme. *Above:* ATP sulfurylase activity (ATP-Sase) *vs.* aspartokinase activity (ASP-Kase), *below:* ATP sulfurylase activity (ATP-Sase) *vs.* homoserine dehydrogenase activity (HSR-Dase); specific activities are expressed in nmol $\times$ min^{-1} $\times$ mg prot^{-1}. *See* legend to Fig. 2.

pathway is subjected to repression by the final amino acid, the observed regulation is coordinate[15]. Such a coordinate response was explained by the association of the structural genes into an operon and by the polycistronic nature of the corresponding mRNA[16]. Although structural genes corresponding to these enzymes are unlinked, it seemed of interest to see if their synthesis was coordinate or not. For this purpose, graphs have been constructed putting together results obtained from different strains and under different growth conditions (*see* Fig. 2 and 3). Owing to the different experimental origin of each point, it is very striking to observe that a rather strong positive correlation exists between specific activities concerned with the different methionine-specific enzymes, and that, in turn, enzymes which do not belong to this group show a lack of coordination[7].

DISCUSSION

Different approaches have led to classification of the enzymes of the threonine − methionine biosynthetic pathway into more than one regulatory group.

Group I comprises three enzymes specifically involved in methionine biosynthesis, namely homoserine acetyltransferase, homocysteine synthase and ATP sulfurylase. Methionine-mediated repression of these enzymes was found to be impaired by mutations of two unlinked genes, eth-2 and ts 296. The mutated allele eth-2r leads to resistance toward ethionine while the mutated allele ts⁻296 leads to a defective methionyl-tRNA synthetase and confers heat sensitivity concomitantly with methionine requirement. On the other hand, the synthesis of two enzymes which are concerned with the part of the pathway common to threonine and methionine biosynthesis, namely aspartokinase and homoserine dehydrogenase, remains unaffected in strains carrying one or the other of the two mutations mentionned.

Involvement of methionyl-tRNA synthetase in such a regulatory system implies that the corepressor is not free methionine but rather a product of this enzyme, more likely methionyl-tRNA itself. However, the existence of more than one tRNA^{Met} in yeast cells[17−20] leads to the hypothesis that only one of these species, which remains to be identified, might have the regulatory function. In any case, the parallel but complementary role of the two unlinked genes, eth-2 and ts 296, indicates that both are concerned with different parts of this regulatory system. Since the final repressor of this system might require the presence of a specific tRNA^{Met}, it seems plausible that the regulatory gene eth-2 might be involved in its synthesis. However, it is difficult to visualize how such a defect could be balanced by an excess of exogenous methionine as it was shown to occur in studies of repressibility of group I enzymes (*see* Fig. 1). Nevertheless, if such an hypothesis proved to be correct, the absence of phenotypic expression of eth-2r allele in minimal medium would provide an additional argument that the regulatory role of methionyl-tRNA is ensured by a minor species of the tRNA^{Met}. On the other hand, the allosteric aspect of the curve obtained with increasing concentrations of methionine and the fact that high levels of repression can still be obtained in eth-2r strains (provided enough exogenous methionine is present), suggest that the eth-2 gene might be concerned with the synthesis of an aporepressor protein. If this were the case it could be assumed that the mutation eth-2s→eth-2r has led to a 5—10 fold lowering of affinity of this aporepressor toward the corepressor methionyl-tRNA. Further genetic experiments (particularly a search for nonsense mutations in the locus eth-2, able to be suppressed by known amber or ochre suppressors) will be undertaken in order to provide more direct evidence for the protein nature of the eth-2 gene product.

The very fact that three structural genes, which are known to be unlinked, show a pleiotropic response to the dual control of two genes involved in the synthesis of a common repressor, favours the concept of a redundant operator-like segment attached to each structural gene concerned. However, slight differences in quantitative response of each of the corresponding enzyme to exogenous methionine, whether in the presence or the absence of the eth-2r allele, seems to indicate that such an operator redundancy might have been subjected to further individual modifications in the course of evolution.

In any case, even slightly diversified, the three operator-like structures have still retained recognition for the same aporepressor + corepressor complex. It seems then worth to remark that coordinate response to a given regulatory signal should not be taken as an indication of clustered arrangement of structural genes into an operon[15,16], but as evidence that even dispersed structural genes respond to the same regulatory system. Besides the pleiotropic methionine regulatory system uncovered here, genetic and biochemical studies of the acetate assimilation pathway in Neurospora have also provided evidence for a coordinate response of unlinked structural genes to the same regulatory factor[21].

Although aspartokinase and homoserine dehydrogenase do not exhibit strictly identical response to high exogenous concentrations of methionine, regulation of their synthesis by this amino acid has been evidenced by cultures in methionine limitation[1]. Nevertheless, (1) the absence of coordination of their synthesis with synthesis of group I enzymes, (2) the absence of response to the regulatory eth-2 gene, and (3) the absence of response to the impairment of methionyl-tRNA synthetase activity, demonstrate that neither aspartokinase nor homoserine dehydrogenase belong to the regulatory group I. The last point also provides evidence that a general regulatory role of acylated tRNA's cannot be assumed. Such a conclusion was already derived from studies in other systems, for example, in the regulation of enzymes involved in tyrosine biosynthesis in Escherichia coli[22] and in the regulation of some enzymes involved in methionine biosynthesis in Salmonella typhimurium[23].

As far as aspartokinase and homoserine dehydrogenase are concerned it seems then that the existence of at least one other corepressor besides methionyl-tRNA has to be postulated. It could of course be assumed that methionine itself might play this role. However, S-adenosylmethionine might also be postulated as a convenient methionine derivative able to assume a regulatory function. In addition, the bivalent control exerted by threonine and methionine over aspartokinase biosynthesis as compared with the monovalent control exerted by methionine alone over homoserine dehydrogenase synthesis, renders it likely that two different regulatory systems might be involved.

Thus, in view of these deductions, the existence of three different types of methio-nine-mediated regulatory mechanisms might be expected in yeast, mechanisms which might differ both by the apo- and the corepressors involved. One should add to those at least one aporepressor and one corepressor for threonine-mediated regulation, and, for each structural gene, one genetic segment, able to recognize the corresponding repressor. It follows that the seven structural genes, the function of which has been found to be regulated so far in threonine and methionine biosynthesis in *S. cerevisiae*, might require no less than 15 genes devoted to a regulatory function.

The authors are thankful to F. de la Torre for her skilled assistance.

This work was supported by grants from the Délégation Générale à la Recherche Scientifique et Technique (66.00.140) and from the C.E.A., France.

REFERENCES

1. DE ROBICHON-SZULMAJSTER H., SURDIN-KERJAN Y., CHEREST H.: Paper at 1st Internat. Symp. Genetics of Industrial Microorganisms, Prague 1970. Abstract Book p. 93; *cf.* this book, p. 149.
2. SURDIN Y., SLY W., SIRE J., BORDES A. M., DE ROBICHON-SZULMAJSTER H.: Biochim. Biophys. Acta **107**, 546 (1965).
3. STEKOL J. A.: Adv. Enzymol. **25**, 369 (1963).
4. MAW G. A.: J. Gen. Microbiol. **25**, 441 (1961).
5. DE ROBICHON-SZULMAJSTER H., CHEREST H.: Genetics **54**, 993 (1966).
6. CHEREST H., DE ROBICHON-SZULMAJSTER H.: Genetics **54**, 981 (1966).
7. CHEREST H.: Thèse de doctorat d'Etat. Paris 1970.
8. TINGLE M., *unpublished results.*
9. MORTIMER R. K., HAWTHORNE D. C.: Genetics **53**, 165 (1966).
10. SCHLESSINGER S., MAGASANIK B.: J. Mol. Biol. **9**, 670 (1964).
11. EIDLIC L., NEIDHARDT F. C.: J. Bacteriol. **89**, 706 (1965).
12. NAZARIO M.: Biochim. Biophys. Acta **145**, 146 (1967).
13. McLAUGHLIN C. S., MAGEE P. T., HARTWELL L. H.: J. Bacteriol. **100**, 579 (1969).
14. McLAUGHLIN C. S., HARTWELL L. H.: Genetics **61**, 557 (1969).
15. AMES B. N., GARRY B. J.: Proc. Nat. Acad. Sci. U.S.A. **45**, 1453 (1959).
16. MARTIN R. G.: Cold Spring Harbor Symp. Quant. Biol. **28**, 357 (1963).
17. GILLAM I., MILLWARD S., VON TIGERSTROM D. B. M., WIMMER E., TENNER G. M.: Biochemistry **6**, 3043 (1967).
18. TAKEISKI K., UKITA T., NISHIMURA S.: J. Biol. Chem. **243**, 5761 (1968).
19. SMITH A. E., MARCKER K. A.: J. Mol. Biol. **38**, 241 (1968).
20. RAJBANDHARY U. L., GHOSH H. P.: J. Biol. Chem. **244**, 1104 (1969).
21. FLAVELL R. B., FINCHAM J. R. S.: J. Bacteriol. **95**, 1063 (1968).
22. RAVEL J. M., SHIVE M. N., SHIVE W.: Biochem. Biophys. Res. Comm. **20**, 352 (1965).
23. GROSS T. S., ROWBURY R. J.: Biochim. Biophys. Acta **184**, 233 (1969).

Specificity and Regulation of the Uptake and Retention of Amino Acids and Pyrimidines in Yeast*

M. GRENSON

*Laboratoire de Microbiologie, Faculté des Sciences, Université libre de Bruxelles, and Institut de Recherches du C.E.R.I.A., Brussels, Belgium***

For optimal functioning, cells need precise regulation of their metabolic pool of low-molecular-weight constituents. This necessity is probably a question of optimal concentrations of metabolic intermediates rather than of real economy. The regulation of the intracellular concentration of metabolites like amino acids must involve a control not only of their biosynthesis and of their eventual degradation, but also of their passage across the cell membrane.

Under certain conditions, yeast cells are able to excrete large quantities of several metabolites into their medium. This phenomenon offers the possibility of interesting practical applications. The purpose of the present paper is to analyze its theoretical aspects: especially its determinism and its physiological significance. The practical interest of this analysis is to make possible a rational choice of conditions where one or more given compounds are excreted into the medium.

This study of the excretion phenomenon has led to the view that excretion is a means of regulating the intracellular concentration of products of cell metabolism in all cases of overproduction. Regulation of uptake of exogenous substances is effected by a completely different mechanism: a specific feedback inhibition of specific uptake systems.

As will appear in what follows, the surrounding membrane of *S. cerevisiae* seems particularly impermeable to most substances, except at the level of specialized transport systems whose properties are similar to those of bacteria (*see* reviews by KEPES and COHEN[1], and by PARDEE[2]) or, possibly, under special conditions which might reversibly alter the membrane configuration.

---

* In this article, which is mainly concerned with studies of the yeast *Saccharomyces cerevisiae*, no exhaustive review of the literature has been made. Additional references can be found in the cited papers.

** Mailing address: Institut de Recherches, C.E.R.I.A., 1 avenue E. Gryzon. B-1070. Bruxelles, Belgium.

AMINO-ACID TRANSPORT SYSTEMS IN *Saccharomyces*

SPECIFICITY

Specific amino-acid permeases. A number of very specific uptake systems have been demonstrated in *Saccharomyces**. They are listed in Table I, which summarizes the criteria used for specificity determination, and where references can be found.

Two criteria were used to delineate the specificity of amino-acid transport systems in yeast.

Good evidence for specificity of a transport system is the absence of competition for uptake with other amino acids, except those which are structurally closely related to the substrate. When inhibition is observed, the inhibitor is considered to be recognized by the active site only as far as inhibition is of a competitive type. Competitive inhibitors were found to fall into two classes: some are transported as well as the substrate (although often with lower affinity), while others are not. For example, the active site of the arginine permease recognizes L-canavanine and L-lysine, since these amino acids competitively inhibit the uptake of arginine although with a lower affinity[3]. That canavanine and lysine are transported by the arginine uptake system is shown by the fact that canavanine toxicity and a part of lysine uptake disappear in an arginine-permeaseless mutant[3,4]. On the other hand, the specific lysine permease, although competitively inhibited by arginine, does not transport it at a detectable rate, since arginine uptake is not altered by the lysine-permeaseless mutation *lsp-1*[4].

TABLE I

Specific Amino-Acid Uptake Systems Demonstrated in *Saccharomyces*[a]

Yeast	Specific transport system	Criteria[b]	Refs
S. cerevisiae Σ1278b	L-arginine	C,M (*agp-1*)	3
	L-lysine	C,M (*lsp-1*)	4
	L-methionine	C,M (*mtp-1*)	5
	L-dicarboxylic amino acids	C,M (*dcp-1*)	6
	L-histidine	C,M (*htp-1*)	7
S. chevalieri	L-proline	C	8

[a] Culture medium used for *S. cerevisiae* was a minimal medium containing ammonium ions as the sole nitrogen source.
[b] C competition experiments (kinetics), M mutation.

---

* All our mutants were isolated from the same wild-type strain of *S. cerevisiae*, Σ1278b (*mat-α*) or from its mating-type mutant 3962c (*mat-a*), and only these strains were used in crosses for genetic analysis. Hence, all the strains used were isogenic, except for the mentioned mutations.

The other test for specificity of uptake was obtained by the study of mutants with specifically impaired uptake of one amino acid (or a small group of amino acids). This seemed to us the most direct and best argument. However, as discussed below, it is not easy to be sure that a mutant is blocked at the level of the uptake system directly. Nevertheless, under certain conditions, mutants can be used to test the specificity of an uptake system, even if their uptake activity is impaired by indirect inhibition, rather than by the loss or modification of the specific binding protein directly. This will also be illustrated below, for the case of pyrimidine uptake.

General amino-acid permease. In addition to these specific amino-acid permeases, *S. cerevisiae* possesses an amino-acid transport system with much broader specificity.

Working under different conditions (namely nitrogen starvation) and with a yeast strain different from ours, HALVORSON and COHEN[9], and SURDIN *et al.*[10] found that all the amino acids enter the yeast cell by one uptake system with broad specificity. On the other hand, we observed that the rate of uptake of several amino acids is increased when cells are grown on a proline rather than an ammonia medium: a 3 to 400-fold increase was observed, depending on the considered amino-acid[3-5,11,12]. In the case of amino acids which are transported by a specific permease, the increased transport activity resulting from growth on proline can be reduced to that of the ammonia medium by several amino acids which have no effect on the specific permease. Specific permeaseless mutants for arginine (*agp-1*), lysine (*lsp-1*) and methionine (*mtp-1*) make the cells resistant to the corresponding toxic analogs canavanine, thiosine and ethionine, respectively. However, sensitivity to the analogs is recovered when the ammonium ions of the medium are replaced by proline as a sole source of nitrogen.

These aspects of amino-acid uptake in yeast were recently reinvestigated and found to be due to the properties of a general amino-acid permease which exists in addition to the specific amino-acid permeases[12].

The general amino-acid permease transports most of the amino acids. Amino acids as different as citrulline, lysine and tryptophan show competitive inhibition. This general amino-acid permease is inhibited by ammonia in its activity, but not in its synthesis. A mutant which has lost this general amino-acid permease activity was isolated: the mutation has been named *gap* (general amino-acid permease). Its phenotype is completely different from that of the *aap* mutant isolated by SURDIN *et al.*[10] and which was thought to have specifically lost the activity of the (general) amino-acid permease. Unlike the *gap* mutation, the *aap* mutation confers resistance to the toxic analogs of a number of amino acids which can enter the cells by two kinds of transport systems, namely the ammonia-sensitive general amino-acid permease and the ammonia-insensitive specific amino-acid permeases. The *gap* and *aap* mutations are not allelic.

A mutation which provokes the same phenotype as the *aap* mutation was obtained in our wild-type strain and shown to be allelic to the *aap* mutation. This mutation, which has been named *apf* (amino-acid permease factor) was found to affect the capacity (maximum rate) of the specific, as well as that of the general amino-acid permeases, but the uptake of pyrimidines remains unchanged. Hence, it is suggested that what is impaired by the *apf* (*aap*) mutation is a factor (V_{max} factor) which is needed for optimal activity of a number of amino-acid permeases[11,13].

The main function of the general amino-acid permease is probably to transport amino acids for katabolic purposes. The simultaneous existence of a general amino-acid transport system and of more specific amino-acid permeases seems to be common in fungi[14,15].

REGULATION OF THE ACTIVITY OF AMINO-ACID UPTAKE SYSTEMS

From the point of view of amino-acid uptake, there is a striking difference between bacteria, such as *Escherichia coli*, and yeast, *viz.* the impossibility of studying amino-acid uptake in yeast under conditions where protein synthesis is inhibited, whereas it is a common practice with bacteria to block amino-acid utilization with chloramphenicol. It was observed in several eukaryotes that when protein synthesis is inhibited by cycloheximide, the uptake of amino acids is almost completely inhibited within a short time[16-21]. This was interpreted by some authors as due to rapid turnover of the corresponding permeases. Another possible interpretation was also considered, and supported by the following observations[21]. In yeast, when protein synthesis was prevented by cycloheximide, the uptake of all the tested amino acids was inhibited. However, when protein synthesis was prevented by histidine starvation in a histidine-requiring mutant, arginine uptake was inhibited, while the rate of histidine uptake remained constant. The simplest interpretation of these facts seemed to be that inhibitions of uptake are due to free amino acids or derivatives, whose synthesis is not interrupted by cycloheximide, and which accumulate as a result of inhibition of protein synthesis. If this is the case, the mentioned observations would indicate that the inhibitions are very specific.

We have now direct observations showing that histidine uptake is regulated by specific feedback inhibition[7]. This, however, does not prove that there is no permease turnover when protein synthesis is inhibited by cycloheximide.

The impossibility of measuring amino-acid uptake in the absence of protein synthesis in yeast is related to the absence of exit and chase of amino acids accumulated from the medium. For instance, a wild-type strain of yeast growing exponentially in the presence of 2 mM ^{14}C-L-histidine accumulates a large pool of free intact histidine. Replacement of the incubation medium by fresh medium containing either no histidine, or unlabelled histidine does not provoke any displacement of the accumulated radioactivity. This seems quite different from the situation which occurs in *E. coli*, where the steady-state concentration level of several exogenous sugars and amino acids is determined by a balance between rates of inflow and outflow[1]. The simplest interpretation of the absence of exit in the case of histidine taken up in yeast seems to be that the cell membrane is impermeable to histidine (except at the level of the transport systems), and that no exit mechanism is functional under our experimental conditions.

In spite of the absence of exit balancing entry, histidine taken up does not accumulate indefinitely, and the limitation of net entry must be due to some kind of feedback control. Fig. 1 shows that preloading cells with histidine leads to rapid inhibition

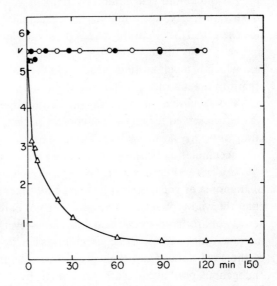

Fig. 1. Effect of preloading cells with several amino acids on the initial uptake rate of 20 μM ^{14}C-L-histidine. Each point represents an uptake rate, v (nmol/min per mg protein). Preloading with 2 mM unlabelled L-histidine (*triangles*), L-lysine (*closed circles*), or L-arginine (*open circles*). Taken from Ref.[7] (*by permission of European J. Biochem.*).

of uptake, whereas preloading with arginine or lysine has no effect on histidine uptake. Similarly, preloading with L-methionine, glycine, L-serine, L-threonine or L-glutamate does not alter histidine uptake significantly. However, these amino acids actually enter the cells (as shown with labelled amino acids used in parallel experiments), and each of the amino acids is accumulated inside the cell at a concentration and in a form able to inhibit its own uptake. These data show that there is a specific feedback inhibition on the process of histidine uptake, and indicate that the uptake of other amino acids is regulated in the same way. Several observations in other systems and organisms can be considered as serious indications of regulation of uptake systems by feedback inhibition (references in Ref.[7]).

REGULATION OF THE SYNTHESIS OF AMINO-ACID UPTAKE SYSTEMS

Some amino-acid transport systems seem to be regulated at the level of their formation. In *S. cerevisiae*, several specific permeases, namely those transporting methionine, leucine, valine and isoleucine appear to be inducible and controlled by a common genetic element[22,23]. The proline permease in *Saccharomyces chevalieri*[24] and possibly the specific dicarboxylic amino-acid permease in *S. cerevisiae*[6] appear to be repressible. The other studied amino-acid permeases seem to be constitutive.

ACCUMULATION OF EXOGENOUS AMINO ACIDS

All the amino-acid uptake systems of yeast that we consider here have a very striking property in common: they accumulate their substrates free and intact inside the cell. This situation makes it easy to demonstrate the existence of specialized uptake systems distinct from the enzymes participating in metabolism, a demonstration which is much more difficult when the activity of the uptake system is limiting in regard to the internal metabolism of the substrate.

The accumulation of large concentrations of free intact substrate inside the cell was not observed for all the amino-acid permeases. Some are typically non-accumulating or poorly accumulating. This makes their study more difficult and explains why only accumulating permeases were used for study so far.

Under conditions where the general amino-acid permease is not active (presence of ammonia or *gap* mutation), amino acids like tryptophan, tyrosine or phenylalanine enter the wild-type strain of yeast Σ1278b at barely detectable intial rates. This reduced activity of aromatic-amino-acid uptake, however, seems to be specific and efficient since mutants which require these amino acids can be supported by normal concentrations of these substances and their growth is not inhibited by other amino acids[25]. The endogenous pool accumulated at the steady state (growth in the presence of 0.1mM ^{14}C-L-tryptophan for instance) is very low, both in the wild-type strain and in a tryptophan-less mutant. Furthermore, the biosynthesis of tryptophan, for instance, is depressed in *S. cerevisiae* grown in the presence of tryptophan[26] (confirmed on our strain). These data indicate that the observed low uptake activity is not due to presence of large pools of aromatic amino acids in the wild-type strain but corresponds to a real strong limitation of the accumulative power of the permeases. This could be due either to a high sensitivity of the permease to feedback inhibition or to some kind of deficiency either in energy coupling for accumulation or in retention of the aromatic amino acids (deficiency of retention might be due for instance to a low threshold level for activity of an exit mechanism). In the last case, however, it should be possible to detect an exchange process between internal and external aromatic amino acids which was not observed.

The absence of accumulation of high concentrations of these amino acids from the medium does not seem abnormal *per se*: the cell activities seem to be adapted to low pools of aromatic amino acids; this means that constituents like the activating enzymes, the feedback-inhibited biosynthetic enzyme, or the specific aporepressor are able to function under conditions of low amino-acid concentrations. What is more astonishing is rather the situation of the accumulated amino acids. In the case of arginine, for instance, it is clear that repression of biosynthesis[27], as well as feedback

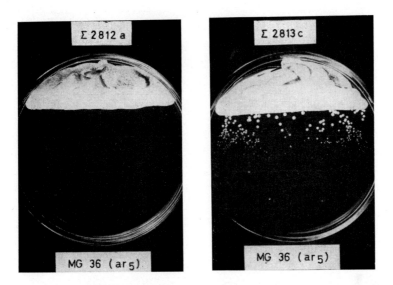

Fig. 2. Feeding of an arginine-less strain by an arginine-permeaseless mutant. Minimal medium containing ammonium ions as a sole source of nitrogen. Receptor strain, MG36 (*arg5*), is plated to a low population density so as to give distinct colonies when growing. Donor strains: Σ2812a, control, contains two mutations, *gap* and *apf*; Σ2813c contains an arginine-permeaseless mutation (*agp-1*) in addition to *gap* and *apf*.

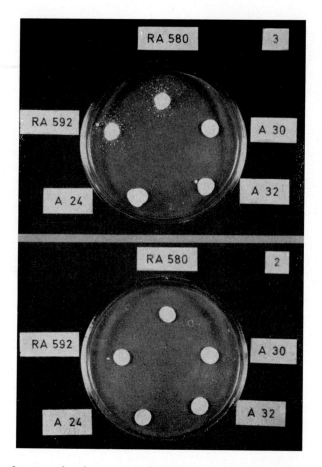

Fig. 3. Feeding of a tryptophan-less yeast strain by two regulatory mutants. The receptor strain, MG779, is able to use indole as a source of tryptophan. Hence it might be either tryptophan itself or indole (or another intermediate of tryptophan biosynthesis) which is excreted. Two excreting strains (RA580 and RA592) can be observed on minimal medium (containing ammonium ions as a sole source of nitrogen; *above*), but feeding is inhibited by adding tyrosine to this medium *below*).

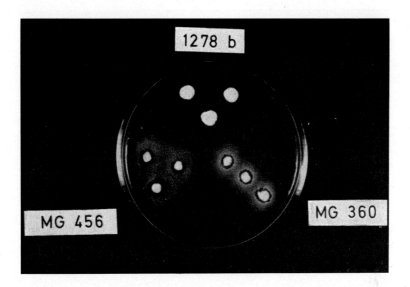

Fig. 4. Feeding of a methionine-less strain by two single auxotrophs. The receptor strain MG331, methionine-less, is able to use homocysteine as a source of methionine. Donor strains: MG456, pyrimidine-less; MG360, lysine-less; control: Σ1278b, wild-type strain. Minimal medium containing ammonium ions as a sole source of nitrogen; since it is unable to support growth of the auxotroph donors, these were inoculated in large quantities.

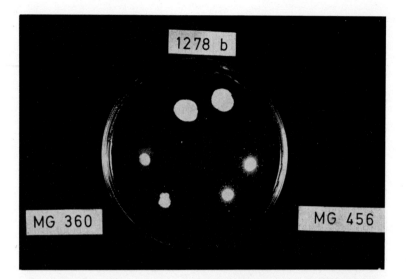

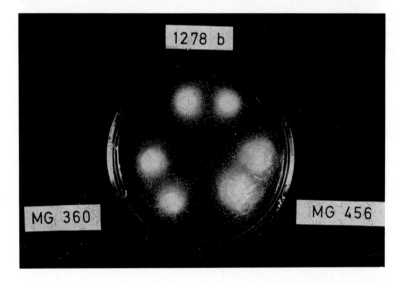

Fig. 5. Feeding of tryptophan-less strains by two single auxotrophs and by the wild-type strain. Donor strains as in Fig. 4. Receptor strain and culture medium: *above:* MG783, *trp* unable to use indole as a source of tryptophan, on minimal medium (containing ammonium ions as a sole source of nitrogen); *below:* same receptor on minimal medium supplemented with indole (250 µg/ml). It can be seen that the two auxotrophs excrete tryptophan on minimal medium, but that the wild-type strain does not (*above*). By adding indole to the medium, a large production of tryptophan is induced, even in the wild-type strain (*below*).

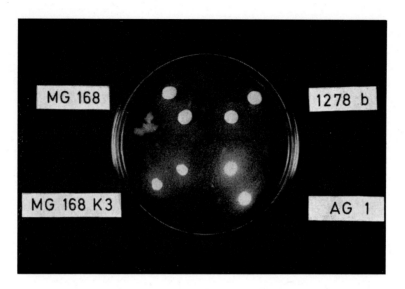

Fig. 6. Feeding of an arginine-less strain in the presence of citrulline. Donor strains: Σ1278b (wild-type strain), MG168 (*agp-1*), MG168K3 (*agp-1 gap*), AG1 (arginase-less). Receptor strain: MG682 (*arg1*, arginine-less unable to use citrulline as a source of arginine). Culture medium: minimal medium with citrulline as a sole source of nitrogen (other nitrogen sources are eliminated in order to obtain good uptake of citrulline). Incubation at 29 °C for 4 d.

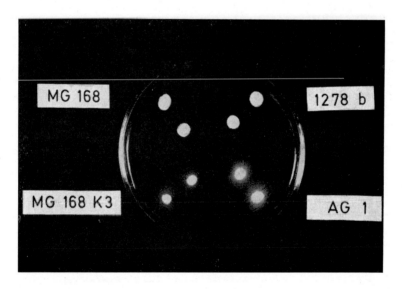

Fig. 7. Feeding of an arginine-less strain in the presence of citrulline. The same experiment as in Fig. 6 was stopped earlier, after incubation at 29 °C for 2 d.

inhibition of uptake[25] occur at high internal arginine pools. This indicates the intervention of a low-affinity element in the regulatory circuit of arginine biosynthesis together with a correlative regulation of permeability. Several hypotheses might be advanced as to the nature of the postulated low-affinity element: aporepressor of arginine biosynthesis, enzyme for formation of the true corepressor, transport system for arginine entry into the nucleus. In the same line, another possibility has to be considered seriously: the role of the cell vacuole.

The vacuole. Yeast, like plant cells, possesses a voluminous organelle whose physiological significance is not completely understood so far. The yeast vacuole seems to represent the lysosomal compartment of the yeast cell since it contains a number of typical hydrolytic enzymes[28]. It is a persistent cell organelle which, during bud initiation, fragments into small vacuoles, some of which pass from the mother cell into the bud[29]. The yeast vacuole seems to accumulate a number of substances of low molecular weight. The isolated vacuoles are sensitive to osmotic shock[30]. As observed microscopically (by UV-absorption), purines and purine compounds (*e.g.* S-adenosylmethionine) are concentrated in the vacuole[31-36]. Variations in vacuolar size and shape, as well as in cell density during the budding cycle appear to be caused by a cyclic alternation in consumption and accumulation of osmotically active molecules[29].

On the basis of these observations, it appears that the yeast vacuole may be considered, at least tentatively, as a storage compartment of the cell in which substances can be reversibly concentrated from the cytoplasm. This, of course, might have important consequences for the permeability and regulatory behaviour of yeast cells.

EXCRETION OF AMINO ACIDS

Some observations on amino-acid excretion by yeast made in our laboratory are presented here and discussed, without any attempt to give a complete review of the question, which is treated by others (*see* also the reviews[36a,b]).

Mutants with impaired permeability. Arginine is excreted by yeast strains bearing the two following mutations simultaneously: *agp-1*, which specifically depresses the activity of the arginine permease[3], and *gap*, which inactivates the general amino-acid permease[12]. This excretion can be detected by feeding an arginine-requiring mutant on minimal medium, as shown in Fig. 2.

This observation shows that the permease activity is somehow needed for arginine retention.

Similar experiments, performed with *lsp-1*, *mtp-1* and *htp-1* mutants (which are affected in lysine, methionine and histidine uptake, respectively) containing a *gap* mutation, indicated much lower excretion rates. This difference might be attributed to several factors, among which the principal are: (1) the existence of a large residual uptake activity; (2) a low internal pool of endogenous substrate unable to trigger the exit process; (3) the properties of the receptor strains which have to be considered when feeding of auxotroph mutants is used as a test for excretion (*e.g.* if several substances are excreted by the donor strain, they might compete for uptake in the receptor strain). It is not excluded that amino acids are excreted by mutants with impaired permeability as soon as the endogenous production reaches a certain level.

Mutants with impaired regulation of amino-acid biosynthesis. The regulation of arginine biosynthesis in yeast involves several regulatory genes controlling distinct parts of the biosynthetic pathway. The present knowledge regarding this question was recently summarized by WIAME[37]. When two regulatory mutations, *cpaO* (leading to derepression of carbamoylphosphate biosynthesis) and *argR* (I or II, leading to derepression of limiting steps of the arginine biosynthetic pathway) are present simultaneously in a strain, arginine is excreted into the medium under certain conditions[27] (growth on ammonia medium supplemented with ornithine). An estimation of the endogenous arginine pool after growth in this medium shows that it is increased from 5 mM in the wild-type strain to 70 mM in the double mutant. This shows that arginine overproduction due to impaired regulation of biosynthesis is accompanied by arginine excretion.

Fig. 3 shows that two mutants, RA580 and RA592, isolated from the wild-type strain Σ1278b of *S. cerevisiae*, excrete a source of tryptophan into their medium, except in the presence of tyrosine. Phenylalanine has no inhibitory effect. Although it is not completely excluded that tyrosine might inhibit tryptophan uptake and hence prevent growth of the receptor strain, it is likely that regulation of tryptophan biosynthesis is altered in both mutants (at the level of anthranilate synthase?), and that the effect of tyrosine is due to inhibition of 3-deoxy-D-*arabino*heptulosonate-7-phosphate synthase, the first enzymic function of the aromatic pathway, which is known to be strongly inhibited by tyrosine[38]. The same phenomenon was observed in *Salmonella typhimurium* and interpreted in this way[39].

Cessation of growth. We recently undertook to isolate excreting mutants systematically. Mutants were found which excrete large quantities of a given amino acid, whereas others excrete a number of different and unrelated amino acids simultaneously. Most of the mutants belonging to this last class were found to be simple auxotrophs. In a minimal medium supplemented with the required component they do not excrete amino acids at a detectable rate. However, when the required substance is omitted, excretion is observed after cessation of growth. Examples are given in Table II and Figs 4 and 5 *above*.

This behaviour could be interpreted in the following way. When growth stops as a result of starvation for a required amino acid, no repression of the other biosynthetic pathways could develop. However, feedback inhibition of key enzymes of the same pathways is possible in spite of the interruption of growth and protein synthesis; nevertheless, as long as feedback inhibition of biosynthesis does not reach 100%,

TABLE II

Amino-Acid Excretion by Auxotroph Mutants: Growth of Receptor Strains on Minimal Medium

Receptor strains		Donor strains		
		Σ1278b wild type	MG360 lysine-less	MG456 pyrimidine-less
MG783	trp[a]	−	+ +	+ +
MG778	trp[b]	−	+ +	+ + +
MG327	met[c]	−	−	−
MG331	met[d]	±	+ +[e]	+ + +[e]
MG682	arg	−	+	+
MG389	his	−	±	±
MG49	thr	−	+ +	+ +·
MG776	tyr	−	+	+ +·

[a] Unable to use indole.
[b] Able to use indole.
[c] Unable to use homocysteine.
[d] Able to use homocysteine.
[e] Inhibition area near the donor (see Fig. 4).

a residual and uncontrolled synthesis of amino acids (or other end products) can occur.

An abnormal behaviour was observed in the feeding experiment shown in Fig. 4: the donor colonies are surrounded by a zone where no growth occurs. This indicates that an inhibitor is being excreted together with a source of methionine. Only some methionine-less receptor strains are sensitive to this inhibitor, and adenine-less strains do not excrete an inhibitor of MG331. These data suggest that the postulated inhibitor might be S-adenosylhomocysteine, which was shown to be inhibitory for some yeast strains[40].

Growth in the presence of intermediates. The addition to the medium of intermediates of amino-acid biosynthesis sometimes leads to excretion of the corresponding amino acids. For instance, we have observed that addition of citrulline to the medium of a wild-type strain of yeast induces excretion of small quantities of arginine (Fig. 6). Addition of indole provokes excretion of very large quantities of tryptophan (Fig. 5 *below*). This can be interpreted readily as due to short-circuiting both feedback regulation and repression of limiting steps of biosynthesis. It is easy to guess why excretion of an end product in the presence of intermediates is not very frequent: the two main reasons are (1) that metabolic intermediates are often poorly or not at all taken up by cells and (2) that regulation of the remaining steps of biosynthesis is often too strong.

Mutation preventing katabolism. The excretion of arginine in the presence of citrulline can be greatly enhanced in the presence of a mutation leading to inactivation of arginase, the first enzyme of arginine katabolism in yeast (Figs 6 and 7).

Mutants unable to use tryptophan as a source of nitrogen were isolated (*e.g.* MG1077): they excrete large quantities of indole.

PYRIMIDINE TRANSPORT IN *S. cerevisiae*

SPECIFICITY AND REGULATION

In order to demonstrate the existence and to evaluate the specificity of pyrimidine transport systems in yeast, we selected mutants with impaired uptake of these compounds.

From a number of observations in several laboratories, it appeared that the uptake of purine and pyrimidine compounds in various types of cells is sharply dependent on their utilization in nucleic acid synthesis or other metabolic transformations (for references, *see* Ref.[41]). In other words, a block in the utilization of nucleic acid precursors could result in a limitation of their uptake rate. This led us to study the pathway of utilization of exogenous pyrimidines in *S. cerevisiae*[41]. The results are summarized in Fig. 8.

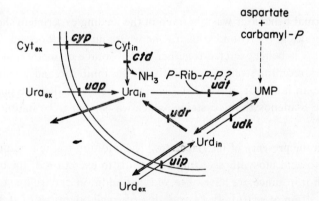

Fig. 8. Tentative scheme of utilization of exogenous pyrimidines, and of recycling of UMP derivatives in *S. cerevisiae*[41]. *Single arrows* represent uptake and anabolic activities; *double arrows* indicate katabolic and exit pathways. The genetic blocks are indicated. Cyt cytosine, Ura uracil, Urd uridine, P-Rib-P-P phosphoribosylpyrophosphate.

Pyrimidines added to the medium to a given concentration (*e.g.* 0.1 mM) enter a wild-type strain of yeast at widely different rates, without any obvious relationship to the molecular structure. Uracil, uridine and cytosine are used as efficient sources of pyrimidines in pyrimidine-requiring strains, while cytidine is used poorly, and thymine and thymidine (0.4 mM) do not support growth at all. The same discrimination between pyrimidines is expressed by *cpa* mutants, which have lost the carbamoylphosphate synthase activity regulated by arginine, with the consequence that growth is inhibited by exogenous uracil[42], cytosine and uridine[41].

The wild-type strain (Σ1278b) is able to use cytosine as a sole source of nitrogen. Mutants affected in this capacity were isolated. They belong to two complementary classes. One class has lost cytosine deaminase activity. In these mutants, cytosine can no longer serve as a source of pyrimidines, although cytosine uptake is normal. The other class of mutants has normal cytosine deaminase activity, but the uptake of cytosine is strongly depressed. ^{14}C-Cytosine taken up by normal cells cannot be detected as such in the soluble pool, whereas it is recovered as intact cytosine in the cytosine-deaminaseless mutant.

These observations show that cytosine can serve as a source of pyrimidines only after deamination to uracil.

The *cyp* mutants are affected in cytosine and cytidine uptake only: they make it possible to define a specific cytosine – cytidine permease.

Since the utilization of cytosine passes through uracil, a mutation affecting uracil utilization, whatever it may be, should affect cytosine utilization as well, except if permeability is concerned. A mutant (RA746), isolated as resistant to 5-fluorouracil, is affected in uracil utilization specifically. The uptake of cytosine, and even that of uridine are normal. Mutations at the same locus were also obtained by selection for specific resistance to uracil in a strain bearing a *cpa* mutation. The *uap* and *cyp* mutations do not show close linkage. The *uat* mutants are unable to use cytosine or uracil as a source of uridine 5′-phosphate but they can use uridine. This indicates that they must have a uridine kinase (which is not used for converting uracil into UMP). They do not transport uracil at a detectable rate, although cytosine and uridine uptake are normal. When a pyrimidine-less mutation is introduced into the same cell as a *uat* mutation, uracil uptake can be detected after starvation for pyrimidines, which shows that there is a feedback inhibition on the uracil permease in the *uat* strains. These observations also show that the pool of uracil which accumulates in these mutants is not able to inhibit uridine uptake.

Two classes of mutants were specifically affected in exogenous uridine utilization: *uip* and *udk*. Uridine kinase activity is absent in *udk*, but normal in *uip* mutants. Uridine uptake is affected in both classes of mutants, but it can be restored by pyrimidine starvation in *udk* mutants, and not in *uip* mutants. This shows that there is a feedback inhibition on the uridine permease in *udk* strains. Uracil uptake is not affected. This shows that the feedback on uracil permease is specifically mediated by uracil, which cannot be replaced by uridine in this function. In fact, in order to demonstrate this specificity, it was necessary to have a block in an enzymic activity (uridine ribohydrolase) which transforms uridine into uracil and ribose. Such a mutant (*udr*) was isolated as described *below*.

This study shows that the uptake of exogenous pyrimidines is mediated by at least three specific permeases, two of which are subject to very specific feedback inhibition. It also underlines the difficulty linked to a rigorous demonstration that a mutation affecting permeability is actually located at the level of the uptake mechanism. The main difficulty, in fact, is due to the existence of a feedback mechanism affecting the activity of the uptake system.

EXCRETION OF PYRIMIDINES AND RELATED SUBSTANCES INTO THE CULTURE MEDIUM

Purines and pyrimidines usually appear to have a tendency to leak out, or to be rejected, from the cells. When guanine and adenine are taken up by *S. cerevisiae*, the deaminated purines are excreted into the medium[43,44]. When nucleosides are taken up in Ehrlich ascites cells[45], as well as in *E. coli*[46-49], the corresponding bases, possibly deaminated, appear in the medium. The excretion of uracil and of a number of other compounds involved in nucleic acid metabolism during thymine starvation of thymine-less strains of *E. coli* has been reported[50-51]. In *S. cerevisiae*, excretion of endogenous uracil has been described for several mutants: overproducers of UMP, mutants blocked in uracil uptake or utilization[52,53]. Our results[41,54] have confirmed and extended these last observations.

Substances able to support growth of pyrimidine-less strains of yeast are released into the culture medium under several conditions. The test used to detect and identify pyrimidines excreted into the medium lies in feeding of pyrimidine-less strains of yeast bearing additional mutations. It is concluded that only uracil is excreted when a pyrimidine-less strain can be fed unless it simultaneously has a *uap* mutation (uracil-permeaseless). Likewise, a *uip* mutation (uridine-permeaseless) and a *ctd* mutation (cytosine-deaminaseless) are used to detect eventual loss of uridine or cytosine, and also the excretion of uracil in the presence of uridine or cytosine, respectively. The results are summarized in Table III. The substances excreted by prototrophic strains were uracil or uridine. Excretion of cytosine was never observed. The pyrimidine-less strains *ura-4*, *ura-1* and *ura-3*[55] excrete ureidosuccinic acid when deprived of pyrimidines (receptor strain: *ura-2*, aspartate carbamoyltransferase-less[55]).

A wild-type strain supplemented with ureidosuccinate or cytosine excretes uracil. It is known that ureidosuccinate is an inducer of the pyrimidine biosynthetic pathway in yeast and leads in this way to an overproduction of UMP[53]. This intermediate of pyrimidine biosynthesis is not taken up by wild-type yeast cells, but mutants can be isolated which can use it[55]. We obtained such a mutation in our strains which was named *usu*[41].

The strongest pyrimidine excretion was observed in *uat-2* mutants where formation of UMP from uracil is prevented. From a *uat* strain, a mutant was isolated which does not excrete uracil (*udr*): it has lost the uridine ribohydrolase activity (*see* Fig. 8). It excretes uridine instead of uracil.

TABLE III

Excretion of Uracil or Uridine by Several Mutants of *S. cerevisiae*[a,b]

Donor strain	Genotype of the donor	Supplement to the medium					
		none		ureidosuccinate 100 µg/ml		cytosine 20 µg/ml	uridine 100 µg/ml
		Receptor strains[c]					
		1	2	1	2	3 or 4	2
Σ1278b	wild type	—	—	+	—	++	—
RA746	uap	+	—	++	—	++	+++
4842d	uap ura-2		pyr	—	—	+	+++
MG887	cyp	—	—	+	—	—	+
4595a	uip	—	—	+	—	0	±
MG911	ctd	—	—	+	—	—	—
RA750	uat-2	+++	—	d	d	d	d
FM57	uat-2 udr	—	±	—	+++	+++	—
4837c	udr	—	—	—	+	++	—
RA848	udk udr	—	—	—	+++	0	—
4949b	udk	—	—	+	—	0	—
MG456	ura-2		pyr	—	—	0	—
FM1	ura-2 usu		pyr	+++	—	0	—
4682a	usu	—	—	+++	—	0	—
5517d	usu udr	—	—	—	+++	0	0
3678a	ura-2 uat-2 usu		pyr	+++	—	0	+++
4897d	uip udr	—	—	—	++	0	—

[a] Minimal medium containing ammonium ions as the sole nitrogen source.

[b] 0 not tested.

[c] **1** 4842d *uap ura-2*, **2** 4596a *uip ura-2*, **3** 4425b *cyp ura-2*, **4** 4654d *ctd ura-2*; For further description of the strains and mutations, *see* the text and Ref.[41].

[d] Spontaneous excretion too great.

CONCLUSION

On the basis of the reported observations, it appears that the internal pool of a number of amino acids, as well as that of pyrimidines is regulated by means of at least two main mechanisms in yeast. Uptake of substances from the medium is controlled by specific feedback inhibition of entry at the level of specific permeases. On the other hand, an exit process, which seems to operate at detectable rates only under certain conditions, appears to be responsible for pool-size regulation in cases of endogenous overproduction.

The study of culture conditions and mutations leading to excretion of amino acids or pyrimidines offers possibilities of practical applications. In this regard, it is likely that production of these substances could be increased by bringing together in the same strain several well-chosen mutations, each one of which individually leads to some excretion.

This work was supported by a grant from the Fonds de la Recherche Fondamentale Collective.

REFERENCES

1. KEPES A., COHEN G. N., p. 179 in Gunsalus & Stanier (Eds.): *The Bacteria*, Vol. IV. Academic Press, New York 1962.
2. PARDEE A. B.: Science **162**, 632 (1968).
3. GRENSON M., MOUSSET M., WIAME J. M., BECHET J.: Biochim. Biophys. Acta **127**, 325 (1966).
4. GRENSON M.: Biochim. Biophys. Acta **127**, 339 (1966).
5. GITS J., GRENSON M.: Biochim. Biophys. Acta **135**, 507 (1967).
6. JOIRIS C. R., GRENSON M.: Arch. Internat. Physiol. Biochem. **77**, 154 (1969).
7. GRABEEL M., GRENSON M.: European J. Biochem. **14**, 197 (1970).
8. MAGAÑA-SCHWENCKE N., SCHWENCKE J.: Biochim. Biophys. Acta **173**, 313 (1969).
9. HALVORSON H. O., COHEN G. N.: Ann. Inst. Pasteur **95**, 73 (1958).
10. SURDIN Y., SLY W., SIRE J., BORDES A. M., DE ROBICHON-SZULMAJSTER H.: Biochim. Biophys. Acta **107**, 546 (1965).
11. GRENSON M.: Rev. Ferment. Ind. Aliment. **22**, 7 (1967).
12. GRENSON M., HOU C., CRABEEL M.: J. Bacteriol. **103**, 770 (1970).
13. GRENSON M., HENNAUT C.: J. Bacteriol **105**, 477 (1971).
14. BENKO P. V., WOOD T. C., SEGEL I. H.: Arch. Biochem. Biophys. **122**, 783 (1967).
15. PALL M. L.: Biochim. Biophys. Acta **173**, 113 (1969).
16. ADAMSON L. F., LAGELLUTING S. C., ANAST C. S.: Biochim. Biophys. Acta **115**, 355 (1966).
17. ELSAS L. J., ROSENBERG L. E.: Proc. Nat. Acad. Sci. U.S.A. **57**, 371 (1967).
18. WILEY W. R., MATCHETT W. H.: Bacteriol. Proc. 106 (1967).
19. WILEY W. R., MATCHETT W. H.: J. Bacteriol. **95**, 959 (1968).
20. YAMADA C., CLARCK A. J., SWENSEID M. E.: Science **158**, 129 (1967).
21. GRENSON M., CRABEEL M., WIAME J. M., BECHET J.: Biochem. Biophys. Res. Comm. **30**, 414 (1968).
22. GITS J., GRENSON M.: Arch. Internat. Physiol. Biochem. **77**, 153 (1969).
23. GITS J., GRENSON M., *unpublished results.*
24. SCHWENCKE J., MAGAÑA-SCHWENCKE N.: Biochim. Biophys. Acta **173**, 302 (1969).
25. HENNAUT C., GRENSON M., *unpublished results.*
26. DOY C. H., COOPER J. M.: Biochim. Biophys. Acta **127**, 302 (1966).
27. RAMOS F., THURIAUX P., WIAME J. M., BECHET J.: European J. Biochem. **12**, 40 (1970).
28. MATILE P., WIEMKEN A.: Arch. Mikrobiol. **56**, 148 (1967).
29. WIEMKEN A., MATILE P., MOOR H.: Arch. Mikrobiol. **70**, 89 (1970).
30. INDGE K. J.: J. Gen. Microbiol. **51**, 441 (1968).
31. ROUSH A. H.: Nature **190**, 449 (1961).
32. SVIHLA G., DAINKO J. C., SCHLENK F.: J. Bacteriol. **85**, 399 (1963).
33. BROWN C., McCLARY D. O.: Bacteriol. Proc. 97 (1963).
34. BALISH E., SVIHLA G.: J. Bacteriol. **96**, 259 (1968).
35. INDGE K. J.: J. Gen. Microbiol. **51**, 447 (1968).
36. WIEMKEN A.: Thesis, *cited in* Ref.[26]

36a. DEMAIN A. L.: Adv. Appl. Microbiol. **8**, 1 (1966).

36b. DEMAIN A. L., BIRNBAUM J.: Current Topics Microbiol. Immunol. **46**, 1 (1968).

37. WIAME J. M.: Paper at 9th Internat. Microbiol. Congress, Mexico 1970; p. 243 in Pérez-Miravete, D. Peláez (Eds.): *Recent Advances in Microbiology*, 1971.

38. DOY C. H.: Biochim. Biophys. Acta **151**, 295 (1968).

39. ZALKIN H.: Biochim. Biophys. Acta **148**, 609 (1967).

40. SPENCE K. D., SHAPIRO S. K.: J. Bacteriol. **94**, 1136 (1967).

41. GRENSON M.: European J. Biochem. **11**, 249 (1969).

42. LACROUTE F., PIERARD A., GRENSON M., WIAME J. M.: J. Gen. Microbiol. **40**, 127 (1965).

43. LAHOU J.: Biochim. Biophys. Acta **27**, 371 (1958).

44. ROUSH A. H., SAEED M.: Biochem. Biophys. Res. Comm. **2**, 43 (1960).

45. JACQUEZ J. A.: Biochim. Biophys. Acta **61**, 265 (1962).

46. LUTWAK-MANN C.: Biochem. J. **30**, 1495 (1936).

47. MANS R. J., KOCH A. L.: J. Biol. Chem. **235**, 450 (1960).

48. PETERSON N. R., KOCH A. L.: Biochim. Biophys. Acta **126**, 129 (1966).

49. KAMMEN H. O.: Biochim. Biophys. Acta **134**, 301 (1967).

50. COHEN S. S., BARNER H. D.: Proc. Nat. Acad. Sci. U.S.A. **40**, 885 (1954).

51. BREITMAN T. R., BRADFORD R. M.: Biochem. Biophys. Res. Comm. **17**, 786 (1964).

52. LACROUTE F., SLONIMSKI P. P.: Compt. rend. **258**, 2172 (1964).

53. LACROUTE F.: Thesis. Faculté des Sciences, Université de Paris, 1966.

54. GRENSON M., *unpublished results*.

55. LACROUTE F.: J. Bacteriol. **95**, 824 (1968).

Properties of Sugar Transport Systems in *Aspergillus nidulans* and Their Regulation

ANTONIO H. ROMANO

Department of Biological Sciences and Graduate Division of Microbiology, University of Cincinnati, Cincinnati, Ohio, U.S.A.

In spite of the progress that has been made in the elucidation of mechanisms of transport in microorganisms, relatively little is known concerning the regulation of uptake systems. The classic work of COHEN and MONOD[1] clearly demonstrated that uptake systems are under genetic control, and are inducible in a large number of instances by specific substances. Nevertheless, there is little information as to whether regulatory mechanisms exist whereby intracellular metabolites control the rate of entry of substrates into cells.

Developments in the field of enzymology have made it clear that the rate of flow of metabolites along metabolic pathways within the cell can be regulated by control of the synthesis or activity of key enzymes in the pathway[2]. Numerous systems have now been described whereby the end product of a metabolic sequence regulates the activity of an enzyme catalyzing an early step in the pathway ("feedback control" or "allosteric control")[3]. Since the uptake of a metabolite into a cell represents the very first step in its metabolism, it seems reasonable that the uptake system itself should be subject to such control. Thus, since sugars represent a major source of energy and carbon for biosynthesis in many heterotrophic organisms, it seems reasonable to expect that mechanisms should exist to regulate the rate of uptake of sugars so that it would be consonant with the rate of utilization for energy and biosynthesis.

KEPES[4] has speculated that ATP, as an important end product of glucose metabolism, may be an allosteric effector of glucose uptake. ROSEMAN[5], on the other hand, has presented arguments in favor of the proposition that phosphoenolpyruvate, as both an end product and as the energy source for active sugar transport in bacteria *via* the PEP phosphotransferase system, should be the regulating metabolite. Evaluation of these hypotheses must await further experimentation.

KRULWICH and ENSIGN[6] have shown that in *Arthrobacter crystallopoietes*, succinate is preferentially utilized over glucose, and moreover succinate both represses the formation of the glucose transport system and inhibits it.

KABACK[7] has reported that the uptake of glucose by isolated membrane vesicles from *E. coli* is inhibited by glucose 6-phosphate and glucose 1-phosphate. Moreover, the uptake of a number of sugars was inhibited by glucose 1-phosphate; thus KABACK has ascribed a central role to this compound in the regulation of sugar transport in general.

The work to be described in this paper is based on an observation that acetate restricts the utilization of certain sugars by *Aspergillus nidulans*. Evidence will be presented that this effect of acetate is on the uptake of these sugars rather than on its subsequent metabolism. Moreover, the hypothesis has been formulated from these studies that this acetate effect is due to acetylcoenzyme A which, as an end product of glycolysis, can regulate sugar utilization by controlling sugar uptake.

An understanding of the mechanism of this possible regulatory system is dependent upon a better understanding of the mechanism of sugar transport in filamentous fungi. And in spite of the significant advances in the elucidation of uptake systems in bacteria[1,5,8,] on the one hand and yeast[9,10] on the other, relatively little is known concerning sugar transport in fungi. Thus, a comprehensive study of the characteristics of sugar transport systems in *A. nidulans* has been undertaken.

Perhaps the chief reason for the paucity of work on uptake systems in fungi is the technical difficulty imposed by their mycelial mode of growth; there is a tendency of cells to clump, making sampling of uniform suspensions difficult. Also, individual cells within an organized mycelium vary in physiological activity due to their varying ages. These technical difficulties have been largely overcome by the use of spore suspensions that have been allowed to germinate and grow for 18 h under conditions of vigorous agitation in baffled flasks. Such suspensions are sufficiently dispersed to allow good sampling, and represent a young, uniformly active population. *A. nidulans* is an advantageous organism for these studies, since dispersed growth on simple defined media is easily obtained, the organism uses a relatively wide range of carbon sources and, as a result of the work of PONTECORVO[11] and his colleagues, genetic systems have been worked out, and mutants are readily obtainable.

GROWTH STUDIES: EFFECT OF ACETATE ON SUGAR UTILIZATION BY *A. nidulans*

When *A. nidulans* R46, a wild-type strain with respect to carbohydrate metabolism, was grown in a synthetic medium[11] containing glucose and acetate as sources of carbon, there was a preferential utilization of acetate, and an inhibition of the utilization of glucose. Results of an experiment in which the rates of growth and glucose utilization were compared in a medium containing only glucose and in one containing glucose and acetate are shown in Fig. 1 *left*. The growth rates in presence and absence of acetate were similar, but the rate of glucose utilization in the presence of 0.1 M acetate was sharply reduced. Moreover, it is clear that in the presence of acetate, there was a reduction in the amount of glucose carbon that was incorporated into cellular material. This is shown in Fig. 1 *right*, where the decrease in sugar content of the medium is plotted against the increase in cell mass. Such a plot gives a straight line, from which can be calculated the rate of utilization of glucose per unit cell material synthesized. It is seen that in the absence of acetate, the synthesis of 1 mg cell material was accompanied by the disappearance of 3 mg glucose; in the presence of 0.1 M acetate however, only 1 mg glucose was utilized per mg cell material synthesized. Thus, there was a 66 % inhibiton of glucose utilization at this concentration of acetate.

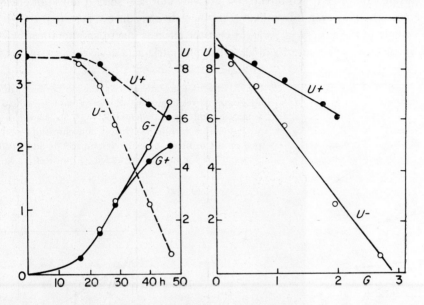

Fig. 1. Effect of acetate on glucose utilization during growth by *A. nidulans* R 46. *Left:* rate of growth and glucose utilization: *right:* relationship between glucose utilization and growth; *G* rate of growth (cell weight, mg/ml), *U* glucose utilization (residual glucose, mg/ml); (−) absence and (+) presence of 0.1 M sodium acetate.

This effect was independent of the concentration of glucose in the medium, as shown in Fig. 2, where the relationships between glucose utilization and cell weight in the absence and presence of 0.1M acetate are shown, at three different initial glucose concentrations. The effect of acetate is virtually identical at all sugar concentrations; thus, this acetate effect is not a competitive one.

Data presented thus far do not reveal the nature of the acetate effect, nor do they reveal whether the effect is due to acetate itself or to a metabolite derived from acetate. To pursue the latter question, the acetate effect was investigated in *A. nidulans* R104, a mutant that cannot grow on acetate because it is deficient in acetylcoenzyme A synthetase. The results of a growth experiment are shown in Fig. 3. It is seen that the wild type (strain R46) and the mutant (strain 104) were identical with regard to the amount of glucose utilized per unit cell mass synthesized in the absence of acetate. However, in the presence of acetate, the mutant impaired in the ability to synthesize acetylCoA from acetate showed a much smaller decrease in glucose utilization than did the wild type. Thus, it is clear that the acetate effect is not due to acetate itself, but to a metabolic product of acetate, and there is strong indication that the active metabolite is acetylCoA.

The possibility that the active substance is a later product of acetate metabolism, such as citrate or some intermediate of the citric-acid cycle cannot be positively ruled out. Citrate cannot be tested directly on growing cultures, since it does not enter the cells efficiently. However, it is unlikely that the active product of acetate is a C_5 or C_4 intermediate of the citric-acid cycle, or that the effect is due to an alteration of ATP levels in the cell as a result of stimulation of the citric-acid cycle by acetate, because

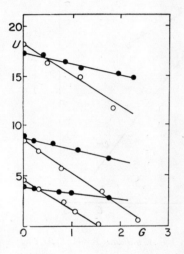

Fig. 2. Effect of acetate on relationship between glucose utilization (U residual glucose, mg/ml) and growth (G cell weight, mg/ml) by *A. nidulans* R 46. Cells grown in synthetic medium with initial glucose concentrations as indicated on the ordinate. *Open circles*, no acetate added; *closed circles*, 0.1 M sodium acetate added. (From Ref.[12])

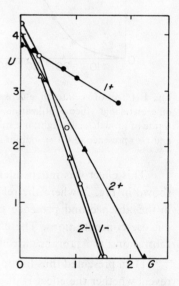

Fig. 3. Effect of acetate on relationship between glucose utilization (U residual glucose, mg/ml) and growth (G cell weight, mg/ml) by *A. nidulans* R 46 (1) and mutant *A. nidulans* R 104 (2). Cells grown on glucose synthetic medium in the absence (−) and presence (+) of 0.1 M sodium acetate. Acetylcoenzyme A synthetase activity (units/mg protein): R 46: 1.13, R 104: 0.11.

neither glutamate nor aspartate have any effect on the amount of glucose utilized per unit cell mass synthesized. Both of these latter compounds are readily utilized by *A. nidulans* and would be expected to give rise directly to C_5 and C_4 intermediates of the citric-acid cycle. A mutant devoid of citrate synthase would be most useful in establishing with certainty that acetylCoA is in fact the active species; unfortunately such a mutant of *A. nidulans* has not been found as yet. Thus, at present, while other substances cannot be unequivocally ruled out, it is considered that acetylCoA is the most probable effector.

The first indication that acetate exerted its inhibitory effect by interfering with the uptake of glucose rather than its subsequent metabolism was provided by studies with a mutant of *A. nidulans* R46 *pdh* deficient in a component of the pyruvate dehydrogenase complex. As a consequence of this metabolic lesion, this mutant requires

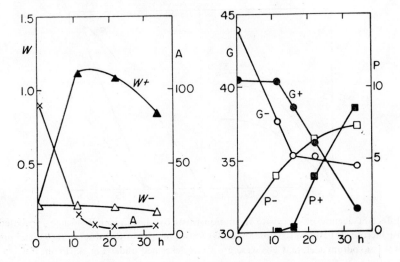

Fig. 4. Effect of acetate on growth, glucose utilization and pyruvate excretion by *A. nidulans* R 46 *pdh*. Cells were grown on synthetic medium containing 100 mм acetate for 44 h, harvested, washed, and resuspended in synthetic medium containing 50 mм glucose with no added acetate (−) and with 100 mм sodium acetate (+). *Left:* W cell weight (mg/ml), A acetate (mм); *right:* G glucose (mм), P pyruvate (mм). (From Ref.[12])

acetate for growth; furthermore, since this mutant cannot oxidize pyruvate, it excretes this compound into the medium when supplied with appropriate precursors, such as sugars. These properties are shown in Fig. 4. Cells were grown on acetate, washed, and resuspended in a growth medium containing 50 mм glucose, in the presence and absence of 100 mм acetate. In the absence of acetate, there was no growth (*left*); however, there was an immediate utilization of the glucose, and a concomitant excretion of pyruvate, which reached a final concentration of 7 mм in the medium (*right*). In contrast, in the presence of acetate, cells grew until the acetate concentration in the medium reached a low level, after 12 hours (*left*). However, the onset of both glucose utilization and pyruvate excretion were delayed in this culture for a time corresponding to that required for the disappearance of the acetate (*right*). Thus, again, acetate inhibited the utilization of glucose.

This mutant offered a convenient method for determining the effect of acetate on a number of different sugars; in each case, the degree of sugar utilization could conveniently be measured by determining the amount of pyruvate formation. Accordingly, the onset of pyruvate excretion by the *pdh* mutant when incubated with fructose, mannose, galactose, mannitol, sorbitol, glycerol, maltose, lactose, and gluconic acid was delayed as long as acetate was present in the medium, but occurred without a lag if acetate was omitted. This is shown in Fig. 5, where fructose and glucose are representative examples. However, Fig. 5 also shows a significant exception from this pat-

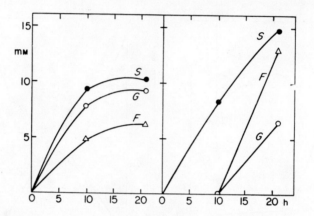

Fig. 5. Effect of acetate on pyruvate formation (mM) by *A. nidulans* R 46 *pdh* incubated with (S) 50 mM sucrose, (F) 50 mM fructose, and (G) 5 mM glucose, in the absence (*left*), and presence (*right*) of 0.1 M sodium acetate. Cells were grown as described under Fig. 4. (From Ref.[12])

tern: pyruvate formation from sucrose was not at all affected by the presence of acetate.

The fact that sucrose utilization was not inhibited by acetate indicated strongly that the inhibition of utilization of the other sugars was not due to an effect on the enzymes of glycolysis, since the formation of pyruvate from sucrose involves the necessary action of the same enzymes on hexose phosphate that are involved in the katabolism of other sugars. Rather, a more specific early effect was indicated, probably transport. Studies on the effect of acetate on sugar transport were undertaken, therefore.

UPTAKE STUDIES: EFFECT OF ACETATE ON THE UPTAKE OF 2-DEOXY-D-GLUCOSE

In order to interpret studies on the uptake of metabolites by cells effectively, it is necessary to establish conditions such that the uptake event can be separated from subsequent metabolism. These conditions can be met by using an analogue of the metabolite under study that will be transported but not metabolized, or by employing a mutant strain that is impaired in an essential early step in metabolism. Two analogues of glucose which have been used to study glucose transport are methyl α-D-glucoside[13] and 2-deoxy-D-glucose[14]. Unexpectedly it was found that ^{14}C-methyl α-D-glucoside, which has been widely used in studies of transport by bacteria, is not transported to a significant degree by *A. nidulans*. However, 2-deoxy-D-glucose, which does not support the growth of *A. nidulans*, is efficiently taken up by this organism. Uptake of this analogue follows Michaelis-Menten kinetics, as shown in Fig. 6, with

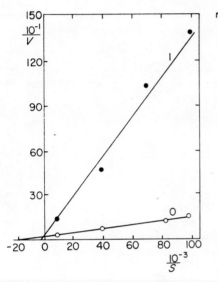

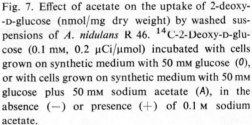

Fig. 6. Lineweaver-Burk plot of the uptake of 2-deoxy-D-glucose (μmol/mg dry cells /30 min) and its inhibition by D-glucose. *A. nidulans* R 46 incubated with ^{14}C-2-deoxy-D-glucose (0.2 μCi/μmol) with no further addition (*0*), and in the presence (*1*) of 1 mM D-glucose.

Fig. 7. Effect of acetate on the uptake of 2-deoxy--D-glucose (nmol/mg dry weight) by washed suspensions of *A. nidulans* R 46. ^{14}C-2-Deoxy-D-glucose (0.1 mM, 0.2 μCi/μmol) incubated with cells grown on synthetic medium with 50 mM glucose (*0*), or with cells grown on synthetic medium with 50 mM glucose plus 50 mM sodium acetate (*A*), in the absence (−) or presence (+) of 0.1 M sodium acetate.

a K_m of 6.1×10^{-5}. This figure also shows that the transport of ^{14}C-2-deoxy-D-glucose is competitively inhibited by D-glucose, thus establishing that these two sugars are transported by the same system.

The conclusion that acetate or a metabolic product thereof inhibits the uptake of glucose by *A. nidulans* was supported by studies of the effect of acetate on the uptake of 2-deoxy-D-glucose. Fig. 7 shows that cells that were grown on glucose alone or on a mixture of glucose and acetate as carbon sources showed virtually the same rate of uptake of ^{14}C-2-deoxy-D-glucose when there was no acetate in the uptake medium; this indicates that acetate did not repress the formation of the glucose transport system. When acetate was added to the uptake medium, however, there was a strong inhibiton of ^{14}C-2-deoxy-D-glucose uptake by cells that had been grown on glucose and acetate, amounting to approximately 66%. This degree of inhibition is quantitatively similar to that seen on the utilization of glucose in the growth experiments described earlier.

Fig. 7 also shows that acetate had a lesser effect on the uptake of ^{14}C-2-deoxy-D-glucose by cells that were grown on glucose alone, and that there was a lag in its action; no inhibition of uptake was manifest until 15 min after the addition of 2-deoxy-D-glucose and a total of 30 min after the addition of acetate, taking into account

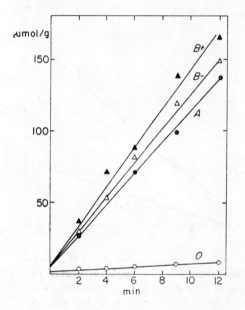

Fig. 8. Uptake and incorporation of sodium ^{14}C-acetate (μmol/g) by *A. nidulans* R 46 grown on 50 mM glucose (*0*), 50 mM glucose plus 50 mM sodium acetate (*A*), or 100 mM sodium acetate (*B*), incubated in the absence (−) and presence (+) of 0.1 M glucose.

the pre-incubation period of 15 min. The reason for this delay in the action of acetate on cells grown in the absence of acetate is that the systems for the uptake and incorporation of acetate are inducible; this is shown in Fig. 8. It is clear that cells grown on glucose alone took up ^{14}C-acetate poorly; on the other hand, acetate was incorporated rapidly by cells grown in the presence of acetate, and glucose had no effect on this incorporation of acetate, whether glucose was present in the growth medium or in the uptake medium. Thus, the lag in the inhibitory action of acetate on the uptake of 2-deoxy-D-glucose is explainable on the basis that in uninduced cells, some time is required for the concentration of the active metabolite derived from acetate to reach an effective intracellular level.

CHARACTERISTICS OF THE SUGAR TRANSPORT SYSTEM; SPECIFICITY OF HEXOSE TRANSPORT

A. nidulans is able to accumulate sugars against a concentration gradient. The concentrative nature of the uptake of the analogue 2-deoxy-D-glucose can be readily discerned from the data shown in Fig. 7. It can be seen that control suspensions (without acetate) took up 0.02 µmol of 2-deoxy-D-glucose per mg dry wt. cells. If live cells are considered to contains 85 per cent H_2O, and all of it is available to the substrate (a high estimate), one mg dry wt. cells would be equivalent to 5.6 mg, or 5.6 µl, H_2O. Thus, 0.02 µmol per 5.6 µl intracellular H_2O would correspond to 0.0035 µmol per µl, or 3.5 µmol per ml cell H_2O. With an external concentration of 0.1 µmol per ml, this would then represent a concentration factor of 35. This represents a case of active transport, therefore. The dependence of this process on energy has been further shown by its susceptibility to inhibition by 2,4-dinitrophenol. This data is in agreement with the demonstrations by ROBERTS[15] that galactose is actively transported by *A. nidulans*, and by LESTER *et al.*[16] that lactose is accumulated against a gradient by *Neurospora crassa*.

Transport systems for the monosaccharides glucose, galactose, and fructose appear to be constitutive. *A. nidulans* R46 took up an equivalent amount of 2-deoxy-D-glucose when cells were grown on glycerol as they did when grown on glucose[17]. Similarly, both galactose and fructose were taken up at the same rate by a galactokinaseless strain R31 (*gal-9*)[15] and a fructokinaseless strain R139, respectively, when grown in the absence or presence of the substrate in question.

TABLE I

Inhibition of ^{14}C-2-Deoxy-D-glucose Uptake in *A. nidulans* R46

Sugar[a]	Changes of the *C1* D-glucopyranose structure	Apparent K_i mM
D-Glucose	—	0.06
1,5-Anhydroglucitol	1-deoxy	1.5
Methyl α-D-glucoside	α-1-O-methyl	N. I.[b]
D-Mannose	2 axial OH	1.3
3-O-Methyl-D-glucose	3-O-methyl	0.14
D-Galactose	4 axial OH	1.1
D-Xylose	5-dehydroxymethyl	3.8
6-Deoxy-D-glucose	6-deoxy	0.38
D-Fucose	6-deoxy, 4 axial OH	N. I.
D-Rhamnose	6-deoxy, 2 axial OH	N. I.
L-Arabinose	5-dehydroxymethyl, 4 axial OH	N. I.
L-Glucose	1-CH₂OH, 5-OH (anomeric)	N. I.
L-Xylose	1-deoxy, 5-OH (anomeric)	N. I.
L-Sorbose	1-deoxy, 5-(CH₂OH + OH) (anomeric)	N. I.
D-Fructose	1-deoxy, 2 axial OH 5-(CH₂OH + OH) (anomeric)	N. I.

[a] All sugars were considered as *C1* chairs except L-glucose, L-xylose, L-sorbose, and D-fructose which were considered in the *1C* conformation.

[b] N. I. = No inhibition. $K_i > 50$ mM. (From Ref.[17])

Since it was shown that acetate was able to inhibit the utilization of number of sugars, the question was asked as to whether all these sugars shared a common transport system of limited specificity, or whether a number of different systems were involved. Thus, a study of specificities was undertaken to answer these questions, and to gain some information on the relationships between sugar structure and affinity for the transport system.

Since it is clear that transport systems are stereospecific, and are subject to competitive inhibition, the relative affinities of various sugars and sugar derivatives for the glucose carrier system can be determined by measuring the degree of competitive inhibition that they exert on the transport of ^{14}C-2-deoxy-D-glucose, which has been established to be a nonmetabolizable substrate of the glucose system. Thus, a comparison of K_i values from competitive inhibition allows a convenient procedure for ordering sugars with respect to their carrier-complexing activity, as has been done by CIRILLO[18] with bakers' yeast and by LEFEVRE[19] with erythrocytes. Table I lists the sugars tested, the change in the structure of glucose which is represented by each (for a discussion of the structure and configuration of sugars in solution, *see* CIRILLO[18]), and the apparent K_i measured against 2-deoxy-D-glucose. The following points can be made from this data: (1) The only single change which completely abolished activity was substitution of the anomeric hydroxyl group at carbon 1 (methyl α-D-glucoside); this is identical with results obtained with yeast[18], and in strong contrast to the situation in many bacteria studied. Any other single change, though leading to decreased activity, is tolerated to some degree. (2) Any two changes completely abolish activity. (3) The hydroxyl group at carbon 1 contributes to activity, but is not essential. (4) The hydroxyl group at carbon 2 can be removed with little loss in activity; a change to the axial configuration (mannose) reduced activity more than removal. (5) The hydroxyl group at carbon 6 contributes to activity, but is not essential (6-deoxy-

TABLE II

Comparative Inhibition of ^{14}C-2-Deoxy-D-glucose and ^{14}C-D-Galactose Uptake by Various Sugars[a]

Sugar	Inhibition of uptake of	
	2-Deoxy-D-glucose	D-Galactose
D-Glucose	6.0×10^{-5} M	5.0×10^{-4} M
D-Mannose	1.3×10^{-3} M	5.9×10^{-4} M
D-Fucose	N. I.[b]	2.9×10^{-4} M
L-Arabinose	N. I.	1.1×10^{-2} M
D-Xylose	3.8×10^{-3} M	N. I.
6-Deoxy-D-glucose	3.8×10^{-4} M	N. I.

[a] Expressed as K_i.
[b] N. I. = No inhibition. $K_i > 50$ mM. (From Ref.[17])

TABLE III

Comparative Affinities[a] of Sugars as Substrates for Glucose Transport Systems in *A. nidulans*, Yeast, and Erythrocytes.

Sugar	*A. nidulans*	Yeast[18]	Erythrocyte[19]
D-Glucose	1	1	1
2-Deoxy-D-glucose	1	1	1.4
D-Fructose	<0.0012	0.2	0.0025
D-Galactose	0.055	0.13	0.2
D-Mannose	0.046	0.1	0.3
1,5-Anhydroglucitol	0.04	0.1	0.2
3-O-Methyl-D-glucose	0.48	0.02	—
D-Xylose	0.016	0.13	0.11
D-Fucose	<0.0012	0.02	0.03
D-Arabinose	<0.0012	0.02	0.05
L-Sorbose	<0.0012	0.05	<0.0025
L-Glucose	<0.0012	<0.0025	<0.0025
Methyl α-D-glucoside	<0.0012	<0.0025	<0.0025

[a] Expressed as a ratio of K_m or K_i of D-glucose to that of the sugar listed.

D-glucose). This is also in strong contrast to the situation in *E. coli*, where ROGERS and YU[20] identified this group as the single most important functional group. This is significant with respect to the phosphorylation of this group during transport by the phosphoenolpyruvate phosphotransferase system[21], to be discussed in the next section of this paper. (6) Fructose, representing more than one change in structure, is not transported by the glucose system. (7) Galactose, representing a single change, can be transported by the glucose system.

The affinity of galactose for the glucose system raised the question as to whether there was a single system for the transport of both of these sugars, or whether there were separate systems with less than absolute specificity. That the latter is the case is shown in Table II, where the inhibition exerted by various sugars on the uptake of 2-deoxy-D-glucose and on D-galactose [measured with galactokinaseless strain R31 (*gal-9*)] are compared. D-Xylose and 6-deoxy-D-glucose, which represent single changes of the D-glucose structure, competed with 2-deoxy-D-glucose but not with galactose. Conversely, D-fucose (6-deoxy-D-galactose) and L-arabinose, which represent single changes of the D-galactose structure, competed with D-galactose but not with 2-deoxy-D-glucose. D-Glucose and D-mannose competed with both systems, but the affinities of these sugars for each system differed by one order of magnitude. Thus, the two systems can be differentiated.

In comparing the pattern of specificities found here with those reported in other classes of organisms, one is struck with the salient differences between *A. nidulans* and bacteria, as represented by *E. coli*. On the other hand, comparison of the fungal system with yeast and mammalian red-blood cells reveals a pattern of broad similarity with respect to specificity of the glucose system, as can be seen in Table III. While there are differences, and *A. nidulans* appears to show stricter specificity than either yeast or erythrocytes, the overall pattern is one of similarity. This may indicate fundamental similarities in the carrier systems of eukaryotic cells.

EVIDENCE AGAINST PHOSPHORYLATION DURING TRANSPORT

The notion that sugars were phosphorylated during their transport was prominent in early studies on sugar transport in mammalian cells (for review *see* WILBRANDT and ROSENBERG[22]). This hypothesis was eventually discarded for lack of convincing evidence, and following the demonstration of CRANE and KRANE[23] in 1956 that neither the C-1 nor the C-6 hydroxyl group of glucose needed to be present and phosphorylated in order for it to be transported by hamster intestine. The phosphorylation hypothesis was revived, however, with the discovery in bacteria of "a novel phosphotransferase system" by KUNDIG, GHOSH, and ROSEMAN[21] in 1964. This phosphoenolpyruvate-dependent system has been shown unequivocally to be involved in the transport of a number of hexoses in certain bacteria; its physiological significance has been emphasized by the isolation of pleiotropic mutants of *E. coli*[24], *Aerobacter aerogenes*[25], *Staphylococcus aureus*[26], and *Salmonella typhimurium*[27], which lack a component of this system, and are concomitantly impaired in their ability to transport a number of hexoses. Also, KABACK[28], working with isolated membrane vesicles from *E. coli*, showed that uphill transport of methyl α-D-glucoside was absolutely dependent on PEP, and that the sugar was phosphorylated during transport.

Also, VAN STEVENINCK[29,30] has supplied evidence that phosphorylation is involved in active transport of D-glucose and 2-deoxy-D-glucose by yeast; in this case, adenosine triphosphate, orthophosphate, or polyphosphate have been suggested as possible phosphate donors. It was of interest, therefore, to determine whether phosphorylation was a necessary event during the active transport of sugars by *A. nidulans*.

When a wild-type strain of *A. nidulans* was allowed to take up ^{14}C-2-deoxy-D-glucose, and the intracellular sugar was then extracted at intervals and chromatographed to separate free 2-deoxy-D glucose from its phosphorylated derivative, an early accumulation of intracellular 2-deoxy-D-glucose phosphate was detected (Fig. 9); during the first 10 min, there was approximately a two-fold excess of sugar phosphate over free sugar. At first glance, this could be interpreted as an indication that phosphorylation was associated with transport; more critical experimentation revealed, however, that this accumulation of sugar phosphate was the result of the action of intracellular kinases rather than the result of the transport event itself.

If phosphorylation were necessary for transport, one would expect that a requirement of the sugar to be transported would be a free C-1 or C-6 hydroxyl group.

Following this line of reasoning one would further expect that sugar analogs not satisfying these structural requirements should neither be transported, nor should they compete with the transport of glucose or 2-deoxy-D-glucose. However, Fig. 10 shows that both 1,5-anhydroglucitol and 6-deoxy-D-glucose, which lack a free hydroxyl group at C-1 and C-6, respectively, competitively inhibit the uptake of 2-deoxy-D-glucose. Furthermore, Fig. 11 shows that 6-deoxy-D-glucose is in fact transported and

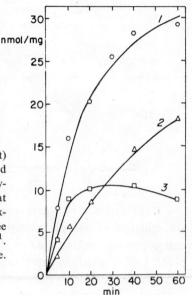

Fig. 9. Uptake of 2-deoxy-D-glucose (nmol/mg dry weight) by *A. nidulans* R 46 and formation of phosphorylated derivatives. Cells incubated in the presence of ^{14}C-2-deoxy--D-glucose (0.1 mM, 0.4 µCi/µmol); samples removed at appropriate intervals, extracted with hot water, and extracts chromatographed to separate and determine free sugar and sugar phosphate by method of WINKLER[31]. *1* total sugar taken up, *2* free sugar, *3* sugar phosphate. (From Ref.[12])

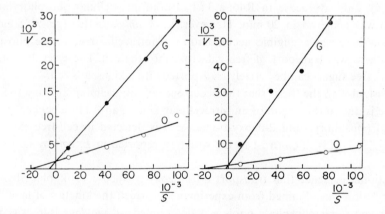

Fig. 10. Lineweaver-Burk plots of the inhibition of 2-deoxy-D-glucose (c.p.m./mg/min) uptake by 6-deoxy-D-glucose and 1,5-anhydro-D-glucitol. *A. nidulans* R 46 incubated with ^{14}C-2-deoxy-D-glucose (0.2 µCi/µmol) with no further addition (O), and with 1 mM 6-deoxy-D-glucose (G; *left*), and 10 mM 1,5-anhydro-D-glucitol (G; *right*). (From Ref.[32])

that the free sugar is concentrated 15 times over that present in the medium. Negligible amounts of 6-deoxy-D-glucose phosphate were found in the cell, presumably because intracellular kinases cannot phosphorylate this sugar. It is clear, however, that this inability to phosphorylate had no adverse effect on the active transport of 6-deoxy-D-glucose.

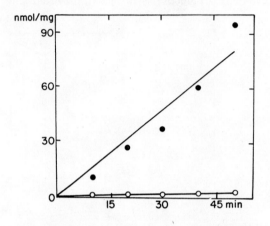

Fig. 11. Uptake of 6-deoxy-D-glucose (nmol/mg dry weight) and determination of intracellular free and phosphorylated sugar. *A. nidulans* R 46 incubated with 1mM 6-deoxy-D-glucose; samples were taken at appropriate intervals and extracted with hot water; free 6-deoxy-D-glucose and its phosphate were determined[33] after separation by barium precipitation[14]. (From Ref.[32])

If the transport of sugars is independent of phosphorylation, and the intracellular accumulation of phosphorylated sugars is due entirely to intracellular kinases, mutant organisms lacking kinases should show no such accumulation of sugar phosphate. Unfortunately, a glucokinaseless mutant was not available to test this directly in the case of 2-deoxyglucose. However, this was tested with galactokinaseless strains. Fig. 12 shows the results obtained. *A. nidulans* R8 (*gal-5*; galactokinase-positive but lacking UDP transferase[15]) phosphorylated galactose in a manner similar to that seen for 2-deoxyglucose in R46; a 3 : 1 ratio of intracellular phosphorylated to free sugar was achieved in 20 min. However, in *A. nidulans* R31 (*gal-9*[15]; galactokinaseless), there was negligible accumulation of intracellular galactose phosphate, although there was transport of free galactose into the cell. These results establish clearly that free sugars can be actively transported into a kinaseless cell.

Consistent with the thesis that the occurrence of intracellular 2-deoxy-D-glucose phosphate is due to the action of an intracellular kinase, an ATP-dependent enzyme capable of phosphorylating 2-deoxy-D-glucose was detected in cell-free extracts of *A. nidulans*[32]. This is in contrast to the situation reported in *E. coli* by GHOSH and GHOSH[34].

Additional evidence that transport od 2-deoxy-D-glucose took place before its phosphorylation was obtained from experiments in which the kinetics of labelling of free sugar and sugar phosphate pools was determined. *A. nidulans* R46 was preincubated in the presence of nonradioactive 2-deoxy-D-glucose for 15 min; at this time, intracellular pools of free sugar and sugar phosphate were approximately equal (0.03 μmol/mg dry weight). The cells were then separated by filtration, resuspended in

^{14}C-2-deoxy-D-glucose, and the rate of labelling of the respective pools were determined. Results of the experiment are shown in Fig. 13. The first sugar detected in the
cell was free; no sugar phosphate was detected for about 1 min (Fig. 13 *left*). Also, the
specific activity of the free sugar pool increased much more rapidly than did the sugar
phosphate pool (Fig. 13 *right*); within several minutes, however, both pools approached

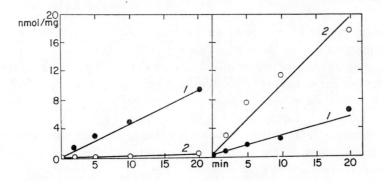

Fig. 12. Uptake and phosphorylation of D-galactose (nmol/mg dry weight). ^{14}C-D-Galactose
(0.1 mM, 0.2 µCi/µmol) incubated with suspensions of *A. nidulans* R 31 (galactokinaseless; *left*)
of *A. nidulans* R 8 (UDP-galactose transferaseless; *right*). 1 intracellular free sugar, 2 intracellular
phosphorylated sugar. (From Ref.[32])

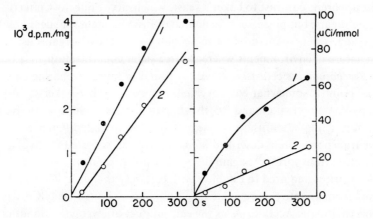

Fig. 13. Kinetics of labeling of intracellular free 2-deoxy-D-glucose and 2-deoxy-D-glucose-
phosphate pools. *A. nidulans* was preincubated with 0.5 mM unlabeled 2-deoxy-D-glucose for
15 min; ^{14}C-2-deoxy-D-glucose (0.5 mM, 0.4 µCi/µmol) was added at 0 time. *Left:* rate of uptake
of isotope (10^3 d.p.m./mg dry weight); *right:* specific activity (µCi/mmol); 1 free sugar, 2 sugar
phosphate pool. (From Ref.[32])

equal specific activities. At 30 s, the specific activity of the free sugar pool was nearly
9 times that of the sugar phosphate pool; this ratio decreased to approximately 2 after
5 min, indicating that the free sugar was subsequently phosphorylated by intracellular
kinases.

Thus it appears that phosphorylation is not a necessary event during the transport of sugars by *A. nidulans*, and again the situation is in sharp contrast to that in the bacteria that have been most intensively studied.

DISCUSSION

Although present knowledge of the mechanism of sugar transport in fungi is scant, and the work reported here has added only a little to this field, it seems clear that the fungal system as represented by *A. nidulans* has a good deal of individuality, and has features that are distinct from bacteria on the one hand, and yeast on the other. The specificity pattern of the glucose transport system is quite different from that reported in bacteria, but quite similar to that reported in baker's yeast[18] and mammalian erythrocytes[19]. On this basis, one is tempted to speculate that there are evolutionary relationships in the carriers involved in both facilitated diffusion and active transport in eukaryotic organisms.

A further difference from the bacterial system is the apparent absence in *A. nidulans* of the PEP phosphotransferase system, or any other system involving phosphorylation during sugar transport.

A. nidulans can accumulate sugars intracellulary against a concentration gradient. This is in apparent contrast to baker's yeast, which, according to CIRILLO[9], does not concentrate sugars, but possesses a facilitated diffusion system for sugar uptake. This is not surprising, when one considers the ecology of the two organisms; *A. nidulans* occurs in nature in environments where the free sugar concentration may be very low, whereas *Saccharomyces cerevisiae* is usually found in more saccharine environments. This latter point is somewhat controversial, however, in that baker's yeast has been reported by VAN STEVENINCK and ROTHSTEIN[35] to possess a system for active transport as well as for facilitated diffusion. These may represent strain differences. At any rate, the active transport system described in yeast has been reported to involve phosphorylation during transport[29,30], again differing from *A. nidulans*.

More work is required to establish with certainty that acetylCoA is the metabolite derived from acetate that regulates sugar uptake, although acetylCoA is the most reasonable candidate. AcetylCoA, as the end product of glycolysis, stands at a branch point of metabolism, between the entrance to the citric-acid cycle and the pathways to the synthesis of lipids and other cellular components. And there are already numerous instances of acetylCoA acting as an allosteric effector, both as a positive effector, as in the case of pyruvate carboxylase[36] and phosphoenolpyruvate carboxylase[37], and as a negative effector, as in the case of pyruvate dehydrogenase[38] and malic enzyme[39].

Also, it has been suggested that acetylCoA regulates glucose katabolism in liver by inhibiting glucokinase and pyruvate kinase[40]. Finally, it is of interest that MORGAN and KORNBERG[41] have shown that acetylCoA can inhibit sugar transport by *E. coli*.

It is surprising that more work has not been done on the regulation of sugar uptake by fungi, when one considers the number of important or potentially important industrial processes that involve the utilization of sugars by fungi. To name a few, these include the production of penicillin, griseofulvin, citric, fumaric, itaconic, oxalic, and kojic acids, and numerous other substances. The rate and manner in which sugars are utilized are of primary importance in all these processes. Perhaps the most dramatic example of this is the well-documented observation that glucose is deleterious to penicillin production, unless added to cultures slowly[42]; a less readily metabolized sugar, such as lactose, has been found to be more advantageous. Production of griseofulvin is also affected adversely by over-feeding of glucose[43]. It is of interest that acetate has been reported to increase penicillin yields[44]. This has been interpreted as a precursor—product relationship[45], but it would be interesting to determine whether this acetate effect was due to its modulating effect on sugar uptake.

The amount of work being done at present stresses the overriding importance of metabolic regulatory mechanisms in the microbial production of industrially important substances. I am confident that future work will give due consideration to the process of nutrient uptake as an important physiological process that is also subject to metabolic regulations.

I wish to express deep gratitude to Professor H. L. Kornberg of the University of Leicester, England, in whose laboratory much of this work was carried out, and with whose collaboration the work on regulation of sugar transport was formulated. Thanks are also due to colleagues in my own laboratory, C. E. Brown, C. G. Mark, and S. L. Dingle.

This work was supported in part by U.S. Public Health Service Grant AI-04739, the University of Cincinnati Research Development Fund, and a National Science Foundation Senior Postdoctoral Fellowship.

REFERENCES

1. COHEN G. N., MONOD J.: Bacteriol. Rev. **21**, 169 (1957).
2. UMBARGER H. E.: Cold Spring Harbor Symp. Quant. Biol. **26**, 301 (1961).
3. STADTMAN E. R.: Adv. Enzymol. **28**, 41 (1966).
4. KEPES A., p. 155 in J. E. Hoffman (Ed.): *The Cellular Functions of Membrane Transport.* Prentice-Hall, Inc., Englewood Cliffs 1964.
5. ROSEMAN S.: J. Gen. Physiol. **54**, 138s (1969).
6. KRULWICH T. A., ENSIGN J. C.: J. Bacteriol. **97**, 526 (1969).
7. KABACK H. R.: Proc. Nat. Acad. Sci. U.S.A. **63**, 724 (1969).
8. KEPES A., COHEN G. N., p. 179 in I. C. Gunsalus, R. Y. Stanier (Eds.): *The Bacteria*, Vol. IV. Academic Press, New York 1962.
9. CIRILLO V. P.: J. Bacteriol. **84**, 485 (1962).
10. KOTYK A.: Folia Microbiol. **12**, 121 (1967).
11. PONTECORVO G., ROPER J. A., HEMMONS L. M., MacDONALD K., BRUFTON A. W. J.: Adv. Genet. **5**, 141 (1953).
12. ROMANO A. H., KORNBERG H. L.: Proc. Roy. Soc. B. **173**, 475 (1969).
13. HAGIHARA H., WILSON T. H., LIN E. C. C.: Biochim. Biophys. Acta **78**, 505 (1963).

14. Augustin H. W., Hofmann E.: Acta Biol. Med. Germ. **11**, 628 (1963).
15. Roberts C. F.: Biochim. Biophys. Acta **201**, 267 (1970).
16. Lester G., Azzena D., Hechter O.: J. Bacteriol. **84**, 217 (1962).
17. Mark C. G., Romano A. H.: Bacteriol. Proc. **70**, 145 (1970).
18. Cirillo V. P.: J. Bacteriol. **95**, 603 (1968).
19. LeFevre P. G.: Pharmacol. Rev. **13**, 39 (1961).
20. Rogers D., Yu B.: J. Bacteriol. **84**, 877 (1962).
21. Kundig W., Ghosh S., Roseman S.: Proc. Nat. Acad. Sci. U.S.A. **52**, 1067 (1964).
22. Wilbrandt W., Rosenberg T.: Pharmacol. Rev. **13**, 109 (1961).
23. Crane R. K., Krane S. M.: Biochim. Biophys. Acta **20**, 568 (1956).
24. Tanaka S., Fraenkel D. G.. Lin E. C. C.: Biochem. Biophys. Res. Comm. **27**, 63 (1967).
25. Tanaka S., Lin E. C. C.: Proc. Nat. Acad. Sci. U.S.A. **57**, 913 (1967).
26. Egan J. B., Morse M. L.: Biochim. Biophys. Acta **97**, 310 (1965).
27. Simoni R. D., Levinthal M., Kundig F. D., Kundig W., Anderson F., Hartman P. E., Roseman S.: Proc. Nat. Acad. Sci. U.S.A. **58**, 1963 (1967).
28. Kaback H. R.: J. Biol. Chem. **243**, 3711 (1968).
29. Van Steveninck J.: Arch. Biochem. Biophys. **130**, 244 (1969).
30. Van Steveninck J.: Biochim. Biophys. Acta **163**, 386 (1968).
31. Winkler H. H.: Biochim. Biophys. Acta **117**, 231 (1966).
32. Brown C. E., Romano A. H.: J. Bacteriol. **100**, 1198 (1969).
33. Dische Z., Shettles L. B.: J. Biol. Chem. **175**, 595 (1948).
34. Ghosh S., Ghosh D.: Indian J. Biochem. **5**, 49 (1968).
35. Van Steveninck J., Rothstein A.: J. Gen. Physiol. **49**, 235 (1965).
36. Utter M. F., Keech D. B.: J. Biol. Chem. **238**, 2603 (1963).
37. Canovas J. L., Kornberg H. L.: Biochim. Biophys. Acta **96**, 169 (1965).
38. Schwartz E. R., Old L. O., Reed L. J.: Biochem. Biophys. Res. Comm. **31**, 495 (1968).
39. Sanwal B. D., Wright J. A., Smando R.: Biochem. Biophys. Res. Comm. **31**, 623 (1968).
40. Weber G., Lea M. A., Stamm N. B.: Life Sci. **6**, 2441 (1967).
41. Morgan M. J., Kornberg H. L.: F.E.B.S. Letters **3**, 53 (1969).
42. Arnstein H. R. V., Grant P. T.: Bacteriol. Rev. **20**, 133 (1956).
43 Rhodes A.: Progr. Ind. Microbiol. **4**, 165 (1964).
44. Calam C. T., Hockenhull D. J. D.: J. Gen. Microbiol. **3**, 19 (1949).
45. Martin E., Berky J., Godzesky C., Miller P., Tome J., Stone R. W.: J. Biol. Chem. **203**, 239 (1953).

5/ Genetic Problems in the Biogenesis of Natural Substances

Antibiotic Production as an Aspect of Secondary Metabolism

H. Boyd Woodruff

Merck Sharp & Dohme Research Laboratories, Merck & Co., Inc., Rahway, New Jersey, U.S.A.

INTRODUCTION

The investigation of secondary metabolites has proved highly productive in advancing fundamental knowledge in microbial biochemistry. Apparently, the science of genetics is also to benefit from this unique attribute of microorganisms. In the context of the genetics of industrial microorganisms, strong emphasis is being given to various aspects of secondary metabolism.

Secondary metabolites are produced with great frequency by microorganisms. They are found rarely in animal cell cultures, possibly because of the greater degree of differentiation and more efficient control mechanisms evolved. Their production is greatly facilitated by the accepted methods of cultivation of microorganisms in the laboratory. Microorganisms are incubated for extended periods of time on agar plates or in liquid media, either stationary or with agitation. As a result, much of the investigational period is during a time of partial or total growth arrest, during which secondary-product formation is the major aspect of metabolism. Had the technology of microbiological research progressed chiefly along lines of continuous culture, taking advantage of this highly developed technique for maintaining optimal growth conditions[1], research on secondary metabolism would have been less productive.

A major expansion of industrial microbiology occurred in the second quarter of this century. Attention was given to secondary metabolites — the antibiotics, the alkaloids, plant growth hormones, vitamins, and various cytotoxins. This is in contrast to earlier emphasis on primary metabolites — the alcohols, organic acids, enzymes and cellular proteins. Scientific curiosity soon led to suggestions for the metabolic basis of secondary metabolism, with the proposal of shunt metabolism most closely approximating today's understanding[2]. The basic observations of many workers have been synthesized by Bu'Lock[3] into a unified thesis for secondary metabolism. With antibiotics as examples, the secondary metabolites have been categorized as simple acid derivatives, fatty acid derivatives, isoprenoid substances, amino-acid and peptide derivatives, sugar derivatives, the phenylpropanes, and metabolites of mixed origin.

The role of the precursor molecule and development of biosynthetic pathways need no further mention in view of the presentations by HASSALL[4] and TAMM[5].

The author has drawn attention to the importance of biological factors in secondary metabolism, the role of the producing microorganism, the importance of the phase of growth as a determinant of primary or secondary metabolism, the effects of secondary metabolites on the producing culture, the varied biological activities of secondary metabolites, and has contributed to the speculations concerning the possible selective advantage granted by secondary metabolism[6]. In view of the extensive literature treatment and of the consideration of specific secondary metabolites in this volume, as well as the more general chapters on gene expression, regulation, and use of mutants in defining pathways of secondary metabolism, it would be repetitious to provide a general review.

The topic, genetic problems in relation to antibiotic production, is considered, therefore, from the point of view of the industrial microbiologist, who is interested in the discovery and efficient production of new secondary metabolites. The presentation must be made as a series of isolated examples, grouped into several sections for convenience. It is not yet possible to synthesize an integrated essay. We start from a statement made at an unpublished symposium on Control of Secondary Metabolism at the 1970 Annual Meeting of the American Society for Microbiology, "*there is yet no real genetics of secondary metabolism*".

ANTIBIOTICS

Most microorganisms, when cultured under appropriate conditions, produce antibiotics. A recent compilation of antibiotics obtained from a single genus of microorganisms, the *Streptomyces*, includes over 1000 named entities[7]. Because antibiotics are secondary metabolites, those factors which promote secondary-metabolite formation are commonly employed in the search for new antibiotics. Thus, cultures are frequently grown on agar plates, where limitation in oxygen supply, in diffusion of nutrients, or nutrient imbalance develops quickly, forcing the cultures into the idiophase which favors antibiotic formation. In practically every screening program, a variety of nutrients is built into the screen, in the hope that a favorable concentration of precursors or initiators will be present at the time antibiotic production begins.

LOSS OF YIELD

The impact of genetic factors on antibiotic screening is primarily negative in nature. Microbiologists are continually plagued with the loss of antibiotic production on the second to third transfer of a culture newly isolated from soil. The practical

Fig. 1. Colonial variation from plating a spore suspension of a streptomycete isolated from soil.

answers to the problem are varied. Some workers simply ignore losses and proceed with screening. Experience proves that there is much redundancy in discovery of anti-biotics (Table I). A culture whose productivity is lost today will surely be replaced by a repeat discovery tomorrow. Other workers follow a stabilization procedure, in which single isolates are made by plating a sporulated culture. A range of morphological variations is usually seen among the individual colonies (Fig. 1), each being tested for productivity. Normally, one or two cultures equal in productivity to the original iso-late will be recovered, but they will differ from the original in having greater produc-tive stability (Table II). A third approach to prevention of culture degeneration is based on the observation that potency loss in molds and actinomycetes is coincident with sporulation. Stability can be achieved by keeping the organism continually in the vegetative state[8]. When it is necessary to preserve the culture, mycelial fragments obtained from submerged culture are lyophilized.

Losses in productivity have been observed in prolonged vegetative or continuous culture, also. In general, non-producing variants grow somewhat more rapidly and this leads to simple overgrowth of the desired antibiotic producer. The growth effect has been described for a number of industrially important fermentations[9].

The genetic basis for loss in productivity of most secondary metabolites has not been established by experiment. The possibility of segregation of heterokaryons has been considered, and has been demonstrated to be a factor in decreased synnematin yields with *Emericellopsis* sp.[10]. Industrial microbiologists hope that continued fun-

TABLE I

Redundancy in Antibiotic Detection with Morphologically Distinct Streptomycetes Isolated from Different Soils in a Two-Month Period[a]

Antibiotic type	Number of isolates
A	3
B	3
C	3
D	3
E	2
F	2
G	2
H	2
J	2
K	1
L	1
M	1
N	1

[a] Data from E. O. Stapley, Merck Sharp & Dohme Research Laboratories.

Table II

Recovery of Antibiotic-Producing Isolates from a Degenerated Soil Streptomycete[a,b]

Isolate[c]	Diameter of antibiotic zone[d] mm	Isolate	Diameter of antibiotic zone mm
Parent	0	11	0
1	18.5	12	0
2	25.5	13	0
3	0	14	0
4	21.5	15	0
5	23.0	16	20.5
6	23.0	17	0
7	23.0	18	15.0
8	23.0	19	18.0
9	18.5	20	16.5
10	0	21	18.0

[a] Data from M. Jackson, Merck Sharp & Dohme Research Laboratories.
[b] Potency loss occurred during four subcultures and lyophilization.
[c] Isolates selected on the basis of morphological variation after plating spore suspensions.
[d] Cultures incubated 2 days in submerged culture at 28 °C with vegetative inoculum. Dextrose yeast-extract medium employed. Agar diffusion assay employed with *Vibrio percolans* inoculum and 13 mm disc.

damental investigation on genetics of the streptomycetes will provide scientific explanations for the problem and lead to rational solutions.

YIELD LOSS THROUGH MUTATION

Even with highly selected antibiotic producers, loss in productivity occurs readily through mutation. Sporulation is not required. For example, a pyocyanine nonproducer of *Pseudomonas aeruginosa* was isolated as early as 1899 by Jordan[11], and the suggestion made that it resulted from mutation. The action of carbobenzoyl-DL-alanine in restoring pyocyanine productivity in apyocyanogenic strains is an interesting observation which has never been explained biochemically[12].

Nonproducing mutants of the cyclic peptide mycobacillin were obtained from *Bacillus subtilis*, following ultraviolet irradiation. No intermediates in synthesis could be demonstrated by growing eleven nonproducers in various combinations, nor did analysis of the cell pool show presence of the nucleotide-linked peptides which are intermediates in biosynthesis[13,14]. Much greater success has been achieved in employing mutants for identification of intermediates in biosynthesis of the tetracyclines[15], prodigiosin[16], and erythromycin[17]. Intermediates have also proved useful as starting materials for formation of chemically modified antibiotics. In addition, mutants blocked at an intermediate stage of antibiotic synthesis can accept varied precursor molecules leading to biosynthesis of families of new antibiotic hybrids[18].

The relation of production of polymyxin to sporulation has been subject to considerable study. (For review *see* Ref.[19]) Mutants of *Bacillus polymyxa* which do not produce polymyxin also do not sporulate, and no asporogenic strain has been found which produced polymyxin. Growth conditions or inhibitors which block sporulation also prevent antibiotic formation. Attempts to find polymyxin as a constituent of spores have failed and the data to date suggest that polymyxin formation and sporulation are subject to joint metabolic control.

REVERSION TO ANTIBIOTIC PRODUCTIVITY

A useful aspect of the mutational loss of secondary-metabolite production has been its application in obtaining superior antibiotic producers with increased frequency. Cultures isolated from nature usually make such low yields of antibiotics that it is very difficult to recover these secondary metabolites in sufficient quantity to study their properties. Mutation can be applied and, as indicated above, primarily leads to strains of decreased productivity. Infrequently, possibly averaging one culture in 3,000 survivors, a stable strain of improved productivity is obtained. DULANEY and DULANEY[20,21] working with *Streptomyces viridifaciens* have shown that a tetracycline-nonproducing mutant, following growth, sporulation, and reisolation, will revert in low frequency to antibiotic production. Five antibiotic producers were recovered from 13,104 single-spore isolates selected at random from a nonproducing strain. Treatment of the nonproducing strains with UV-light greatly increased the frequency of recovery of antibiotic producers. At 0.15% survival, 993 antibiotic producers were recovered from 16,983 cultures evaluated.

The productivity observed for 100 isolates of the control tetracycline-producing culture of *S. viridifaciens*, when grown in shake flasks, is presented diagramatically in Fig. 2. The comparative shift in quantity of tetracycline produced in survivors following exposure to a mutagen is also shown in Fig. 2. Fig. 3 demonstrates productivity of 100 revertants from UV-treatment of a nonproducing strain. The revertants were preselected for antibiotic production by overlayering colonies with a bacterium sensitive to tetracycline, and *S. viridifaciens* colonies surrounded by a zone of inhibition were chosen as revertants for production trials. The high frequency of strains producing yields significantly superior to individual strains of the control population is evident.

Mutation is an important aspect of the reversion. None of 100 isolates recovered from a nonproducer by natural selection proved superior to the parent culture from which the nonproducer was derived, compared with the situation recorded in Fig. 3. Correlation of tetracycline productivity with pigmentation in sporulating hyphae of

S. viridifaciens may be a factor, since pigmented strains often are more resistant to UV-radiation. For example, an increase in the number of stable mutants producing the red-colored antibiotic pigment prodigiosin has been reported by LABRUM and BUNTING[22].

Unfortunately, the favorable yield response from reversion is not predictable. Some nonproducers are incapable of reversion to production, others do so infre-

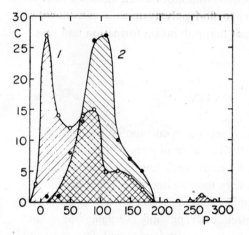

Fig. 2. Comparison of productivity of natural isolates with ultraviolet survivors. C % of cultures, P productivity, % of parent; *1* ultraviolet survivors, *2* natural isolates. (*Data from E. L. Dulaney, Merck Sharp & Dohme Research Laboratories.*)

Fig. 3. Comparison of productivity of natural isolates with reversion mutants. C % of cultures, P productivity, % of parent; *1* reversion mutant, *2* natural isolates. (*Data from E. L. Dulaney, Merck Sharp & Dohme Research Laboratories.*)

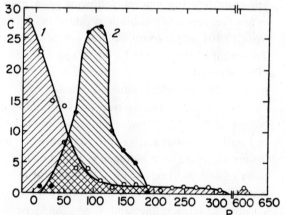
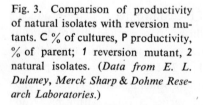

quently. Therefore, although the method works well, especially with strains freshly isolated from soil, much labor is required and many isolates must be evaluated. The technique is much less successful with high-producing industrial strains. Total nonproducers are seldom found in the surviving cultures after treatment with a mutagen, and when they do occur they are such physiological cripples that they cannot be reverted to high or excess productivity with any known mutational agent.

NON-ANTIBIOTIC SECONDARY METABOLITES

Antibiotics are the most frequently encountered secondary metabolite because their growth-inhibiting property makes them easy to detect. Non-antibiotic secondary metabolites also accumulate in fermentation broths and such substances having varied physiologic activity were mentioned in the *Introduction*. Diatretyne I[23], crystalline products B-73 and C-73[24], 3-benzyl-1,6-propano-2,5-dioxopiperazine[25], and gentisic acid[26] are non-antibiotic secondary metabolites accompanying the antibiotic diatretyne II, cycloheximide, gramicidin J, and patulin, with no known pharmacologic value.

The primary controlling factor in production of a secondary metabolite is the genetic character of the producing culture. The early suggestions that peptide antibiotics are direct translation products of messenger RNA have proved unfounded[27]. Some peptide antibiotics have been produced as products of cell-free enzymatic reactions[28,29]. The amino-acid sequences of isofunctional enzymes of different microbial species vary considerably as is evident from the wide variety of DNA compositions of the genomes[30]. That such varied enzymes should shunt accumulating metabolites into varied secondary products during periods of growth arrest is not surprising. One would not expect, however, that these secondary metabolites would necessarily be antibiotic in action. Non-antibiotic secondary metabolites should be as organism-

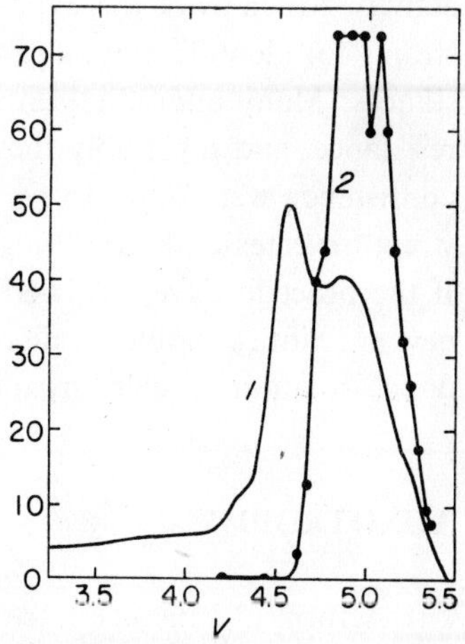

Fig. 4. Distribution of antibiotic and non-antibiotic secondary metabolites in a typical antibiotic purification system. Bio-Gel P-2 chromatograph of amphoteric secondary metabolites previously purified by batch elutions from Dowex 1×2 and Dowex 50×2, followed by chromatography over Dowex 1×2. V effluent volume, l; *1* refractometer response, *2* bioassay, ml broth equivalent. (*Data from T. W. Miller, Merck Sharp & Dohme Research Laboratories.*)

specific as are the antibiotics. This has proved to be the case. In an antibiotic selection program in the author's laboratory, it has frequently been noted that the procedure being used for final purification of an antibiotic deliniates several additional non-antibiotic substances of fermentation origin accompanying the antibiotic (Fig. 4). A different culture, however, with an antibiotic not recoverable by the above isolation method, will also show few non-antibiotic fermentation products by the first separation procedure, but its antibiotic and accompanying non-antibiotic fermentation products are recoverable by a secondary separation procedure.

Even though the appropriate genetic complement must be present to permit a culture to produce a specific secondary metabolite, it is, of course, obvious that the amount of an antibiotic formed and which one of a series of antibiotics within the genetic capability of a culture will accumulate are factors determined by the fermentation conditions. The observation of BRADLEY[31] that enzymes which produce the antibiotic of *Streptomyces violaceoruber* are adaptive in nature and only appear when induced specifically by substances generated in the cytoplasm of nutritionally imbalanced cells is an interesting extension of the importance of fermentation conditions.

ANTIBIOTIC INDUCTION BY MUTATION

Mutation can be used to force expression of secondary metabolites in organisms which are nonproducers. KELLNER[32] proposed the application of the mutation technique to nonantibiotic-producing cultures as a mechanism for discovery of new antibiotics. The importance of the genetic make-up of the producing culture became apparent during attempts in the author's laboratory to apply the procedure. Exposure of lavender-spored, antibiotic-nonproducing streptomycetes isolated from soils to UV-treatment did lead to recovery of a few antibiotic producers of relatively low activity. Much work was expended to isolate sufficient antibiotic materials from a few strains for identification, and they proved to be streptothricin or closely related substances. Many additional antibiotic-producing isolates were then checked for cross resistance, and invariably they proved cross-resistant with streptothricin. The same consistency was found in recovery of tetracyclines from yellow-pigment-producing streptomycetes and streptomycin from cultures indentified as *Streptomyces griseus*. If the procedure were carried broadly through different species of microorganisms, new antibiotics should be discovered, but at much greater labor expenditure than by direct isolation of microorganisms of varied morphology from nature.

AUXOTROPHY

A form of genetic control of antibiotic biosynthesis, through mutation to auxotrophy, has proved useful in producing variants of known antibiotics. The 6-demethyltetracycline, originally recognized because of its greater stability in fermentation broth, was a discovery resulting from a standard mutation program directed at yield improvement of tetracycline[33]. When a directed mutational approach was applied aimed at recovery of methionine-deficient auxotrophs, the majority produced demethyltetracycline, and a later check of 6-demethyltetracycline producers isolated at random showed that they also were methionine auxotrophs[34]. Mutant cultures deficient in methyl transfer reactions are candidates for substitution reactions. For example, with appropriate mutants, ethyl has been substituted for methyl on an N of tetracycline[35], an O of griseofulvin[36], and an S of lincomycin[37].

YIELD IMPROVEMENT THROUGH AUXOTROPHY

Auxotrophs of antibiotic-producing microorganisms have a more general value for increasing the efficiency of yield improvement. As indicated earlier, the major deficiency in obtaining organisms with increased capacity for production of antibiotics is the low frequency with which desirable genetic alterations are induced. Great progress over early studies with X-rays, UV-light, or nitrogen mustards has been accomplished through the use of mutagens with extreme mutation capacity at low lethality, for example, nitrosoguanidine. Greater success could be envisioned if one was certain that every strain selected for antibiotic evaluation was, in fact, a mutant. The data of DULANEY and DULANEY[21] prove the value of the approach. Of 12 auxotrophs, each requiring 1 of 7 different essential growth substances, 6 produced yields of tetracycline in excess of the parent culture. The improved yields were particularly apparent when an excess of the required growth substance was added to a normally adequate medium, the maximum being over 11 times the control value.

The obtaining of auxotrophs is somewhat simplified by the penicillin selection procedure of DAVIS[38], but it is not applicable to all antibiotic-producing microorganisms. The labor involved in large measure negates the advantages in greater frequency of obtaining improved mutants, and many growth factors are too expensive to be added to an industrial fermentation. An alternate approach, also suggested by DULANEY and DULANEY[21], is dependent upon producing reversion from auxotrophy. Selection is easily accomplished on minimal-medium agar plates; hundreds, even thousands, of revertants can be recovered from a single auxotroph for trial of antibiotic productivity, and one is guaranteed that every culture studied is a mutant. The results have proved unpredictable. Revertants from some auxotrophs are stable in antibiotic productivity. Others are highly variable and show greatly increased frequency of high-producing strains.

RECOMBINATION

The various recombinational mechanisms in microorganisms should be mentioned in a paper falling within the general category of genetic control of biosynthetic pathways. Recombination and exchange of genetic information have so far proved of limited value, however, to the industrial microbiologist interested in improving yields of antibiotics. The processes have great scientific interest, even when limited to examples involving antibiotics. Plasmids and episomes are receiving particular attention at the present time because of the characteristics of resistance to antibiotics which they carry. It is interesting that, in the course of episomatic inactivation, the antibiotically active product of one organism is converted through phosphorylation or acetylation by a second organism into a new metabolite lacking antibiotic potency. Had the proper acetylating or phosphorylating enzymes been present in the primary producer, no antibiotic would have been observed. Such a secondary metabolite would be overlooked completely in most fermentations.

Colicinogenic factors, the episomatic elements which control formation of the antibiotic colicins, have been particularly valuable in fundamental understanding of induction of mating and in integration. The colicins are too limited in spectrum to have practical value as antibiotics. Many antibiotic-producing cultures are lysogenic and carry temperate bacteriophages. This applies especially to species of *Streptomyces*[39]. The strains with improved antibiotic productivity, which are sometimes found during lysogeny, are more likely to have resulted from phage as a selector of mutants than from phage-DNA incorporation into the genome. Certain literature statements that actinophages transfer genetic information directing streptomycin and erythromycin production have not been fully evaluated[40].

Recognition of the parasexual cycle in filamentous microorganisms, occurring as it did following the great successes in yield improvement with conventional mutation, was received with enthusiasm by the industrial worker. As applied to penicillin production by *Penicillium chrysogenum*[41,42] and synnematin B production by *Emericellopsis* sp., some cultures with improved yields were obtained, and the recombinants appeared to be stable on transfer. Most recombinants were inferior to the parent culture, however. In the streptomycetes, recombination has been observed with several antibiotic producers, including *Streptomyces violaceoruber*[43], *Streptomyces aureofaciens*[44], *Streptomyces rimosus*[45], and *Streptomyces erythreus*[46]. Yields of antibiotic are influenced, much more often leading to inferior than superior strains. ALIKHANIAN[47] has clearly pointed out that recombination should not be expected to yield desirable recombinants in high frequency, and that selection must be imposed on a large population as in mutation or other forms of culture improvement, if improved cultures are to be obtained.

At the present state of the science, recombination may be looked upon more favorably as a means of combining several desirable properties in a single strain, rather than for yield improvement. With this modification in goal, not only recombination by the parasexual process, but genetic transformation employing isolated DNA deserves serious consideration. The widespread success with many Gram-positive and -negative bacteria with transfer of characteristics other than antibiotic production indicates that experimentation on transfer of antibiotic production is desirable. When the work is directed at combination of two separate characteristics, the selection of desirable recombinants becomes technically feasible, compared with the inefficient selection methods when increased yield is the sole objective.

RECAPITULATION

In view of the many isolated examples presented above in which highly skilled workers have employed secondary metabolites as tools in elucidating the principles of microbial genetics, it is surprising that so few practical applications have been found. In part, this arises from lack of familiarity of the industrial microbiologist with the principles of genetics. Some geneticists have attempted to meet the challenge of practical application, and they too have failed. In the author's opinion, the right mix of skills has not yet been found in a single laboratory. Progress cannot be made without the background of microbial geneticists, and they are few in number. An important adjunct is skill in selection techniques for improved productivity. Microbiologists have become so skilled in the design of random mutation experiments, so adept in detecting improved cultures, and in stabilizing them, that genetic manipulations appear inefficient in contrast, especially when inefficient evaluation procedures are employed. It is hoped that an assemblage of research workers with complementary skills will accept antibiotic yield improvement as a suitable objective in order that a clear evaluation of the potential of genetic control of antibiotic biosynthesis can be made. Until that time, the genetics of antibiotic secondary-metabolite biosynthesis will remain the uncoordinated programs reviewed in this chapter, *i.e.*, research on loss of strain productivity, on mutation, reversion and auxotrophy.

REFERENCES

1. MÁLEK I., BERAN K., HOSPODKA J. (Eds.): *Continuous Cultivation of Microorganisms*. Publ. House Czech. Acad. Sci., Prague 1964.
2. FOSTER J. W.: Bacteriol. Rev. **11**, 167 (1947).
3. BU'LOCK J. D.: Adv. Appl. Microbiol. **3**, 394 (1961); BU'LOCK J. D., p. 61 in Z. Vaněk, Z. Hošťálek (Eds.): *Biogenesis of Antibiotic Substances*. Publ. House Czech. Acad. Sci., Prague 1965.
4. HASSALL C. H.: Paper at 1st Internat. Symp. Genetics of Industrial Microorganisms, Prague 1970. Abstract Book, p. 98.
5. TAMM C.: Paper at 1st Internat. Symp. Genetics of Industrial Microorganisms, Prague 1970. Abstract Book, p. 100.
6. WOODRUFF H. B., p. 22 in B. A. Newton, P. E. Reynolds (Eds.): *Biochemical Studies of Antimicrobial Drugs*. Cambridge University Press, New York 1966.
7. UMEZAWA H.: *Index of Antibiotics from Actinomycetes*. University Park Press, State College, Pennsylvania 1967.
8. WHIFFEN A. J., SAVAGE G. M.: J. Bacteriol. **53**, 231 (1947).
9. REUSSER F., KOEPSELL H. J., SAVAGE G. M.: Appl. Microbiol. **9**, 342 (1961).
10. FANTINI A. A.: Genetics **47**, 161 (1962).
11. JORDAN E. O.: J. Exp. Med. **4**, 627 (1899).
12. AZUMA Y., WITTER L. D.: J. Bacteriol. **87**, 1254 (1964).
13. BANERJEE A. B., BOSE S. K.: J. Appl. Bacteriol. **27**, 93 (1964).
14. BANERJEE A. B., MAJUMDAR S. K., BOSE S. K., p. 271 in D. Gottlieb, P. D. Shaw (Eds.): *Antibiotics*, Vol. II. *Biosynthesis*. Springer-Verlag, New York 1967.

15. McCormick J. R. D., p. 73 in Z. Vaněk, Z. Hoštálek (Eds.): *Biogenesis of Antibiotic Substances.* Publ. House Czech. Acad. Sci., Prague 1965; McCormick J. R. D., p. 113 in D. Gottlieb, P. D. Shaw (Eds.): *Antibiotics,* Vol. II. *Biosynthesis.* Springer-Verlag, New York 1967.
16. Williams R. P., Hearn W. R., p. 410 in D. Gottlieb, P. D. Shaw (Eds.): *Antibiotics,* Vol. II. *Biosynthesis.* Springer-Verlag, New York 1967.
17. Martin J. R., Rosenbrook W.: Biochemistry 6, 435 (1967).
18. Shier W. T., Rinehart K. L., Jr., Gottlieb D.: Proc. Nat. Acad. Sci. U.S.A. 63, 198 (1969).
19. Paulus H., p. 254 in D. Gottlieb, P. D. Shaw (Eds.): *Antibiotics,* Vol. II. *Biosynthesis.* Springer-Verlag, New York 1967.
20. Dulaney E. L., p. 93 in *Genetics and Breeding of Streptomyces.* Proc. Internat. Symp., Dubrovnik 1968.
21. Dulaney E. L., Dulaney D. D.: Trans. N. Y. Acad. Sci. 29, 782 (1967).
22. Labrum E. L., Bunting M. I.: J. Bacteriol. 65, 394 (1953).
23. Anchel J.: Science 121, 607 (1955).
24. Rao K. V., Cullen W. P.: J. Am. Chem. Soc. 82, 1127 (1960).
25. Kurahashi K.: 5th Internat. Congr. Biochemistry, Sect. 2, p. 95. Moscow 1961.
26. Brack A.: Helv. Chim. Acta 30, 1 (1947).
27. Mach B., Tatum E. L.: Proc. Nat. Acad. Sci. U.S.A. 52, 876 (1964).
28. Yukioka M., Tsukamoto Y., Saito Y.: Biochem. Biophys. Res. Comm. 19, 204 (1965).
29. Kurylo-Borowska Z., Tatum E. L.: Biochim. Biophys. Acta 113, 206 (1966).
30. Marmur J., Doty P.: J. Mol. Biol. 5, 109 (1962).
31. Bradley S. G.: Developments Ind. Microbiol. 3, 362 (1961).
32. Kelner A.: J. Bacteriol. 57, 73 (1949).
33. McCormick J. R. D., Sjolander N. O., Hirsch V., Jensen E. R., Doerschuk A. P.: J. Am. Chem. Soc. 79, 4561 (1957).
34. Hendlin D., Dulaney E. L., Drescher D., Cook T., Chaiet L.: Biochim. Biophys. Acta 58, 635 (1962).
35. Dulaney E. L., Putter I., Drescher D., Chaiet L., Miller W. J., Wolf F. J., Hendlin D.: Biochim. Biophys. Acta 60, 447 (1962).
36. Jackson M., Dulaney E. L., Putter I., Shafer H. M., Wolf F. J., Woodruff H. B.: Biochim. Biophys. Acta 62, 616 (1962).
37. Argoudelis A. D., Mason D. J.: Biochemistry 4, 704 (1965).
38. Davis B. D.: J. Am. Chem. Soc. 70, 4267 (1948).
39. Bradley S. G.: Adv. Appl. Microbiol. 8, 29 (1966).
40. Alikhanian S. I., Teteryatnik A. F.: Mikrobiologiya 31, 54 (1962).
41. Alikhanian S. I., Kameneva S. V.: Sci. Reports Ist. Super. Sanità 1, 454 (1961).
42. Macdonald K. D., Hutchinson J. M., Gillett W. A.: J. Gen. Microbiol. 33, 365, 375, 385 (1963).
43. Bradley S. G.: Developments Ind. Microbiol. 6, 296 (1965).
44. Borisova L. N., Konyoukhova M. V., Ivakina N. S.: Antibiotiki 7, 685 (1962).
45. Alikhanian S. I., Mindlin S. Z.: Nature 180, 1208 (1957).
46. Huang-Lo L.: Mikrobiologiya 31, 61 (1962).
47. Alikhanian S. I., Chernosvitova V. I., Lubinskaya S. I.: Antibiotiki 7, 491 (1962).

Improvement of Penicillin Productivity in *Penicillium chrysogenum* by Recombination

C. BALL

Glaxo Laboratories Ltd., Ulverston, Lancashire, Great Britain

INTRODUCTION

In attempting titre improvement genetically there are hypothetical reasons for using recombination methods as well as mutation methods[1]. It is important to decide whether recombination methods are essential for strain improvement or merely a useful alternative to mutation. Evidence in support of the essential need for recombination is difficult to obtain, but any demonstration that such methods can aid titre improvement has importance because they could be essential in certain situations.

An earlier communication[2] illustrated and discussed the relevance of haploidization analysis in *Penicillium chrysogenum* to penicillin yield improvement. The studies to be described here are a continuation of this work, with particular emphasis on the penicillin yields of haploid segregants from crosses. It was shown previously that such titres could be more variable than the parental titres, even in crosses that segregated parentally for spore colour and auxotrophic markers. As the strains used were immediately related in lineage, the question posed was whether mutation rather than recombination could contribute to this variability. The work since undertaken has extended these titre studies to a point at which recombination between linkage groups can be detected because master strains (*i.e.* strains marked on all linkage groups) have been used.

On the question of titre variability the influence of medium batch variation and strain mutation due to storage is rarely mentioned in the literature. The testing of parents and segregants on a medium of given age compares segregants with the stored parent. Similarly, a comparison of segregants with the unstored parent must be based on titre data obtained from media that differ. In the past, the problem has been discounted by assuming that such variability is not large. In the present work, however, we have attempted to estimate the extent of such variability when assessing whether segregants have higher titres than the parents.

MATERIALS AND METHODS

Media. The media, *i.e.* complete medium (CM) and minimal medium (MM), were similar to those used previously[2]. All platings were carried out on CM except when otherwise stated.

Haploid and diploid recognition. Haploids and diploids were distinguished in the first instance on the basis of spore colour, spore size and prototrophy. However, in classifying the ploidy of segregants, the spore size criterion becomes all important[2]. The method of PONTE-CORVO and SERMONTI[3] is entirely adequate for this purpose.

Haploidisation analysis. Analysis of haploidisation was carried out in conjunction with a modified *p*-fluorophenylalanine technique. When there was segregation of any two mutants, emphasis was put on those crosses that showed either two parental classes only or two parental and two recombinant classes. If three classes, *i.e.* two parental and one recombinant, are each recovered in approximately similar frequency, interpretation can be ambiguous. Either linkage or nonlinkage can be deduced, depending on whether mitotic crossing-over or selection against a recombinant genotype by PFA is postulated.

Mutagenic treatments. Diploids were treated with physical agents (UV- and γ-rays) and chemical agents [N-methyl-N′-nitro-N-nitrosoguanidine (NTG) and ethyl methanesulphonate (EMS)]. Spore survivals were of the order of 1%. Following chemical treatments the spores were removed by centrifugation and washing. All treatments were carried out with shaking at 25 °C for a short time. Treatment with EMS was for 45 min at a final concentration of 1% in 0.1M phosphate buffer (pH 7.5). NTG treatment was for 2 h at a final concentration of 0.1% in 0.2M acetate buffer of pH 5. γ-Irradiation was carried out on CM sloped cultures, the dose being 50,000 roentgens from a ^{60}Co-source. UV-irradiation was carried out with a mercury vapour lamp (2567 Å) for 2.5 min, on spores suspended in water in a Petri dish.

Only surviving colonies with apparently normal morphology were selected for genetic analysis.

Strains. An unstable strain was derived from Q 176 (to which it was in many ways analogous[4]) after serial UV-irradiation and selection for titre increases. From this, a 'stable' strain was isolated by serial selection for dense sporulation and fast linear growth rate[2]. This 'stable' strain then became the parent of all the ensuing studies.

The mutant alleles of the stable strain used in this work will be identified as follows:
bgn bright green spore colour
brw brown spore colour
whi white spore colour
ylo yellow spore colour
cho choline requirement
his histidine requirement
nic nicotinamide requirement
thi thiamine requirement
rbs inability to use ribose as sole sugar source.

Data obtained previously[2] and based on various segregations of at least 20 independently isolated mutants, indicated that there were not less than three linkage groups in *P. chrysogenum*. Consequently, strains marked on all three linkage groups have been termed 'master strains'. Master strains such as *whi his rbs* and *brw his nic* have been used in this work. *whi* and *brw* are located on group I; *his* is located on group II; *rbs* and *nic* are group III markers. In addition, strains *bgn thi* and *whi lys* have been used. *bgn* and *thi* have been allocated to group I, and *lys* is a group III marker.

The markers *his* and *lys* drastically modify titre. However, titre increases were UV induced in a single step for strains *bgn thi* and *brw nic* as the markers in these strains did not modify titre. In this way titre-increasing determinant t_1 was induced in *bgn thi*, and determinants t_2 to t_5 were induced in strain *brw nic*.

In addition, all strains were stored as freeze-dried soil samples.

Testing of strains for yield. In various tables, the titre indicated for a given strain is based on at least four replicate fermentations. Data is presented on a yield group basis using yield groups A to M; *e.g.* group A contains all titres of 0 to 500 units per ml and group M all titres of 6,001 — 6,500 units per ml. International units per ml of penicillin G have been used throughout.

RESULTS

SEGREGATION OF AUXOTROPHIC AND SPORE-COLOUR MARKERS

Three general types of cross have been carried out. These are crosses of general type *bgn thi* × *whi nic* and *brw nic* × *whi his rbs* and *bgn thi* × *brw nic*. The first two types involved crossing strains that carried determinants for titre increase with strains that did not carry such determinants. The remaining type of cross involved on the other hand crossing strains which each carried such determinants.

In Table I it can be seen that, in keeping with previous data[2], *bgn* and *thi* and *whi* and *thi*+ can be allocated to linkage group I. Furthermore, the data clearly indicates that *nic* segregates independently, as expected, because *nic* had previously been allocated to linkage group III. Results of crosses of general type *bgn thi* × *brw nic* showed a very similar type of segregation pattern, given that *brw* is one of the chromosome I markers instead of *whi*.

Table II, however, illustrates that difficulty was experienced with certain crosses, other than the control cross, in recovery of *brw* and more specifically of *brw his* genotypes. The reason for this could be that, in those crosses, the *brw nic* parent carried

TABLE I

Haploid-Segregation Data from Class I[a]

	Cross[b]	Segregant genotypes			
		bgn thi	*bgn thi nic*	*whi*	*whi nic*
	bgn thi × *whi nic*	8	2	4	5
a	*bgn thi* (t_1) × *whi nic*	7	2	4	6
b	*bgn thi* (t_1) × *whi nic*	2	7	3	3
c	*bgn thi* (t_1) × *whi nic*	2	2	7	1

[a] Crosses of general type *bgn thi* × *whi nic*.
[b] Cross *bgn thi* (t_1) × *whi nic* carried out in triplicate.

TABLE II

Haploid Segregation from Crosses of Class II[a]

Haploid segregant genotypes	Cross[b]									
	a		b		c		d		e	
	his$^+$	his	his$^+$	his	his$^+$	his	his$^+$	his	his$^+$	his
brw rbs	5	2	1	0	1	0	3	0	0	0
brw nic	2	2	1	0	3	0	3	0	1	0
whi rbs	9	5	2	4	0	4	0	11	5	8
whi nic	8	4	4	4	5	3	0	17	3	3

[a] General type brw nic × whi his rbs.

[b] a whi his rbs × brw nic
  b whi his rbs × brw nic (t$_2$)
  c whi his rbs × brw nic (t$_3$)
  d whi his rbs × brw nic (t$_4$)
  e whi his rbs × brw nic (t$_5$).

a determinant for improved titre. Induction of such mutations was usually accompanied by a slight change of morphology, and determinants for such a morphological change might prevent ready recovery from PFA. Such determinants could be located on the brw marked linkage group I. Indeed morphological change could be a pleiotropic expression of improved titre determinant (see next section).

With cross brw nic (t$_4$) × whi his rbs the absence of brw his segregants is paralleled by the absence of whi his$^+$ segregants, unlike any of the other crosses in Table II. It is likely that a translocation involving linkage groups I and II is probably present in strain brw nic (t$_4$).

TITRES OF PARENTS AND SEGREGANTS

The understandable lack of an absolutely reliable comparison between parents and segregants has been discussed earlier (see Introduction). However, the extent of variability due to medium batch variation coupled with strain mutation on storage can be estimated from data shown in Table III. The titre data presented refers to tests using parent strains of major interest in this work and carried out on various medium batches at different times during a period of about one year. In estimating whether crossing such strains has improved titre, reference has been made to Table III which includes the titre of the parent strains of each cross, tested on the same batch of medium as the segregants.

TABLE III

Yield Group Titres of Various Sister Strains

Strain	Yield group[a]												
	A	B	C	D	E	F	G	H	I	J	K	L	M
Original parent	—	—	—	—	—	2	5	—	—	—	—	—	—
brw nic	—	—	—	—	—	3	3	—	—	—	—	—	—
whi nic	—	—	—	1	5	1	—	—	—	—	—	—	—
whi rbs	—	—	—	—	1	2	1	1	—	—	—	—	—
bgn thi	—	—	—	—	—	4	4	1	—	—	—	—	—
bgn thi (t_1)	—	—	—	—	—	—	—	1	2	2	1	—	—
brw nic (t_2)	—	—	—	—	—	—	—	—	—	2	7	18	—
brw nic (t_3)	—	—	—	—	—	—	13	11	—	—	—	—	—
brw nic (t_4)	—	—	—	—	—	—	—	2	9	13	—	—	—
brw nic (t_5)	—	—	—	1	2	—	—	2	9	5	—	—	—

[a] See Materials and Methods.

Table IV shows the titres of segregants of crosses of general type bgn thi × whi nic. (The segregation of other markers has been given in Table I.) A highly significant feature of this data is that, in certain crosses, all whi segregants had reduced titres. However, in one of the crosses carried out in triplicate, i.e. cross c, the titre was not found to be reduced. Another outstanding feature of the data in Table IV is the reduced titres of bgn thi nic segregants compared with those of bgn thi segregants. This would suggest that an interaction has taken place between a determinant on linkage group I as marked by bgn thi and a determinant on linkage group III in the strain marked by nic. The determinants could in fact be the auxotrophic markers themselves. In addition, the titre-increasing determinant t_1 would appear to be located on group I

TABLE IV

Yield Group[a] Titres of Segregants of Crosses of Class I[b]

Haploid segregant genotypes	Cross whi nic ×			
	× bgn thi	× bgn thi (t_1) u	× bgn thi (t_1) b	× bgn thi (t_1) c
bgn thi	2F, 3G	2G, 5H	1G, 1H	1G, 1H
bgn thi nic	1B, 1C	1C	5D, 2E	1C, 1D
whi	4C	4C	3C	1E, 2F, 3G, 1H
whi nic	2B, 3C	3B, 3C	3B	1F

[a] Numbers given before yield group letters indicate those in the respective group.

[b] General type bgn thi × whi nic.

TABLE V

Yield Group[a] Titres of Segregants from Crosses of Class II[b]

Haploid segregant genotypes	Cross whi his rbs ×				
	× brw nic	× brw nic (t_2)	× brw nic (t_3)	× brw nic (t_4)	× brw nic (t_5)
brw nic	2G	1I	1H, 1I, 1J	1J, 1L[c]	1I
brw rbs	5G	1I	1H	2J	—
whi nic	2C	1E, 3F	1E, 1F	—	2E, 1H
whi rbs	2C	1E, 1F	—	—	4E, 1H

[a] Numbers given before the yield group letters indicate those in the respective group.
[b] General type brw nic × whi his rbs.
[c] Titres significantly in excess of parental and original parental titres (Table III).

as marked by bgn thi, or on group II. Allocation to linkage group III as marked by nic^+ is considered less likely on the grounds that, in cross c, nic^+ recombinants are detected with titres less than t_1. Previous data[5], involving crossing bgn thi strains to master strains, indicated that strain bgn thi did not show free recombination between groups I and II and, as with strain brw nic (t_4), is considered translocated between such groups.

The titres of segregants from crosses of type brw nic × whi his rbs and of bgn thi × brw nic are shown in Tables V and VI.

Table V shows that, although one can obtain segregants carrying whi with very reduced titre, this is found only in the control cross. Furthermore, in other crosses, segregants with improved titres could be detected. However, the scarcity of data has to be taken into account when considering allocation of titre-increasing determinants to linkage groups. It would appear, even so, that t_2, t_3 and t_4 can be allocated to linkage group I as marked by the brw marker, but allocation of t_5 is more ambiguous and

TABLE VI

Yield Group[a] Titres of Segregants from Crosses of Type bgn thi × brw nic

Haploid segregant genotypes	Cross bgn thi (t_1) ×	
	× brw nic (t_3)	× brw nic (t_5)
bgn thi	3G	2J
bgn thi nic	1D, 2E, 2F	1 G, 1H, 1I
brw	3H, 1I, 1L[b]	—
brw nic	1F, 3G, 2H, 1I	1J

[a] Numbers given before the yield group letters indicate those in the respective group.
[b] Titres significantly in excess of parental and original parental titres (Table III).

could be attributed to the possible high instability of the parent strain *brw nic* (t_5) (*see* Table III).

Strains carrying high-titre determinants are currently being crossed. Initial data for two such crosses is shown in Table VI. In cross *bgn thi* (t_1) × *brw nic* (t_5), the strain carrying t_5 was a high-titre recombinant from an earlier cross (*see* Table V). The results in Table VI indicate that indeed t_5 could reside on the linkage group III, marked by *nic*, as *bgn thi nic* segregants all had much higher titres than expected on the basis of results from cross *bgn thi* (t_1) × *whi nic* (*see* Table IV) which is effectively a control for the above cross.

The cross *bgn thi* (t_1) × *brw nic* (t_3) is outstanding because of the brown prototrophic segregants with enhanced titre. Mitotic crossing-over, to generate ultimately a genotype *brw* t_1 t_3, cannot be ruled out as a possible explanation.

MUTAGEN-INDUCED RECESSIVE LETHAL AND TRANSLOCATION FREQUENCIES

This topic could not only have a bearing on the possible mechanism of spontaneous mutability of diploids (*see below*) but could also indicate the relative ease with which a balanced lethal system can be induced in the organism. MACDONALD[6] and AZEVEDO and ROPER[7] pointed out that balanced lethals could make possible the use of diploids in commercial fermentations because without such systems a diploid might be unstable in generating recombinants spontaneously.

One purpose of the present studies was to discover whether any of the mutagens used was efficient at producing point mutations as opposed to translocations, as the latter are known to restrict viable recombinant recovery (*see above*). The diploid selected for this work was *whi lys* × *brw his nic*. On group I *whi* and *brw* are in repulsion; on group III *lys* and *nic* are in repulsion, and *his* marks group II. Because of markers in repulsion on certain linkage groups it is possible to discern, through subsequent segregation data, whether the inability to recover a certain allele following mutagen treatment is due to mitotic crossing-over or to non disjunction rather than to a recessive lethal on the group initially marked by the missing marker. In addition, chromosome translocation is inferred if markers that were previously segregating independently now show linked segregation.

Analysis in duplicate of the untreated population of spores from this diploid gave expected segregation data. The allele ratio approximated 1 : 1 in each case, and the various possible genotypes were recovered in approximately equal frequency. The results of mutagen treatment are shown in Table VII. The outstanding feature of these results is that, with all the mutagens, double lethals involving homologous linkage groups exceed both those involving single linkage groups and those involving nonhomologues. Another interesting feature is that the frequency of double lethals involving homologues also exceeds translocations even when γ-rays are used.

TABLE VII

Recessive Lethal and Translocation Frequencies Induced by Mutagen

Mutation		Mutagen				Total
		UV	γ	EMS	NTG	
Single lethals on	group I	3	1	1	2	7
	group II	1	1	0	0	2
	group III	3	4	1	2	10
Double lethals	nonhomologues	0	0	0	1	1
	homologues	2	6	8	11	27
Translocations	I + II	0	1	0	0	1
	I + III	1	1	0	0	2
	II + III	0	0	0	0	0
	multiple	1	1	0	0	2
Translocation-lethals		0	0	2	2	4
Normal		8	4	7	2	21
Total		19	19	19	20	77

It is tempting to speculate that chromosomes in diploids of *P. chrysogenum* are paired before mutagen action and this pairing, which may be unequal, might make such regions more accessible to mutagen attack. Consequently, non-allelic lethals may be preferentially induced for "homologous" pairs. Indeed ROBINOW and CATEN[8] have suggested on the basis of cytological evidence in *A. nidulans* that in the diploid condition homologous chromosomes could be paired.

DIPLOID INSTABILITY

Diploid instability is relevant to possible mechanisms generating segregants with various titres. The term "diploid instability" has been used by various workers when referring to a variety of properties such as high frequency of parental spore-colour segregants in the population or diploids in the population with variable titres. In the present work, however, the term is being used to indicate variation in morphology between diploid colonies produced by plating spores on solid agar. The main distinguishing characteristic used was degree of sporulation, *i.e.* dense, intermediate, and poor sporulation.

The previous communication[2] (*see* also *Materials and Methods*) described how an unstable haploid strain could be diploidised with the stable derivative. The unstable determinant was dominant; the diploid colony exhibited the same range of morphological variability as that of the unstable haploid parent. A more careful examination of the diploid population has revealed, however, that there are at least twice as many poor-sporulating types in the population as in the unstable parent. The frequency of dense sporulating types is at least ten-fold higher. These observations invite comparison with those of NGA and ROPER[9] who found, with *A. nidulans*, diploid instability analogous to that described above. These workers showed, on the basis of enhanced instability in the diploid condition, that a region of genetic material in duplicate was the cause of mutant induction. The mutants induced were in the same region or in regions adjacent to the duplication.

A microorganism in industrial use could well have numerous regions of the genome in duplicate, many of which do not have a readily detectable effect on morphology particularly as the duplications are small. The effects on titre, however, could be much more significant. The mechanism by which these mutations are achieved must remain as elusive for the time being as it is for *A. nidulans*[9,10]. It has been suggested that unequal chromatid exchange, generating further duplication as well as deficiency, could be the mechanism. In this connection, any study that indicated pairing in the diploid condition could be relevant. Indeed, if unequal chromatid exchange is between non-sister strands in the diploid, the mechanism that generates enhanced titre variability could be considered as one of mutation due to recombination.

DISCUSSION

On the basis of the results described here it can be concluded that a fundamental approach to crossing in *P. chrysogenum* is useful for titre improvement purposes and aids interpretation of segregation data to a greater degree than do empirical methods alone. Such problems as strain instability, parental genome segregation and selection against alleles can be largely overcome. The selection problem is reduced by use of PFA, and the use of master strains enables detection of free recombination between linkage groups[2]. However, the solution to the instability problem encounters a difficulty in that the titres of segregants are more variable than those of the parents or original parent of crosses, even if the parent strains used are relatively stable and closely related in lineage.

Various possible mechanisms* could be responsible for producing segregants with different titres. The possibility of mutation in a haploid parent before diploidisa-

---

* A recent cross, carried out on the assumption that t_2 is linked to *brw* on linkage group I and that t_5 is linked to *nic* on linkage group III, has generated *brw nic* recombinants with titres in yield groups N, O, P. As these values are in excess of any recorded previously it is concluded that the most likely cause is improvement of penicillin productivity by recombination of whole chromosomes.

tion must be considered, as well as the more interesting possibilities of mutation due to diploidisation or recombination in the residual genotype. The latter would involve either mitotic crossing-over or segregation of a fourth linkage group. This is not favoured, as sister strains have been used throughout this work and thus scope for residual genotypic variability has been reduced to a minimum. Also, the frequency of mitotic crossing-over is not high, and no evidence exists for a fourth linkage group.

The hypothesis that mutation in parent strains has taken place before diploidisation might explain certain decreases of titre, especially when one linkage group determines low titre amongst all segregants carrying it. Even so, this does not sufficiently explain the increases in titre that occur amongst certain segregants of a given genotype, and not amongst others of the same genotype. Moreover, although clonal propagation has been observed to produce decreases in titre, increases have never been detected.

The hypothesis of mutation due to diploidisation (which could involve recombination) is favoured. Indeed MACDONALD[11] has described diploids, formed between the same sister strain parents, which had differing titres. A hypothesis of mutation was favoured there, but it was not known whether mutation had occurred before or after diploidisation. Also, though such mechanisms could contribute to the titre variability of segregants from crosses between strains of divergent lineage, there is a strong possibility in these cases that orthodox recombination could contribute to such variability. This may occur because of the high frequency, in such crosses, of non-allelic differences in the diploid.

The enhanced titre variability of segregants has advantages because titre increases are produced. And, although the unequivocal allocation of titre-increasing alleles to linkage groups is hindered by such variability, the data presented here indicates that this can be approached on a large scale. It is possible to obtain indications as to which linkage groups in various strains one should recombine by haploidisation. So far, we have evidence that titre increases can be produced in a strain on either groups I or III. Subsequent work is extending this study with a view to achieving controlled manipulation of titre-increasing alleles by mitotic crossing-over. So far, we have a strain *ylo cho* which has a number of initially desirable properties. Neither marker modifies titre drastically, and they locate on opposite arms of linkage group I.

In conclusion it should be said that, on the question of mutation due to diploidisation and in other aspects of this study, reference to work with *A. nidulans* has been invaluable[11]. Studies carried out in the last decade with that organism have encouraged the present work with *P. chrysogenum* directed at its controlled breeding with the object of increasing penicillin productivity. It can be claimed that some success has already been achieved.

The author wishes to acknowledge the expert assistance of Mrs. M. Burge and Miss J. M. Weatherburn.

REFERENCES

1. SERMONTI G., in *Genetics of Antibiotic Producing Microorganisms*. John Wiley & Sons, London 1969.
2. BALL C.: J. Gen. Microbiol. **66**, 63 (1971).
3. PONTECORVO G., SERMONTI G.: J. Gen. Microbiol. **11**, 94 (1954).
4. STAUFFER J. F., BACKUS M. P.: Ann. N. Y. Acad. Sci. **60**, Art 1 (1954).
5. BALL C., *unpublished results*.
6. MACDONALD K. D.: Nature **204**, 404 (1964).
7. AZEVEDO J. L., ROPER J. A.: J. Gen. Microbiol. **49**, 149 (1967).
8. ROBINOW C. F., CATEN C. E.: J. Cell Sci. **5**, 403 (1969).
9. NGA B. H., ROPER J. A.: Genet. Res. **14**, 63 (1969).
10. BALL C.: Genet. Res. **10**, 173 (1967).
11. MACDONALD K. D.: Antonie van Leeuwenhoek **32**, 431 (1966).

The Use of Parasexual Genetics in an Industrial-Strain Improvement Program with *Penicillium chrysogenum*

R.P. ELANDER, M.A. ESPENSHADE, S.G. PATHAK, C.H. PAN

Wyeth Laboratories, Inc., West Chester, Pennsylvania, U.S.A.

INTRODUCTION

Although classical methods of mutation and selection have yielded the majority of improved strains utilized for the production of penicillin, a number of industrial laboratories are now utilizing parasexual techniques for the selection of strains with improved fermentation yields. The use of parasexuality for isolating potentially improved industrial strains of *Penicillium* was suggested by PONTECORVO and SERMONTI[1] nearly two decades ago. Additional reports by SERMONTI[2], ALIKHANIAN[3], MACDONALD, HUTCHINSON and GILLET[4] and MACDONALD[5] were disappointing due to the divergence of opinions concerning the practicality of parasexual genetics to industrial -strain improvement programs. Most of the diploid strains were reported to have barriers restricting recombination, thereby leading to strains which were highly unstable and which segregated back to their respective poor-yielding auxotrophic parents.

ELANDER[6] summarized an intensive large-scale industrial-strain development program with *Penicillium* and described a highly stable diploid strain which synthesized greater than parental yields of phenoxymethylpenicillin. The present report summarizes the activities of another large-scale industrial program utilizing both mutation-selection and parasexual techniques to generate high-yielding strains for commercial penicillin production.

MATERIALS AND METHODS

STRAINS

The genealogy of improved Wyeth strains of *Penicillium chrysogenum* utilized for the commercial production of penicillin G is shown in Fig. 1. The M-designation refers to actual production usage. These strains are maintained as lyophilized ampoules in order to obviate back-mutation and degeneration with respect to penicillin fermentation. The M-strains show profuse green (wild-type) conidiation on slant culture and are prototrophic with respect to their nutritional requirements.

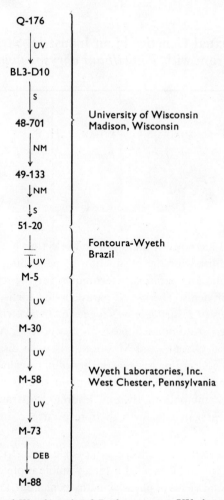

Fig. 1. Lineage of improved Wyeth strain of *P. chrysogenum*. **UV** ultraviolet radiation, **S** spontaneous, **NM** methyl-bis-(β-chloroethyl)amine, **DEB** 1,2,3,4-diepoxybutane.

Parasexual recombination was observed in heterokaryotic cultures synthesized from complementary auxotrophic mutants derived from a wild-type commercial strain designated M-30. The lineage of several mutants and recombinants derived therefrom is summarized in Fig. 2.

MUTAGENIC TECHNIQUES

Ultraviolet radiation. The source of ultraviolet (253 nm) radiation was a low-pressure mercury arc lamp, Model PCQ-X1*. A dose rate of $300-400 \, \mu W/cm^2/s$ was used for all UV-radiation experiments. The intensity measurements were determined with the aid of an ultraviolet intensity meter*.

---

* Ultra-Violet Products, Inc., San Gabriel, California, U.S.A.

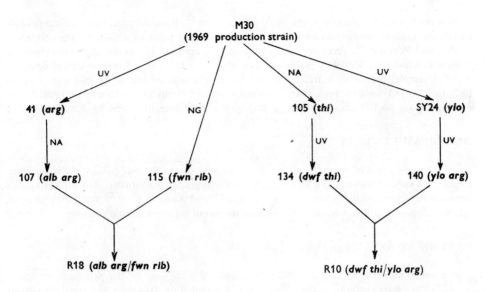

Fig. 2. Lineage of Wyeth diploid strains of *P. chrysogenum*. **UV** ultraviolet radiation (253 nm), **NA** nitrous acid, **NG** N-methyl-N'-nitro-N-nitrosoguanidine.

Chemical mutagens: *Nitrogen mustard (NM)*: The procedure of STAHMAN and STAUFFER[7] was followed for mutation induction with methyl-bis-(β-chloroethyl)amine obtained as mechlorethamine hydrochloride*.

Nitrosoguanidine (NG): N-Methyl-N'-nitro-N-nitrosoguanidine* in a concentration of 0.025 M was routinely used and the spores were exposed for periods ranging from 0 to 20 min. The details of the mutation protocol were adopted from ADELBERG *et al.*[8] and MOORE[9].

Nitrous acid (NA): The procedure described by SIDDIQUI[10] was used. The conidia of *P. chrysogenum* were exposed to a concentration of 0.017 M in Tris buffer for a time period ranging from 0 to 30 min.

Diepoxybutane (DEB): A concentration of 0.05 M of 1,2,3,4-diepoxybutane** was used for a maximum exposure period of 60 min.

SHAKE-FLASK PENICILLIN G AND V FERMENTATIONS

The shake-flask fermentations were carried out in 500 ml or 250 ml Erlenmeyer flasks (wide-mouth) utilizing three milk filter discs*** for closure. The fermentation medium was composed of corn-steep liquor, $(NH_4)_2SO_4$, lactose, cerelose, $CaCO_3$ and sperm oil. Sodium phenylacetate and sodium phenoxyacetate were used as "precursors" for either penicillin G or penicillin V, respectively. Aliquots were aseptically removed from the fermentor flasks at 72, 120 and 168 h and assayed for penicillins G or V chemically by the method of BOXER and EVERETT[11] or biologically according to the procedure described in the Federal Register[12]. Residual carbo-

---

* Aldrich Chemical Co., Inc., Milwaukee, Wisconsin, U.S.A.

** K & K Laboratories, Inc., Plainview, New York, U.S.A.

*** Perfection Milk Filters, Schwartz Manufacturing Co., Two Rivers, Wisconsin, U.S.A.

hydrate levels were determined by the method of SUMNER[13]. Residual precursor (sodium phenyl-acetate or sodium phenoxyacetate) levels were determined colorimetrically by the procedure of PAN and PERLMAN[14]. Precursor levels were also determined by gas-liquid chromotography using a Perkin-Elmer Model N° 900 chromatograph equipped with a flame ionization detector. A 0.2 µl sample, previously extracted with benzene, was injected into a six foot glass column [0.25 inch (O.D.) by 0.08 inch (I.D.)] packed with 10% diethyleneglycol adipate treated with 2% H_3PO_4 on 80/100 diatoport-S according to the procedure described by NIEDERMAYER[15].

DNA MEASUREMENTS

A concentration of approximately 10^9 spores was obtained from an inoculated one liter Erlenmeyer flask containing 75 g of moistened medium-grade cracked corn. The diphenylamine colorimetric procedure described by ISHITANI *et al.*[16] and ARRIGHI *et al.*[17] for DNA analysis was used. Spore densities were determined with the aid of a Levy counting chamber.

ENZYME ASSAYS

Alkaline protease. Alkaline protease was assayed according to the procedure of IKEDA *et al.*[18]. The various strains of *Penicillium* were propagated on a medium described by DWOR-SCHACK *et al.*[19]. Total protein was determined with Folin phenol reagent according to the procedure of LOWRY *et al.*[20]. Bovine serum albumin was used as standard.

β-Galactosidase. β-Galactosidase was determined according to the procedure outlined by NIEDHARDT and BOYD[21].

Glucose oxidase. Glucose oxidase was determined according to the procedure developed by UNDERKOFLER[22]. The mycelial extracts were compared to a purified glucose oxidase enzyme standard*.

RESPIRATION STUDIES

Standard Warburg manometric methods described by UMBREIT, BURRIS and STAUFFER[23] were used. Potassium hydroxide (0.2 ml of 20%) was added to the center well of all flasks to absorb carbon dioxide. A Gilson Model G-8 differential respirometer was employed.

PARASEXUAL TECHNIQUES — COMPOSITION OF MEDIA

Complete medium (CM — in %):

Corn-steep liquor (Corn Products)	0.5
Bacto-Peptone (Difco)	0.5
Yeast extract (Difco)	0.5
Casamino Acids (Sheffield N-Z Case)	0.25
Corn dextrin	3.0
$FeSO_4.7H_2O$	0.001
KCl	0.05
$MgSO_4.7H_2O$	0.05
KH_2PO_4	0.1
NH_4NO_3	0.3
Agar (Difco)	2.0

pH 5.5

* Nutritional Biochemical Corporation, Cleveland, Ohio, U.S.A.

Minimal medium (MM — in %):	NH_4NO_3	0.3
	KH_2PO_4	0.1
	$MgSO_4.7H_2O$	0.05
	KCl	0.05
	$FeSO_4.7H_2O$	0.001
	Sucrose (reagent grade)	3.0
	Ionagar*	1.0

pH 5.5

NOMENCLATURE

The system of abbreviations for nutritional, conidial color, and colonial-growth mutants was adopted according to the proposals of DEMEREC et al.[24] and SERMONTI[25].

ISOLATION OF HETEROZYGOUS DIPLOIDS AND SEGREGANT STRAINS

Homogenized spore suspensions of the two complementary biochemically-deficient parental strains were inoculated to complete medium (CM) broth and incubated at 25 °C for 7—10 days. The mycelial mass was then removed, washed 3—5 times with minimal medium (MM), and teased apart on the surface of minimal agar. Diploid conidia were isolated from green diploid sectors in heterokaryotic mycelial fans or from green conidiating colonies derived from spores from heterokaryotic growth which were previously diluted and plated out on either CM or MM agars. Segregants were selected as sectors formed in diploid mycelium.

PLOIDY ESTIMATION

Cultures propagated on complete medium (CM) were used as sources of conidia for microscopic measurements. Conidia were examined under high-dry magnification (430×) and measured to the nearest 0.25 μm. Conidial volumes were calculated using the formula, $V = 1/6 \pi d^3$. DNA content and enzyme level were also used to differentiate haploid and diploid conidia and/or mycelium.

RESULTS

GENESIS OF IMPROVED WYETH MUTANTS OF *P. chrysogenum* BY STEP-WISE MUTATION AND SELECTION

The lineage of improved Wyeth strains of *P. chrysogenum* is presented in Fig. 1. Recent production strains utilized for the commercial production of penicillin G in large 20,000 gallon fermentors begins with strain M-5, an improved UV mutant generated in the strain bank of our South American facility, Fontoura-Wyeth, S. A. This strain was first utilized in our West Chester plant in 1967 and served as the progenitor strain for a number of improved mutants developed in our strain bank during

---

* Consolidated Laboratories, Inc., North Chicago, Illinois, U.S.A.

the ensuing three years. With the exception of strain M-88, all of the improved mutants adopted in our Production Fermentation Department were selected from UV (253 nm) survivor populations. The strains are excellent sporulators and exhibit extremely uniform colony-population patterns. The mutants are maintained as lyophilized pellets stored in sterile vacuum-sealed glass ampoules. The capacity of various strains to synthesize benzylpenicillin (penicillin G) and phenoxymethylpenicillin (pe-

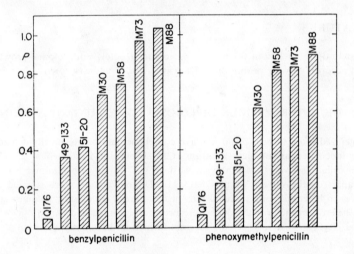

Fig. 3. Penicillin accumulation by improved Wyeth mutants in 168 hour shake-flask fermentation. *P* relative penicillin potency.

nicillin V) in shake-flask fermentation is presented in Fig. 3. A consistent step-wise improvement can be observed. An approximate twenty-six-fold improvement is observed for benzylpenicillin comparing the relative performance of Q-176 and the most recent mutant to be added to the Wyeth Lineage, strain M-88. A similar dramatic improvement in productivity is also observed for phenoxymethylpenicillin (Fig. 3).

GENESIS OF IMPROVED WYETH MUTANTS OF *P. chrysogenum* BY PARASEXUAL RECOMBINATION

The lineage of two high-yielding heterozygous diploid strains of *P. chrysogenum*, R10 and R18, is summarized in Fig. 2. The diploids were obtained as diploid sectors in heterokaryotic cultures and derived from matings between complementary auxotrophic-conidial color mutants.

PROPERTIES OF HETEROZYGOUS DIPLOID STRAINS

Conidial volumes and DNA content. A number of parental wild type, conidial color, biochemically-deficient strains and heterozygous diploid strains were compared with respect to spore volume and DNA content (Table I). The diploid strains R9, R10 and R18 characteristically exhibited larger spore volume and increased DNA content when compared to their respective wild-type haploid grandparents (M-30 or N°2) or their complementary biochemically-deficient parents. Strain 105, a thiamine-requiring mutant had an unusually large 76.93 μm^3 spore volume and a greater DNA content (1.26×10^{-7} μg) compared to its wild-type parent strain (M-30). When the *dwarf* marker was introduced into the mutant, the conidial volume and DNA content were reduced by nearly one-half.

Spore germination. Table II shows the results of a comparative spore-germination study between a particularly vigorous diploid strain (WC-9) and two haploid strains (N°2 and M-30). Equivalent numbers of conidia (adjusted to approximately 5×10^7 spores/ml) were obtained from slants of similar age (7 days at 25 °C) and inoculated as suspensions to Erlenmeyer (250 ml) flasks containing 50 ml of sterile Czapek-Dox solution. The inoculated flasks were then incubated at 25 °C on a rotary shaker (250 r.p.m.). After a period of 24 h, aliquots were removed and placed on a Levy hemocytometer for microscopic observation (430×). The average number of spores counted

TABLE I

Spore Volume and DNA Content of Selected Haploid and Diploid Strains of *P. chrysogenum*

Strain	Genotype	Ploidy	Spore volume μm^3	DNA per spore $\mu g \times 10^7$
M30	wild type	N	54.38	0.26
41	arg	N	53.00	0.35
107	alb arg	N	27.54	0.26
115	fwn rib	N	22.43	0.33
R18	alb arg/fwn rib	2N	81.10	0.79
105	thi	N	76.93	1.26
134	dwf thi	N	36.08	0.69
SY24	ylo	N	69.77	0.88
140	ylo arg	N	42.1	0.66
R10	dwf thi/ylo arg	2N	65.5	1.15
2	wild type	N	34.99	0.44
WC-2	ylo met	N	36.78	0.57
WC-8	alb ade	N	29.75	0.45
R9	ylo met/alb ade	2N	61.49	0.83

TABLE II

Spore Germination in Haploid and Diploid Strains of *P. chrysogenum*

Strain	Ploidy	Per cent germination[a]
2	N	37.1
M30	N	33.9
WC9	2N	47.5

[a] In Czapek-Dox solution, average of 200 spores, 24 h incubation at 25 °C.

was 200 for each strain. The diploid WC-9 strain showed nearly 50% germination at 24 h in the chemically-defined medium. In contrast, the two haploid strains showed only 33.9% and 37.1% germination, respectively.

Respiration and glucose utilization. Strains N°2 and M-30 were compared to the WC-9 diploid on the basis of oxygen uptake and glucose oxidation. Washed mycelial suspensions of the three strains were added to standard Warburg flasks containing 0.4% glucose solution. The flasks were then placed on the Gilson differential respirometer apparatus and submerged in a 25 °C water bath. Table III shows the average results of three separate experiments with flasks run in duplicate for each strain. The diploid culture (WC-9) respires at a rate nearly 45% greater than its parent (N°2) and approximately 28% greater than the M-30 haploid production strain. At the completion of the experiments, the broths were examined for residual glucose levels. The diploid strain oxidized considerably more glucose than either of the haploid strains — 1.41 mg/ml *vs.* 0.90 mg/ml and 1.09 mg/ml for the two haploid cultures (Table III).

Enzyme studies. Table IV summarizes the comparative mycelial levels of β-galactosidase, glucose oxidase, and alkaline protease for the two haploid strains (N°2 and M-30) and the vigorous diploid WC-9 culture.

TABLE III

Comparison of Oxygen Uptake and Glucose Oxidation of Haploid and Diploid Strains of *P. chrysogenum*

Strain	Ploidy	Oxygen uptake $\mu l\ O_2$ /h per mg mycelium	Glucose oxidized mg/ml		
			supplied	residual	oxidized
2	N	5.03	4.0	3.10	0.90
M30	N	5.06	4.0	2.91	1.09
WC9	2N	7.34	4.0	2.59	1.41

TABLE IV

Comparison of Enzyme Activities in Haploid and Diploid Strains of *P. chrysogenum*

Strain	Ploidy	β-Galactosidase units/h per mg protein	Glucose oxidase units/g CEP[a]	Alkaline protease	
				total units	units per mg protein
2	N	80.26	1,275	700.4	5.58
M30	N	89.68	1,575	347.2	2.52
WC9	2N	106.40	1,950	1341.5	8.92

[a] Crude enzyme preparation; activity of purified enzyme: 27,150.

Mycelia of the various strains were obtained from 250 ml Erlenmeyer flasks containing a growth medium consisting of corn-steep liquor (6%) and lactose (2%). The flasks were inoculated with spores of the various strains and incubated at 25 °C on a New Brunswick rotary shaker for 72 h. Lyophilized mycelium, further desiccated and ground with coarse quartz sand in a mortar, was suspended in 0.1 M phosphate buffer (pH 6.0) and assayed according to the method of NIEDHARDT and BOYD[21]. Protein determinations using Folin phenol reagent were carried out according to the procedure described by LOWRY et al.[20].

The mycelium utilized for alkaline protease activity was harvested from 250 ml Erlenmeyer flasks containing 50 ml of a high protein medium consisting of soybean meal (1.0%), corn meal (2.0%), and $CaCO_3$ (0.5%). Alkaline protease levels were determined also on filtered broth since it was established that most of the enzyme was extracellular.

The diploid strain (WC-9) possessed considerably more β-galactosidase activity than the two haploid strains (Table IV). WC-9 had 32.5% more β-galactosidase compared with its haploid parent (WC-2) and 18.6% more activity when compared with M-30. Air-dried ethanol extracts of WC-9 diploid mycelium contained 52.74% and 23.8% more glucose oxidase when compared to strains N°2 and M-30. The alkaline protease levels of the diploid WC-9 strains were considerably higher than the two haploid strains. The diploid culture produced 58.8% more alkaline protease than its parent haploid strain (N°2). Similarly, WC-9 produced nearly three-fold more enzyme than the M-30 strain.

PHENOXYMETHYLPENICILLIN (PENICILLIN V) PRODUCTION BY TWO RELATIVELY STABLE WYETH DIPLOID STRAINS OF *P. chrysogenum*

The majority of heterozygous and "homozygous" diploid strains synthesized thus far in our West Chester Laboratories have been extremely unstable and poor

TABLE V

Relative Phenoxymethylpenicillin Accumulation by Selected Stable Wyeth Diploid Strains of
P. chrysogenum

Strain	Genotype	Relative yield	Residual carbohydrate %	Residual precursor %	pH	Relative growth %
R10	*dwf thi/ylo arg*	0.79	1.11	0.030	8.1	65
R18	*alb arg/fwn rib*	0.92	1.71	0.055	6.7	66
M30	wild type	1.00	1.80	0.009	7.9	40

in either penicillin G or V productivity. Two relatively stable heterozygous diploid strains, R10 and R18, were compared in shake-flask fermentation with a recent production fermentation strain (M-30). Although the diploids were inferior in productivity to the haploid production control (Table V), strain R18 produced excellent penicillin V levels. The diploids were rapidly growing cultures and produced considerably more mycelial matte (centrifuged mycelial solids) than the haploid production culture. They were also more efficient with respect to carbohydrate utilization. The diploids consumed less precursor (sodium phenoxyacetate) than the haploid (M-30) production strain. Strain R18 had an unusually low pH plateau suggesting that this culture may have potential for longer fermentation cycles. The diploid strains were vigorous mycelial producers and were, possibly, limited by dissolved oxygen levels, expecially in shake-flask fermentors. The strains were now utilized as parental stocks for standard mutation-selection programs.

COMPARISON WITH LILLY HETEROZYGOUS DIPLOID STRAINS OF *P. chrysogenum* DERIVED FROM MUTANT E-15

Several years ago, the senior author reported the discovery of a highly stable, high-yielding heterozygous diploid strain synthesized at another large industrial laboratory[6]. In a mating between a white-spored adenine-less mutant and a yellow-spored methionine-less mutant, an excellent penicillin V-producing diploid strain was discovered (Fig. 4). In fact, it produced more penicillin than its wild-type parent (the production control) and was utilized for commercial penicillin production for a short period of time. However, the strain was soon replaced with a higher yielding haploid prototroph. Although the (*ylo met/whi ade*) diploid was never utilized again for actual production, several excellent natural clonal derivatives were later discovered (Fig. 5). Unfortunately, these selections never gained production status.

A population pattern study of the diploid revealed a remarkably stable colony-population pattern (Table VI). The total variation observed in a large population of

TABLE VI

Population Pattern in a Stable *Penicillium* Diploid and its Wild-Type Haploid Parent[a]

Strain	Colony type	Number	Untreated population %	Total varia-tion %	Number	Treated popula-tion[b] %	Survival %	Total vari-ation %
Hetero-zygous diploid (*ylo met/ whi ade*)	Green[c]	264.5	90.7	9.2	152.0	88.3	59.0	11.6
	Yellow	24.5	8.4		9.5	5.5		
	White	1.2	0.4		4.5	2.6		
	Pale green	1.2	0.4		6.0	3.5		
Wild-type haploid E-15	Light green[c]	176.0	68.9	31.9	109.0	58.9	72.4	41.1
	Dark green	51.3	20.1		47.5	25.7		
	White	30.3	11.8		28.5	15.4		

[a] From Refs[6,26].
[b] After 2 min UV-irradiation.
[c] Predominant type.

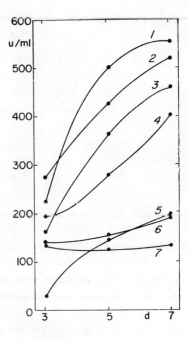

Fig. 4. Penicillin V production (arbitrary units/ml) by parent, biochemical mutants, heterokaryon, heterozygous diploid and diploid segregants. (From Ref.[6])
1 heterozygous diploid (*ylo met/whi ade*), 2 starting prototroph (*whi*+), 3 mitotic segregant PGMS (*ylo met/whi ade*), 4 heterokaryon (*ylo met + whi ade*), 5 mitotic segregant YMS (*ylo met*+), 6 albino auxotroph (*whi ade*), 7 yellow auxotroph (*ylo met*).

colonies was only 9.2% compared to 31.9% for the production E-15 control. The strain was also more UV-sensitive and its population pattern remained essentially unchanged following 2.0 min. of UV-radiation. In contrast, the production haploid increased in variation by nearly 10% following the UV-radiation exposure.

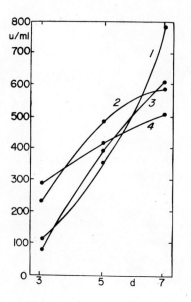

Fig. 5. Penicillin V production (arbitrary units/ml) by parent, diploid and diploid selection. (From Ref.[6]) **1** selection N°2, **2** diploid parent (*ylo met/ whi ade*), **3** selection N°3, **4** starting prototroph (*whi*[+]).

DISCUSSION

The screening of large numbers of minor variants induced by low dosages of mutagens is probably the most important method for improving the efficiency of fermentation processes. Despite the randomness of the procedure, this approach is still the predominant one employed in industrial-strain improvement laboratories throughout the world.

Examples of gradual step-wise improvement in antibiotic production through mutation and selection are numerous. These may be seen in studies with *Penicillium* and penicillin[6,27] and with *Cephalosporium* and cephalosporin[28,29].

The step-wise selection program described herein for *P. chrysogenum* constitutes a greater than twenty-six-fold improvement over strain Q-176 in penicillin G production based on shake-flask fermentation data. Far greater increases in productivity could be recorded, if comparisons could be made in large-scale production vessels, utilizing early Wisconsin strains with the last strain to be added to the Wyeth lineage — strain M-88. Similar shake-flask comparisons with improved Lilly strains showed a dramatic fifty-five-fold improvement over the original Fleming culture[6].

The discovery of parasexuality in asexual molds by PONTECORVO and ROPER in 1952 has provided a means for planned breeding which resulted in the synthesis of vigorous diploid strains with increased capacity to synthesize economically-important fermentation products. The technique consists of combining two strains with particularly interesting properties to form a diploid with more desirable properties. For enhanced penicillin production, notable desirable properties include more rapid, early synthesis of secondary metabolites generating penicillin, more abundant conidiation to facilitate long-term preservation, more efficient sugar and precursor utilization, greater synthesis of 6-aminopenicillanic acid (APA) and low levels of penicillin acylase.

The studies described herein on the conidial volume and DNA content of various haploid and diploid strains of *Penicillium* clearly indicate that strains R10, R18 and R9 possess higher ploidy levels than their respective wild-type and related auxotrophic parents. The data also points out that considerable variation in spore size and DNA content may occur when major mutants are utilized for genetic recombination studies. It appears that both conidial volume and DNA content should be correlated for proper ploidy determination. The present study also suggests that the diploid strains should be compared with immediate parents in the lineage rather than with distantly related great-grandparent strains. In the present study, two wild-type strains of differing lineage were utilized. Although the strains have common ancestry in the Wisconsin series, there is considerable variation in conidial size and DNA content (Table I). The introduction of biochemical markers into strains leads to auxotrophy and may also manifest other unusual pleiotropic effects. In this study, the introduction of the thiamine marker (thi) into strain 105 resulted in large conidium formation compared to its wild-type parent — strain M-30. This particular auxotrophic mutant had a greater conidial DNA content than any of the diploid strains analyzed thus far in our laboratories. However, the introduction of the dwarf marker (dwf) into the strain (N°134) reduced the DNA content by nearly one-half.

The diploid strain R9 appears to be a particulalry vigorous culture compared to the two wild-type haploid production cultures — N°2 and M-30. This culture, when inoculated to fresh agar medium, completes its asexual life cycle (spore to spore) and exhibits abundant green diploid conidiation following 3—4 days incubation. In contrast, the N°2 and M-30 haploid strains generally require an incubation period of 5—6 days before new green conidiation is observed. With this background, it is not surprising that the WC-9 culture exhibits more vigor with respect to more rapid spore germination, oxygen uptake and glucose oxidation.

According to PARDEE[30], a method for improving the rate of enzyme synthesis in industrially important microorganisms is to increase the number of structural genes per given cell. When extra copies of the gene are in the cell, the enzyme can usually be produced in proportionately larger amounts. In *Escherichia coli*, cells with increased numbers of structural genes can be selected by isolating rare mutants possessing several phenocopies, by introducing extra-chromosomal segments known as episomes or by phage-escape synthesis where bacteria can be infected by lysogenic phage which carry structural genes for specific bacterial enzymes in place of phage genetic material.

In filamentous fungi, including those with a predominant yeast phase, parasexuality provides an excellent technique for obtaining cells with increased numbers of structural genes. The utilization of diplophase yeast strains with increased vigor with respect to ethanol synthesis and starch hydrolysis has been common. The Japanese have made similar improvements in the production of soy sauce and other fermented foods through diploid hybridization of selected strains of the heterothallic haploid yeast *Saccharomyces rouxii*[31]. WICKERHAM[32] recently described an unusual diploid isolate of *Candida lipolytica* (*Endomycopsis lipolytica*). He reported that haploid strains of this organism are outstanding for their ability to produce lipase and extracellular protease. This particular yeast also has great potential for converting hydrocarbons to foods and animal feeds. WICKERHAM suggested that hybridization of selected lines and production of large-celled diploids may greatly improve the economic yields of technological processes with this organism.

The WC-9 diploid strain of *Penicillium* described in this report has considerable vigor with respect to the formation of β-galactosidase, glucose oxidase and alkaline protease (Table IV). It would appear that the larger diploid cells of this clone possess a larger number of structural genes regulating the formation of these three particular enzymes and that the diploid state of the organism is probably responsible for the enhanced enzyme formation.

Asexual filamentous fungi are utilized in many important industrial processes from antibiotic syntheses to steroid hydroxylations and reductions. The use of parasexual genetic techniques to generate stable diploid strains would appear to be a simple logical approach to generate more efficient strains. Despite the large volume of literature pertaining to the use of filamentous fungi in bio-conversions of important steroid moieties, the authors know of no published reports concerning the application of artificially induced diploid strains for more efficient steroid conversion.

Contrasting the two parasexual genetic programs described in this report, the recent Wyeth program has not yielded, to date, a stable diploid strain suitable for large-scale commercial penicillin production. Of the two Wyeth heterozygous diploid strains described, strain R18 shows a more homogeneous population pattern than does R10. Undoubtedly, this factor plays an important role in its capacity to accumulate relatively large quantities of penicillin. These two diploid strains have significantly more uniform population patterns than the majority of the diploids synthesized in our laboratories. The remarkably stable diploid (*ylo met/whi ade*) synthesized at the Lilly Laboratories remains unique with respect to its ability to generate large quantities of phenoxymethylpenicillin. Its homogeneous colony population pattern and its limited parental genome segregation were probably significant factors culminating in high levels of penicillin accumulation.

The authors wish to acknowledge the excellent technical assistance of Mr. Vincent Karetsky, Mr. Vincent Cullen and Mr. Edward Ambush who carried out numerous phases of the genetic, fermentation, biochemical and analytical studies. Gratitude is also expressed to Mrs. Virginia Merrion and associates for the chemical and biological penicillin assays and to Mr. Ernest Tyler for performing gas-chromatographic analyses.

REFERENCES

1. PONTECORVO G., SERMONTI G.: Nature **172**, 126 (1953).
2. SERMONTI G.: Ann. N. Y. Acad. Sci. **81**, 950 (1959).
3. ALIKHANIAN S. I.: Adv. Appl. Microbiol. **4**, 1 (1962).
4. MACDONALD K. D., HUTCHINSON J. M., GILLETT W. A.: Antonie van Leeuwenhoek **30**, 209 (1964).
5. MACDONALD K. D.: Antonie van Leeuwenhoek **32**, 431 (1966).
6. ELANDER R. P., p. 403 in *Induced Mutations and Their Utilization*. Abh. deut. Akad. Wiss., Berlin 1967.
7. STAHMANN M. A., STAUFFER J. F.: Science **106**, 35 (1947).
8. ADELBERG E. A., MANDEL M., CHEN G.C.C.: Biochem. Biophys. Res. Comm. **18**, 788 (1965).
9. MOORE D.: J. Gen. Microbiol. **55**, 121 (1969).
10. SIDDIQUI O. H.: Genet. Res. **3**, 303 (1962).
11. BOXER G. E., EVERETT P. M.: Analyt. Chem. **21**, 670 (1949).
12. Federal Register: Antibiotic and Antibiotic-Containing Drugs; Reissuance of Regulations. Vol. 27 (251), Part II, p. 13003 (1962).
13. SUMNER J. B.: J. Biol. Chem. **65**, 393 (1925).
14. PAN S. C., PERLMAN D.: Analyt. Chem. **26**, 1432 (1954).
15. NIEDERMAYER A. O.: Analyt. Chem. **36**, 938 (1964).
16. ISHITANI C. Y., IKEDA Y., SAKAGUCHI S.: J. Gen. Appl. Microbiol. **2**, 401 (1956).
17. ARRIGHI F. E., BERGENDAHL J., MANDEL M.: Exp. Cell. Res. **50**, 47 (1968).
18. IKEDA Y., NAKAMURA K., UCHIDA K., ISHITANI C.: J. Gen. Appl. Microbiol. **3**, 93 (1957).
19. DWORSCHAK R. S., KOEPSELL H. J., LAGODA A. A.: Arch. Biochem. Biophys. **41**, 48 (1952).
20. LOWRY O. H., ROSEBROUGH N. J., FARR A. L., RANDALL R. J.: J. Biol. Chem. **193**, 265 (1951).
21. NIEDERHARDT F. C., BOYD R. F., p. 90 in *Cell Biology: A Laboratory Text*. Burgess Publ. Co., Minneapolis 1965.
22. UNDERKOFLER L. A., p. 486 in Proc. Internat. Symp. Enzyme Chem., Tokyo 1958.
23. UMBREIT W. W., BURRIS R. H., STAUFFER J. F., p. 305 in *Manometric Techniques: A Manual Describing Methods Applicable to the Study of Tissue Metabolism*. Burgess Publ. Co., Minneapolis 1964.
24. DEMEREC M., ADELBERG E. A., CLARK A. J., HARTMAN P. E.: Genetics **54**, 61 (1966).
25. SERMONTI G., p. 389 in *Genetics of Antibiotic-Producing Microorganisms*. Wiley-Interscience, London 1969.
26. ELANDER R. P., p. 89 in D. Perlman (Ed.): *Fermentation Advances*. Academic Press, New York 1969.
27. BACKUS M. P., STAUFFER J. F.: Mycologia **47**, 429 (1955).
28. ELANDER R. P., STAUFFER J. F., BACKUS M. P.: Antimicrob. Agents Ann. **1**, 91 (1960).
29. STAUFFER J. F., SCHWARTZ L. J., BRADY C. L.: Developments Ind. Microbiol. **7**, 104 (1966).
30. PARDEE A. B., p. 3 in D. Perlman (Ed.): *Fermentation Advances*. Academic Press, New York 1969.
31. MORI H., ONISHI H.: Appl. Microbiol. **15**, 928 (1967).
32. WICKERHAM L. J., KRUTZMAN C. P., HERMAN A. I.: Science **167**, 1141 (1969).

Genetics of Penicillin Production in *Penicillium chrysogenum* and *Aspergillus nidulans*

K.D. MACDONALD

Microbiological Research Establishment, Porton Down, Salisbury, Wiltshire, Great Britain

INTRODUCTION

The discovery of parasexuality in the mould *Aspergillus nidulans*[1] began a new era in the genetics of asexual fungi and it was visualised, with the development of parasexual methods of genetical analysis in asexual microorganisms of industrial importance, that suitable crosses would lead to the isolation of recombinant progeny with superior yields of commercial products.

The demonstration that the parasexual mechanism occurred in *Penicillium chrysogenum*[2] was followed by a number of studies on the genetics of penicillin production in this asexual fungus[3-6]. Diploids synthesized between certain strains yielded more penicillin than their parents[3,6] suggesting that it could be beneficial to maintain the heterozygous diploid state either by selecting against segregants as they were formed[7] or by isolating stable diploids[6,8,9]. Attempts to isolate segregants with increased penicillin titres relative to the parent strains forming a diploid have not been encouraging probably due in some cases at least to differences in chromosomal morphology between haploid parents restricting the emergence of recombinants[10]. Crosses were made between strains of divergent lineage produced by separate and repeated mutagenic treatments which could have induced different chromosomal rearrangements in each parent. Probably useful practical results could only be obtained if careful attention was paid to the type of mutagenic agent employed. Parents in parasexual crosses should have a history of mutation treatment in which mutagens inducing gross chromosomal damage were avoided[10].

About the time that *P. chrysogenum* began to be used to produce penicillin on an industrial scale, surveys of several fungi showed that strains of *A. nidulans* were among those yielding penicillin-like substances. Since then the formal genetics of this mould has been the subject of extensive analyses[11-13]. It has the advantage of possessing both a sexual and a parasexual process; a genetical analysis utilising the latter mechanism complements that involving true sexuality by facilitating the positioning of centromeres and allowing genetic markers to be assigned to specific linkage groups. *A. nidulans* therefore seemed a more suitable choice of organism than *P. chrysogenum* for studies on the formal genetics of penicillin production after it was shown that the

255

antibiotic activities of a number of strains of *A. nidulans* were due to the elaboration of a metabolite which by a number of criteria, including chromatographic and electrophoretic analyses, could not be distinguished from penicillin[14].

Preliminary work indicated that penicillin production was a character under nuclear rather than cytoplasmic control in *A. nidulans*[14], as was found in *P. chrysogenum*[3,15]. Crosses were then made between different wild-type isolates of *A. nidulans* and progeny isolated which yielded about twice the amount of penicillin given by their parents[16]. A number of mutants with increased penicillin titres relative to their parents have been isolated after UV-light treatment and these are in the process of being positioned on the chromosome map of *A. nidulans*. Our work indicates that strains with superior yields of penicillin to their parents can be isolated following genetical recombination processes and that a genetical analysis of penicillin production can be performed in *A. nidulans*.

THE GENETICS OF PENICILLIN PRODUCTION IN *P. chrysogenum*

EFFECT ON PENICILLIN YIELD OF LABELLING MUTATIONS

One of the earlier problems in studies on the genetics of penicillin production was that the labelling of strains with genetic markers often resulted in reductions in penicillin yield particularly when auxotrophic mutations were induced. The loss of penicillin titre was usually a pleiotropic effect of the auxotrophic mutation[17]. Nutritional mutations involving more fundamental blocks in metabolism, such as an adenine deficiency, lowered penicillin yields further than those mutations which were more specific in their effects, such as a vitamin deficiency; possibly because the ramifications of pleiotropy following the former sort of mutations would be greater than the latter. The production of penicillin in deep culture under specific environmental conditions is finely balanced and slight alterations in temperature, aeration and media constituents can have rather drastic effects on the final penicillin yield. It is perhaps not too surprising therefore that certain genetic as distinct from environmental changes can also unbalance such a system. Sometimes, the addition of extra amounts of specific nutrients above that required for growth by auxotrophs can go some way to redress the loss of titre[17,18] and possibly more involved environmental adjustments would reduce this loss further, if not completely. However, normally a standard environment is established for the estimation of penicillin titre and the selection of genetic markers with little or no effect on penicillin yield under the chosen conditions would seem best to avoid marker interference with the segregation patterns of genes affecting increases in penicillin production.

HETEROZYGOUS DIPLOIDS

After the formation of a heterokaryon between parent strains labelled, for example, with complementary nutritional deficiencies and different mutant spore colours, heterozygous diploids are normally isolated either as green sectors on the growing heterokaryotic colony or by plating spores from the heterokaryon on minimal medium and isolating green colonies. Since continued nuclear division as well as prolonged storage under certain conditions may result in loss of penicillin titre[19] it would seem desirable to isolate a heterozygous diploid as soon as possible after a heterokaryon has been synthesized between two parent strains which, themselves, should be freshly isolated cultures from single conidia and maintain their accepted penicillin yields. Sometimes when two auxotrophic parents are grown together on a complete medium to allow heterokaryon formation and the resulting mycelial felt is fragmented and dispersed on minimal medium, as well as heterokaryotic colonies growing, heterozygous diploid colonies also appear presumably because of nuclear fusion during heterokaryotic growth on complete medium[20]. The isolation of heterozygous diploids in this manner would avoid the further nuclear divisions necessary if heterokaryons were subcultured and diploids isolated from these either as sectors or following the plating of spores from the heterokaryon. Increasing the number of replicas at the stage of a parasexual cross when the component strains were grown on complete medium would enhance the possibility of isolating heterozygous diploids directly following the plating of fragmented mycelium.

Reports that diploids yield more penicillin than either of their parent strains[3,6,8] suggest the possibility of using diploids in industrial fermentations. The problem then would be to avoid the consequences of segregation during growth in deep culture and the concomitant titre loss which might result. If the parent strains of a heterozygous diploid were not closely related, and during mutation and selection programmes to increase penicillin yield had been subjected to a separate series of treatments with mutagenic agents causing different chromosomal rearrangements, it is likely that because of restrictions on recombination most segregants from the heterozygous diploid would be of one or other parental type. Then, before synthesizing a heterozygous diploid, each of the parent strains could be labelled with different recessive auxotrophic or morphological mutations which reduced growth rate in complex media but did not affect penicillin titre when present with normal alleles in the heterozygous state. From such a diploid most segregants, since they had parental genomes, would be selected against in fermentation media because of their poor growth[7]. If a heterozygous diploid did not exhibit parental genome segregation, consideration could be given to ensuring such a segregation pattern by isolating genetic markers causing restricted growth in the parent haploids after treatment with X-rays or other mutagens known to cause chromosomal rearrangements[7]. Another method of preserving the heterozygous diploid condition would be to introduce recessive lethals by mutagenic treatment of the diploid[9]. Recessive non-allelic lethals linked in the *trans* configuration would largely eliminate haploid segregants and the effectiveness of such a system of balanced lethals would be complete if parental genomes only segregated. In this context it is interesting to note that the isolation of stable diploids which has been reported[6,8] could be due to the occurrence of spontaneous recessive lethal mutations.

Heterozygous diploids synthesized independently between the same two parent strains can vary substantially in penicillin yield[21]. In breeding experiments designed to exploit the parasexual recombination system the problem then is to select those diploids which have retained, intact, the genomes of each of the original haploids. A series of heterozygous diploids were made between two relatively high-yielding parents with penicillin titres of 3,000 and 3,500 u/ml; the mutations in each parent concerned with increasing penicillin yield relative to that of their original wild-type ancestor were shown to be recessive[22]. Two sorts of diploids were found, those with penicillin titres around 3,000 u/ml and those with less than a quarter of this penicillin yield[21]. When the segregation patterns of the diploids were examined, high-yielding diploids gave segregants with the genetic markers of one parent and segregants with the genetic markers of the other parent, all with the original high penicillin yields of the parent strains. The low-yielding diploids were of three kinds. Firstly, those in which segregants carrying the genetic markers of one parent were high, like the original haploid parent, and where segregants with the markers of the other parent were low in penicillin yield, like the diploid. Secondly, the reverse of this, where segregants of the previously high type were low and the previously low type were high in penicillin titre. Thirdly, those where all segregants were low in penicillin yield. It was argued that these results agreed with a hypothesis which attributed the low yields of diploids to the occurrence of spontaneous dominant mutations in either or both parental genomes which reduced penicillin titre. The parent strains were tested for penicillin yield immediately prior to synthesizing the heterozygous diploids and confirmed as high producers. These dominant mutations therefore could have been present only in low numbers in the original haploid parental cultures or taken place in either parental genome during the process of heterozygous diploid synthesis. Of 68 heterozygous diploids tested about a third had high and the remainder low penicillin yields so that a high rate of selection for such mutants would have to be postulated. However, there is evidence that spontaneous mutations to low titre can occur in cultures of high penicillin productivity and that such mutants have advantageous survival rates[19].

Another possibility considered to account for the isolation of diploids with different penicillin yields was that segregation processes had occurred prior to testing the diploids, resulting in either homozygosity of recessive genes concerned with increasing penicillin yield or of their dominant alleles. If the former were taken to explain the isolation of high-yielding diploids then the frequency of segregation was remarkably high since a third of the diploids were of this sort. There could have been a high rate of apparent segregation immediately following heterozygous diploidisation due to an initial lack of synchrony in the division of chromosomes of each parental genome leading to homozygosity of one or more chromosomes by non-disjunction. On the other hand, if the diploids with low penicillin yield had to be explained by the segregation of dominant alleles then it would be difficult to account for the occurrence of segregants of high penicillin titre from such diploids.

It is concluded that spontaneous mutations reducing penicillin titre which are known to occur[3,19], and not nondisjunction processes, led to the titre variation in heterozygous diploids. Different haploid strains of *P. chrysogenum* vary in their stability with regard to penicillin titre[23] and the selection of stable haploid parents could avoid or reduce variability in synthesized diploids. It follows from the mutation hypothesis that it would be the high-yielding diploids which had retained the original genomes of each parent and that these should be investigated if the recombination system of the parasexual cycle is being examined for the segregation of genes concerned with increasing penicillin titre. In the studies which have been reviewed these diploids appeared to have higher penicillin titres than the common ancestor of both haploid parents[5,21] so either similar mutants had been selected during their derivation by independent mutation and selection programmes or there was noncomplementation between different recessive mutations in each parent responsible for increasing the penicillin titre.

SEGREGATION FROM HETEROZYGOUS DIPLOIDS

The lack of success in isolating recombinants from a heterozygous diploid with increased penicillin yields relative to those of the diploid's parents could be explained by the occurrence of parental genome segregation[20,24]. This might be avoided if haploid parents were employed which had not been subjected to mutagenic treatments resulting in chromosomal rearrangements. It would be unnecessary to return to wild-type strains of *P. chrysogenum* in industrial breeding programmes. A current strain used for the industrial production of penicillin could be used as starting material although its chromosomal morphology was substantially different from its original wild-type ancestor. Isolates with increased penicillin yield produced from such a strain after treatment with a mutagen not causing gross chromosomal damage, although having the same chromosomal rearrangements as their parent, would be homozygous for these. Presumably therefore there would be no interference with free recombination between different isolates due to variations in chromosomal morphology. Without a detailed knowledge of the formal genetics of *P. chrysogenum* it is envisaged that from parasexual crosses between strains, isolated as described above, random samples of segregants would be tested for favourable recombinations of mutations increasing penicillin yield. The disadvantage that parasexuality has over sexuality in requiring two independent segregation processes rather than one before recessive non-allelic genes linked in the *trans* configuration can be expressed might be obviated to some extent by a higher rate of parasexual segregation in asexual fungi[25].

Whether the use of the parasexual mechanism as just described would be of any more benefit than mutation and selection would depend on practical experience. It might be easier, for example, to combine two independent mutations by parasexual crossing rather than attempt further mutation induction in either singly mutated strain, if both mutations were relatively rare events.

THE GENETICS OF PENICILLIN PRODUCTION IN *A. nidulans*

ANTIBIOTIC YIELDS OF DIFFERENT WILD-TYPE STRAINS OF *A. nidulans*

Strains of *A. nidulans* were first reported to produce penicillin-like substances about 25 years ago[26,27]. When we began our studies we were fortunate in that Professor J. L. Jinks and his associates at Birmingham University in England were studying natural variability among wild-type isolates of *A. nidulans* and kindly made available to us 100 different strains isolated from various parts of the United Kingdom. We also obtained a number of strains from Glasgow University for which we are indebted to Mr. E. Forbes.

As a first step we assayed the Birmingham strains biologically against a strain of *Bacillus subtilis* sensitive to penicillin with the results shown in Fig. 1 (Ref.[14]). The Glasgow wild-type strain NRRL 194 when assayed under similar conditions gave a titre equivalent to 6 u penicillin/ml. We then selected two Birmingham strains yielding titres equivalent to 20 u/ml, and two, 12 u/ml. These, together with the Glasgow

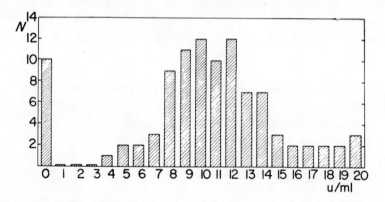

Fig. 1. Antibiotic yields of 100 wild-type strains of *A. nidulans* after biological assay. N number of strains, u/ml titre equivalent to u/ml of penicillin.

wild-type strain and two derivatives of this we intended to use in heterokaryon tests were examined to discover if their antibacterial acitivity was due to the production of penicillin. In no case was it possible to distinguish the antibiotic produced by any of these strains from penicillin by a number of criteria including sensitivity to penicillinase and chromatographic and electrophoretic analyses.

THE GENETICAL ANALYSIS OF PENICILLIN PRODUCTION IN *A. nidulans*

As was found in *P. chrysogenum*[15,28] the inheritance of penicillin production was shown by heterokaryon tests[29] to be under nuclear rather than cytoplasmic control in *A. nidulans*[14].

The two Birmingham strains selected with penicillin titres of 20 u/ml were genetically labelled with auxotrophic and spore-colour markers not affecting antibiotic yield and a suitably labelled strain was also isolated after crossing two Glasgow deri-

TABLE I

Genetic Markers on Parent Haploid Strains and Mean "Allele" Penicillin Yields of 216 Haploid Segregants Isolated after *p*-Fluorophenylalanine Treatment of Synthesized Heterozygous Diploid[a]

Parent strain	Chromosomes							
	I	II	III	IV	V	VI	VII	VIII
Tester strain	sup-1 ade-20 ylo ade-20	ACR-1	gal-1	pdx-4	facA303	sfi-3	nic-8	rib-2
Strain with pen-01	bio-1	pen-01						

Chromosome	Genetic marker segregating	Segregants with mutant allele			Segregants with wild-type allele			Ratio mean mutant titre/mean wild-type titre
		number	mean titre u/ml	standard error of mean u/ml	number	mean titre u/ml	standard error of mean u/ml	
I	ylo	120	10.22	0.52	96	11.10	0.60	0.92
II	ACR-1	112	10.66	0.55	104	10.55	0.57	1.01
III	gal-1	119	10.31	0.53	97	10.97	0.59	0.94
IV	pdx-4	107	10.49	0.56	109	10.72	0.55	0.98
V	facA303	113	9.97	0.55	103	11.30	0.57	0.88
VI	sfi-3	86	10.13	0.62	130	10.94	0.51	0.93
VII	nic-8	108	10.85	0.55	108	10.36	0.60	1.05
VIII	rib-2	103	5.58	0.20	113	15.20	0.37	0.37
I	bio-1	96	11.10	0.60	120	10.22	0.52	1.09

[a] *sup-1 ade-20*, suppresses adenine requirement at *ade-20* locus; *ylo*, yellow spores; *ade-20*, adenine requirement; *ACR-1*, resistance to acriflavine; *gal-1*, inability to utilize galactose; *pdx-4*, pyridoxin requirement; *facA303*, resistance to fluoroacetate; *sfi-3*, sulfite requirement; *nic-8*, nicotinic acid requirement; *rib-2*, riboflavin requirement; *bio-1*, biotin requirement; *pen-01*, increased penicillin yield.

vatives; this latter strain carried a mutation (*pen-01*) produced after UV-light treatment which increased penicillin titre from 6 u/ml to 20 u/ml. The two Birmingham derivatives and the Glasgow strain with *pen-01* were selfed and crossed in the three possible combinations, and 200 progeny isolated from each self and cross were tested for penicillin yield. The titre variation among progeny isolated after crossing was in every case greater than that in progeny from any of the selfs suggesting that the three strains had different genes responsible for increasing penicillin yield which recombined favourably or unfavourably in the progeny from crosses, raising or lowering penicillin titre relative to that of the parents. Three high-yielding recombinants giving titres of 31, 38 and 39 u/ml were examined in detail and it was found impossible to distinguish the antibiotic they produced from penicillin after a number of analytical tests[16].

These studies demonstrated that breeding for increased penicillin titre through processes of genetical recombination was feasible and suggested that a genetical analysis of penicillin production could be undertaken. With this in view we searched for further mutants giving higher yields of penicillin than their parents. The mutant already mentioned, *pen-01*, was from among a series of single conidial isolates tested after UV-light treatment of the biotin-less Glasgow strain, *bio-1* (Ref.[13]). From a white sporing derivative of the latter strain also yielding 6 u/ml of penicillin, two further mutants were independently isolated after UV-light treatment, *pen-02* and *pen-03* giving titres of 12 u/ml and 20 u/ml, respectively. It is hoped to position all three mutants on the chromosome map of *A. nidulans* and work on this has begun.

A heterozygous diploid synthesized between *bio-1*, *pen-01* and a strain labelled on all eight chromosomes[30] had a penicillin titre of 5 u/ml similar to the latter strain indicating that *pen-01* was a recessive mutation (Table I). A technique for assigning a gene of unknown location to a specific linkage group in *A. nidulans* has been described where a heterozygous diploid is first synthesized between a tester strain genetically labelled on all 8 chromosomes and a strain carrying the unlocated marker[30], and then haploidisation of the diploid is induced by *p*-fluorophenylalanine treatment[31]. Haploid segregants show free recombination between the unlocated marker and all the markers of the tester strain except that marker on the chromosome carrying the allele of the unlocated gene. From the diploid made between the strain *bio-1*, *pen-01* and the tester strain, 216 haploid segregants were isolated after *p*-fluorophenylalanine treatment and checked for the presence of genetic markers introduced in the parents (Table I). Assuming free recombination of whole chromosomes after haploidisation then, for a gene of the tester strain not on the homologue of the chromosome carrying *pen-01*, the ratio of the mean penicillin titre of haploids with the mutant allele to the mean titre of haploids with the wild-type allele should be approximately unity providing the mutation did not reduce penicillin yield. On the other hand, if a mutant gene of the tester strain was on the homologue of the chromosome carrying *pen-01*, this ratio should approximate to the titre ratio of the parent haploids, *i.e.* 5 : 20 or 0.25. Only the titre ratio of *rib-2* : *rib-2*$^+$, which was 0.37 differed significantly from unity showing that *pen-01* was located on chromosome VIII[32] (Table I). A series of crosses were then set up between strains with *pen-01* and strains with mutants at different loci on chromosome VIII in the hope of positioning *pen-01* on this chromosome following perithecium analysis[33] (Fig. 2). The results of two crosses showed free recombination between *pen-01* and *orn-8* and between the former and *rib-2*. In a third cross some evidence was obtained of linkage between *pen-01* and *cha*; unfortunately, in this cross one of the strains carried the marker *phoB7* which reduced penicillin yield. However, *pen-01* was

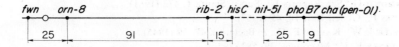

Fig. 2. An outline map of chromosome VIII. The centromere is shown as an open circle. *fwn*, fawn conidia; *orn-8*, requirement for ornithine; *rib-2*, requirement for riboflavin; *hisC* requirement for histidine; *nit-51*, requirement for ammonium; *phoB7*, reduced alkaline phosphomonoesterase activity; *cha*, chartreuse conidia; *pen-01*, increased penicillin yield.

tentatively assigned to a position distal of *cha* on chromosome VIII[34]. Further crosses are in progress to position *pen-01* more accurately on this chromosome.

Apart from locating *pen-02* and *pen-03* on the chromosome map of *A. nidulans* it will also be of interest to attempt the synthesis of recombinant progeny carrying two or more mutants which increase penicillin titre, to discover if the effects on penicillin yield are additive or otherwise.

REFERENCES

1. PONTECORVO G., ROPER J. A., HEMMONS L. M., MACDONALD K. D., BUFTON A. W. J.: Adv. Genet. **5**, 141 (1953).
2. PONTECORVO G., SERMONTI G.: Nature **172**, 126 (1953).
3. SERMONTI G.: Ann. N. Y. Acad. Sci. **81**, 950 (1959).
4. ALIKHANIAN S. I., KAMENEVA I.: Sci. Reports Ist. Super. Sanità **1**, 454 (1961).
5. MACDONALD K. D., HUTCHINSON J. M., GILLETT W. A.: Genetica **36**, 378 (1965).
6. ELANDER R. P., p. 403 in Erwin-Baur-Gedachtnisvorlesungen, Vol. IV. Akademie Verlag, Berlin 1967.
7. MACDONALD K. D.: Nature **204**, 404 (1964).
8. ELANDER R. P., p. 89 in D. Perlman (Ed.): *Fermentation Advances*. Academic Press, New York 1969.
9. AZEVEDO J. L., ROPER J. A.: J. Gen. Microbiol. **49**, 149 (1967).
10. MACDONALD K. D.: Mutation Res. **5**, 302 (1968).
11. PONTECORVO G., KAFER E.: Adv. Genet. **9**, 71 (1958).
12. KAFER E.: Adv. Genet. **9**, 105 (1958).
13. DORN G. L.: Genetics **56**, 619 (1967).
14. HOLT G., MACDONALD K. D.: Antonie van Leeuwenhoek **34**, 409 (1968).
15. MACDONALD K. D., HUTCHINSON J. M., GILLETT W. A.: J. Gen. Microbiol. **33**, 375 (1963).
16. HOLT G., MACDONALD K. D.: Nature **219**, 636 (1968).
17. MACDONALD K. D., HUTCHINSON J. M., GILLETT W. A.: J. Gen. Microbiol. **33**, 365 (1963).

18. ALIKHANIAN S. I., MINDLIN S. Z., GOLDAT S. U., VLADIMIZOV A. V.: Ann. N. Y. Acad. Sci. **81**, 914 (1959).
19. MACDONALD K. D.: Nature **218**, 371 (1968).
20. MACDONALD K. D., HUTCHINSON J. M., GILLETT W. A.: Antonie van Leeuwenhoek **30**, 209 (1964).
21. MACDONALD K. D.: Antonie van Leeuwenhoek **32**, 431 (1966).
22. MACDONALD K. D., HUTCHINSON J. M., GILLETT W. A.: J. Gen. Microbiol. **33**, 385 (1963).
23. CALAM C. T.: Prog. Ind. Microbiol. **5**, 3 (1964).
24. SERMONTI G.: Sci. Reports Ist. Super. Sanità **1**, 472 (1961).
25. LHOAS P.: Genet. Res. **10**, 45 (1967).
26. FOSTER J. W., KAROW E. O.: J. Bacteriol. **49**, 19 (1945).
27. DULANEY E. L.: Mycologia **39**, 582 (1947).
28. SERMONTI G., MORPURGO G.: Ricerca sci. **29**, 171 (1959).
29. JINKS J. L.: Compt. rend. Trav. Lab. Carlsberg Ser. Physiol. **26**, 183 (1956).
30. McCULLY K. S., FORBES E.: Genet. Res. **6**, 352 (1965).
31. MORPURGO G., *quoted in* LHOAS P.: Nature **190**, 744 (1961).
32. HOLT G.: *M. Sc. Thesis*. London University 1968.
33. HEMMONS L. M., PONTECORVO G., BUFTON A. W. J.: Adv. Genet. **5**, 194 (1953).
34. DITCHBURN P., MACDONALD K. D., *unpublished results*.

Hybridisation Experiments with *Penicillium patulum* and *Fusarium moniliforme*

C.T. Calam*, L.B. Daglish, W.S. Gaitskell

I.C.I. Ltd., Pharmaceuticals Division, Alderley Park, Macclesfield, Cheshire, Great Britain

Abbreviations used:

brw	brown
dpu	dark purple
grn	green
mog	Mars orange (013/1)
org	orange
pgn	pale green
pvb	pale violet blue (Lobelia blue 41/2—3)
rpu	reddish purple (Plum purple 934)
vpu	violet purple
vrd	violet red (Erythrite red 0027)
whi	white
ylo	yellow

The names in parentheses are from Ref.[13].

INTRODUCTION

The discovery of parasexual hybridisation in fungi aroused considerable interest when it was first announced. The original description[1] stressed the potential value of the method for industrial strain improvement. Since then numerous papers have been published on the subject. The field has been reviewed by one of the present authors[2,3] and by other writers. On the whole the approach to the parasexual process has rightly been from the standpoint of genetics, and most of the work reported has been aimed at obtaining evidence as to the genetic processes involved. For this purpose it is obviously necessary to carry out well defined experiments, using cultures which are carefully selected for their suitability for the process.

On the whole, attempts to use the method for strain improvement have revealed the existence of a number of complications, perhaps more than the original authors expected. The papers on *Penicillium chrysogenum*[4-6] are of interest in showing the

---

* Present address: Biological Department, Liverpool Polytechnic, Liverpool, L3 3AF, Great Britain.

wide range of work that has been involved. At the same time, useful advances in productivity have been claimed, and it can reasonably be argued that the parasexual
process remains of great interest to the industrial worker.

The present paper is intended to record some experimental work done in an industrial laboratory, where the object was to assess the utility of the method for strain
improvement, rather than to study genetics. Here the choice of strains is limited to
those giving maximal yields, often highly selected mutants, and this is likely to make
hybridisation difficult in a number of ways, ways which it is difficult to forecast. The
experiments about to be described raise a number of points of interest, and it is hoped
that these will prove of value to others working in the field.

Penicillium patulum

Our experiments with *Penicillium patulum* were begun in the hope of obtaining
strains giving increased yields of griseofulvin. They followed a programme of work
with *P. chrysogenum* in which biochemical deficients were used from widely differing
lines as well as from the same strain. All but four of the many crosses attempted
yielded diploids with large spores, from which parental types were readily obtained on
haploidisation. Crossing-over was also observed. Usually a mutagen such as nitrogen
mustard was used to facilitate recombination. *p*-Fluorophenylalanine (PFA) was used
on one occasion, with some success.

In starting work with *P. patulum* the same methods were used, but the results
obtained were somewhat different. It is proposed to describe three typical experiments
which were made during successive stages of the work, which illustrate the problems
encountered.

EXPERIMENTAL METHODS

The procedures used were those already described by MACDONALD *et al.*[7], and the same
minimal medium (MM) and complete medium (CM) were employed. The strains used and the
crosses to be described are given in Table I.

As will be seen later, from the description of the experiments, it proved difficult or impossible
to obtain satisfactory production of diploids from the deficient strains. Methods for the enhancement of diploid induction included setting up the crosses on both liquid limiting medium and
solid complete medium, the incubation of the plates at normal and increased temperatures and
subjecting hyphal tips of likely heterokaryons to camphor vapour.

Camphor treatment was tried in both the second and third experiments, and in both cases
was followed by the isolation of a few green prototrophic strains. This involved the removal of the
extreme edges of putative heterokaryotic colonies; these were transferred to fresh MM and, in the
first experiment, exposed to camphor for 5 to 24 h. It required nearly a month's incubation
before any green prototrophs arose, and the longest camphor treatment resulted in five crosses
yielding prototrophs compared with three crosses from a treatment of only seven hours. Only

Table I

Parent Strains of *P. patulum*, Biochemical Deficients and Crosses Attempted

Parent strains	Description
1) 766 2) G	High-yielding mutants of *P. patulum* originating from a common ancestor but obtained *via* two widely separated lines of mutation, involving numerous mutation steps. The letters and numbers added, *e.g.* G1, GH3 indicate particular strains.

Experiment N°	Cross N°	Cross	Parental lines
1	—	766 grn thi × G1-20 brw met	different
2	11 38	766 ylo-4 amm-2 × GH3 whi-3 met-9 766 ylo-1 bio-1 × GH3 whi-4 amm-8	different same
3	2 9	766 ylo-1 bio-1 gly-1 × 766 whi-2 thi-1 arg-2 G4-110 pgn acr-1 cys-2 amm-12 × GH3 whi-3 met-10 pab-1	same same

three of the crosses yielding prototrophs after camphor treatment had failed to do so without treatment. In the other experiment camphor treatment was for five, twenty-four and forty-eight hours. After about seven weeks, three crosses yielded green prototrophic strains, none of which had done so without camphor treatment.

To improve the possibilities of recombinant formation some of the cultures were stored for considerable periods at 2 °C, and UV-light was also used in some experiments. PFA was also used.

EXPERIMENTAL RESULTS

Experiment 1. A cross was made between singly marked brown and green strains from different lines, one brown, requiring methionine and the other green, requiring thiamine. The general scheme of the experiment is given in Table II, including subsequent experiments. Six days after plating the macerated mixed culture, some growth appeared on the minimal-medium plates, and after a further five days sporulation appeared on raised areas of the crinkly growth; there was also a considerable amount of background growth. When spores were plated on minimal medium many brown, one or two green and one yellow prototroph were obtained. Only the yellow strains had spores larger than the parent, the ratio of the diameters being 1.2 : 1.

As the yellow strain appeared to be a possible diploid it was examined in some detail over a considerable period of time. This work is summarised in the lower half of Table II.

PFA treatment was applied to the yellow strain shortly after isolation in various experiments using agar cultures or liquid shaken cultures with PFA levels from 0.01 to 0.9 g/l. On plating, sectors were not obtained though a few brown colonies appeared with small spores (2.9 — 3.3 μm).

TABLE II

First Experiment

766 *grn thi* × G1-20 *brw met*

brown prototrophs | 2 green prototrophs

yellow prototroph[a]

Storage months	Treatments	Colonies obtained
4	1. CM + PFA	6 browns
	2. CM alone	3 green
10	3. CM alone	2 brown
14	4. UV mutation	27% browns
	controls	7% browns
18	5. reisolated colony, PFA	8 browns[b]
26	6. *ditto* CM plates	2 green[c]
	CM + PFA	3 green[c]

[a] $3.5 \,\mu m$.
[b] 6 prototrophs, 2 recombinants *brw thi*.
[c] prototrophs.

Platings on CM gave very small numbers of brown or green colonies. After a total of 14 months storage of the original culture at $2\,°C$ under oil, fresh slopes were inoculated by mass transfer of spores. These showed 7% of brown prototrophs, or with UV-mutation 27%. Later, on reisolation *via* a single colony, the yellow strain showed a spore size of $3.6 \,\mu m$ and behaved like the original culture. On treatment with PFA eight brown colonies were isolated, six prototrophs and two deficients (*brw thi*) which appeared to be recombinants.

The status of the numerous brown prototrophs which appeared as a result of the original cross is not clear. The most obvious explanation is that they were revertants but they had a rather higher productivity than expected and our view is that some change had occurred during the hybridisation—reversion process, though precisely what was not clear. When the same two parents were again crossed five weeks later only one brown prototroph was obtained.

Experiment 2. Fifteen crosses were set up between single deficients from the same parental line: two of these will be described (Table III). Spores from heterokaryons in Cross 11, plated after 20 days, gave green prototrophs with small spores. In Cross 38 putative heterokaryotic colonies were produced which showed areas of green spores after thirteen days' growth. The numerous green prototrophs obtained from them had slightly larger spores than the parents ($3.8 \,\mu m$ instead of $3.3 \,\mu m$). (Six weeks later most of the other crosses had produced green prototrophic strains but most of them had small spores and were not vigorously prototrophic.) In this experiment crosses from solid complete medium gave only feeble growth on MM plates. In three

TABLE III

Second Experiment[a]

Cross N°	Cross
11	766 *ylo-4 amm-2* × GH3 *whi-3 met-9* ↓ green prototrophs[b]
38	766 *ylo-1 bio-1* × GH3 *whi-4 amm-8* ↓ green prototrophs[c]

Treatments	Colonies obtained
1. plated 2,000 colonies	4 parental *whi amm* 1 parental *ylo bio* 1 recombinant *whi bio* 1 recombinant *ylo*[d]
2. mutated EI + UV	1 parental *whi amm*
3. PFA	no auxotrophs

[a] Same parental lines.
[b] Small spores, no variation on plating.
[c] Large spores (3.8 μm).
[d] Prototroph.

cases green strains arose from crosses on liquid medium. Growth at higher temperature (29 °C instead of 24 °C) was slower but several of the crosses gave green prototrophic strains. The only large-spored prototrophs, however, were isolated from mixed growths and plates grown at 24 °C.

On plating no variants were obtained from Cross 11. With Cross 38 plating or mutation treatment gave a few auxotrophs of which one or two appeared to be recombinants.

Experiment 3. In view of the unexpected results obtained in the previous crosses, considerable attention was given to the third experiment so as to obtain clear-cut results. Ten crosses were set up using doubly deficient coloured mutants, one being acriflavine-resistant. They were carefully checked for reversion before use. They were from the same or from different parental lines.

Mixed growths from complete medium were macerated and plated on minimal medium, where a few rather feeble colonies grew. Only two crosses were successful. After five weeks a green prototrophic strain was isolated from a putative heterokaryotic colony of Cross 9, and after a further two weeks one green prototroph had been isolated from colonies on Cross 2. The results of the two crosses are given in Table IV. Cross 2 gave one green and one yellow prototroph with medium-sized spores. These showed virtually no variation on plating. The green prototroph from Cross 9, which showed the expected intermediate acriflavine resistance and had spores of increased size, was also stable and only one parental strain was recovered from it. Mutagens

TABLE IV

Third Experiment[a]

Cross 2[b]	766 *ylo-1 bio-1 gly-1* × 766 *whi-2 thi-1 arg-2*
	1 green prototroph[c] 1 yellow prototroph[c]

Treatments	Colonies obtained
1. green prototroph plated on CM 2. yellow prototroph plated on CM	2 white deficients yellow prototrophs only

Cross 9	G4-110 *pgn acr-1 cys-2 amm-12* × GH3 *whi-3 met-10 pab-1* ↓ 1 green[d]

Treatments	Colonies obtained
1. plated with PFA 2. plated on CM 3. on acriflavine	no coloureds 1 parental *pgn cys amm* *acr* resistance intermediate between parents

[a] Successful crosses.
[b] Same parental lines.
[c] 3.4 μm.
[d] 3.9 μm.

were not employed in this experiment since their use introduces doubts as to the origin of any recombinant strains which may be produced.

DISCUSSION OF RESULTS

The parasexual process involves the production of heterokaryotic colonies from which prototrophic diploid strains arise. Formation of heterokaryons should be relatively easy and diploidisation not infrequent. As will be seen, even after the application of various processes designed to facilitate diploidisation, the formation of diploids in *P. patulum* was difficult, in fact the five crosses described in the three experiments gave only three large-spored prototrophs; only in Cross 38 was a considerable number of these obtained. It must be said also that whereas with other moulds the formation of heterokaryons showed up clearly with the growth of characteristic colonies, and diploidisation was also obvious when it occurred and the diploids easily recognisable, this was not the case with *P. patulum*. Formation of heterokaryons was slow. Although when diploids were apparently obtained they were usually green and grew well on minimal medium, at best the spore sizes were not much larger than those of the parent strains, the ratio of increased size being about 1.2 : 1 instead of the expected 1.4 : 1 which was found with diploids of *P. chrysogenum* and which corresponds to a doubling in spore volume. Considerable doubt must therefore be felt as to whether true diploids were obtained.

The other criterion for diploid formation is the ready production of recombinants from the diploids. In the case of *P. patulum* this was also uncommon. From the three possible diploids investigated, which showed some increase in spore size, only three or four recombinants were obtained, in addition to which a small number of parental types were recovered. This was in spite of the application of various treatments including the use of mutagens and PFA to induce recombination.

It is difficult to account for the poor results obtained in this work with *P. patulum*.

We would very much like to acknowledge the help of Professor J. A. Roper and Mr. C. A. Boam of Sheffield University who have carefully been through the experimental work with us to try and avoid our overlooking any points of importance.

Fusarium moniliforme

F. moniliforme (or *Gibberella fujikuroi*) does not appear to have attracted much attention as regards the parasexual method of hybridisation. SPECTOR and PHINNEY[8] have described sexual crosses and investigated the genetic control of gibberellin formation. Soviet workers (*e.g.*[9]) have studied the effects of mutation. Parasexual crosses have been fairly extensively studied in the plant pathogen *F. oxysporum, e.g.*[10,11].

In some ways *F. moniliforme* is not an ideal fungus for hybridisation work. The colours of the strains are hard to describe and often shade into each other; they are also affected by age and conditions of growth. The spores are often bicellular. The experiments to be described were exploratory, aimed at discovering whether crosses were possible and whether potentially useful results could be obtained.

EXPERIMENTAL METHODS

The two media used have the following recipes:

Complete medium (in g/l):

Cerelose	60
Malic acid	5
Ammonium tartrate	2.2
KH_2PO_4	1

	MgSO$_4$.7H$_2$O	0.5
	Peptone	1
	Yeast extract	1
	Minor elements solution	10 ml
	Water	1 l
	pH 5.3—5.5	

Minimal medium (in g/l):	Glucose	60
	NH$_4$NO$_3$	2.5
	MgSO$_4$.7H$_2$O	1
	KH$_2$PO$_4$	5
	Minor elements solution	2 ml
	Water	1 l
	pH 5.3—5.5	

Minor elements solution (in g/l):	FeSO$_4$	1
	ZnSO$_4$	1
	CuSO$_4$	0.15
	MnSO$_4$	0.1
	(NH$_4$)$_2$MoO$_4$	0.1
	Water	1 l

Tests for gibberellic acid production were carried out in shaken culture and the yields are given as trace, low, medium, intermediate, high and very high.

Hybridisation was carried out basically in the same way as with *P. patulum*. In the case of the nonsporing strains crosses were carried out using mycelial suspensions obtained by rubbing from agar slopes. A mixture of the two cultures was placed on the surface of a 10 ml CM agar block in a test tube. The mixed culture was allowed to grow for a week in the incubator at 24 °C. The resulting mycelial mass was removed from the tube, placed in a McCartney bottle with beads plus 10 ml of water; the mixture was then shaken to give a mycelial suspension. About 0.1 ml of this suspension was then spread on minimal-medium agar plates, along with this a ten-fold dilution was also plated. The plates were incubated for 14 days at 24 °C. The plates were then examined for growth and pieces of heterokaryotic colonies were transferred to minimal agar plates, with two or three pieces per plate. The cultures which arose from these plates were called Stage I cultures. The plates were again incubated and the type of growth arising was observed. Under these conditions slow but fairly extensive growth took place; after two to three weeks incubation portions from the edges of the colonies were removed and again transferred to minimal medium plates with one colony per plate. The colonies arising from these plates were referred to as Stage II colonies. By this time, consistent growth of the heterokaryons was beginning to take place and occasionally signs of diploidisation were beginning to occur. The subculturing from Stage II colonies was repeated at intervals giving a series of types of growth which will be described in the individual experiments which are reported later. At each stage the colonies were tested for prototrophy by removing a sample with wire and plating on MM. In initial experiments tests were carried out to identify the two parental strains in the heterokaryons. When after two, three or four stages of subculture had taken place and the colonies showed signs of having become prototrophic, some of the culture was removed with a wire from the colonies, streaked on minimal agar plates and single colonies were isolated. In this process the colony was touched lightly with a loop so as to remove spores rather than mycelium. From the monospore colonies which arose, spores were again removed and passed through the streak procedure. In this way pure cultures were obtained from each of the prototrophs arising from the cross. Individual isolates were maintained on minimal agar slopes.

TABLE V

Parent Strains of *F. moniliforme*

Parental code	Strain		Colour	Microspores	Length µm	Gibberellic acid production
	Type	N°				
A	mutant	419[a]	white	—[b]	—	high
B	wild	1001	white	one-celled	8.1	medium
C	wild	NRRL 1252	violet-blue	one-celled	11.5	medium
D	wild	1952 (Mexican)	pale grey-blue	one-celled	10.6	very low

[a] Derived from 917.
[b] Asporogenous strain.

During the development of the heterokaryon throughout the stages sectors appeared in a number of cases; these were treated as separate strains and suitably investigated. In some cases they proved to be prototrophic and to represent new forms, in other cases they were of one of the parental types and in one case a recombinant type was obtained. In the course of our experiments a total of 45 crosses were attempted and hybridisation of one kind or another took place in a total of 30 instances.

It is proposed to describe here 6 typical cases. The following Tables, N^{os} V, VI and VII, describe the parental strains used, the derivation of the biochemically deficient parents and the crosses which were made. Characteristically, the heterokaryon plates showed a very sparse background growth from which arose piled-up, wrinkled and corrugated colonies; sometimes saucer-shaped heterokaryotic colonies developed. When the heterokaryotic colonies were transferred to fresh agar plates there was considerable formation of aerial growth around the original colony and from this outgrowths of embedded dendroid habit developed. From this there often developed a thin velvety surface growth which was formed very slowly. In some cases dendroid growths appeared around the heterokaryotic colonies even on the original heterokaryon plates. Frequently the dendroid outgrowths on the subculture plates were of a different colour from the

TABLE VI

Derivation of Biochemically Deficient Parents

Original parent	Intermediate parent			Biochemical deficient		
	code	mutagen	gibberellic acid production	code	mutagen	gibberellic acid production
A	A1 *vrd*	UV	—	A11 *vrd trp*	UV	medium
A	A2 *org*	multiple	very high	A22 *org ade/nic*	MNNG	inter
B	B1 *mog*	UV	low	B11 *mog met*	UV	very low
B	B2 *dpu*	UV	very low	B21 *dpu his*	UV	trace
C	C1 *pvb nic*	UV	medium	C11 *whi nic lys*	UV	trace
D	D1 *vpu dpu*	UV	medium	D11 *rpu his*	UV	trace

TABLE VII
Crosses Described

Cross N°	Strains crossed
1	A11 *vrd trp* × B11 *mog met*
2	A11 *vrd trp* × B21 *dpu his*
3	A11 *vrd trp* × D11 *vpu his*
5	A11 *vrd trp* × C11 *whi nic lys*
9	B11 *mog met* × C1 *pvb nic*
16	A22 *org ade/nic* × B11 *dpu his*

main colony, and sometimes spread extensively across the plate even when the central aerial colony remained relatively small in size. In the later stages, if part of the dendroid growth was transferred to another plate, the normal type of flat aerial growth developed. The various changes in growth habit which were observed between the original heterokaryotic colony and the final flatish aerial growth resembled the stages in diploid formation which occurs in *P. chrysogenum*.

Attempts were made to obtain recombinants from the prototrophs. To do this, filtered culture suspensions were exposed for a short time to UV-light resulting in a kill of from 50 to 99.0%. The treated suspensions were plated and isolates picked off. Replica-plating was used to obtain auxotrophs.

EXPERIMENTAL RESULTS

It will be noted that in most of the crosses several prototrophs or diploids were produced, which usually differed among themselves, and that most of these were mutated so as to produce recombinants. Table VIII summarises the spore sizes of the parental and deficient strains used, and also of the diploids and prototrophs which were mutated. It will be noted that in a number of instances there were considerable increases in size of spores: spores which were approximately double the size of the parents were regarded as "large". Table IX records the results of the mutations, giving types and numbers of recombinants and the amounts of gibberellic acid produced by them.

Description of Individual Crosses

Cross 1. Orange and red-violet cultures requiring methionine and tryptophan respectively were crossed giving colonies which were orange in colour; these were flat and spreading with a little dendroid growth at the edges. Prototrophs were obtained from the central sporulating area. Six were mutated. Out of 59 recombinants retained

TABLE VIII

Lengths of Microconidia from Parent Strains and Diploids

	Source	Diploid strains examined	Conidia					
			monocellular			bicellular		
			length μm	number measured	range μm	length μm	number measured	range μm
	Ref.[12]	—	8.4	—	5.0—12	17	—	12—22
Parents	419		≈4[a]	—	—	—	—	—
	1001	—	8.1	14	6.6—10			
	1252		11.5	11	5.5—16.5			
	1952		10.6	12	6.6—15.4			
Deficients	A1		0	—	—	—		—
	A2		0	—	—	—		—
	B11		10.5	13	7.0—19	0		0
	B21	—	9.9	12	7.0—19	0	—	0
	C1		8.9	12	5.5—14	0		0
	C11		10.9	11	9.0—14	0		0
	C12		11.8	13	5.5—20	0		0
	D11		11.6	10	5.5—18	0		0
Cross N°	1	1	12.1	6	7—20	25.3	6	19—31
		2	14.9	6	10—19	24.2	6	18—23
		3	10.4	7	8—13	0	—	0
		4	12.4	9	6—15	16.5	2	15—18
		5	11.0	10	7—17	21.7	4	20—26
	2	1	17.1	20	8—24	23.9	4	22—25
		2	17.4	20	9—27	—	—	—
		3	18.0	19	7—25	22.0	6	12—29
		4	16.4	19	9—30	—	—	—
		5	15.6	20	9—27	—	—	—
	3	1	4.2	11	2—7	0		
		2	3.7	11	2—6	0		
		3	3.6	10	2—7	0	—	—
		4	4.2	10	3—6	0		
		5	4.1	10				
	5	1	3.7	10	2—6	0		
		2	3.7	6	2—7	0	—	—
		3	4.5	10	2—10	0		
	9	1	14.1	6	10—17	—	—	—
	16	6[b]	≈2	—	—	—	—	—

[a] Based on a few spores obtained by growth on *Fragmites* stems.
[b] Prototrophs.

TABLE IX

Summary of Crosses and Mutations

Diploids mutated

Cross N°	Strains used	number	colours[a]	spore size	gibberellic acid production
1	AII vrd trp × BII mog met	2	Marigold 113	large	medium
		4	Carrot Red 612	large	medium
2	AII vrd trp × B2I dpu his	1	Crimson	large	low
		7	Purple	large	low
		1	violet red	large	low
		1	dark purple	large	low
		1	violet	large	low

Isolates from diploids

Cross N°	parental type	colour	parental type	requirement	number	gibberellic acid production majority	gibberellic acid production best
1	A	violet red	—	nil	0	—	—
	A	violet red	A	Trp	0	—	—
	A	violet/violet red	B	Met	6	low	—
	B	Mars orange	B	Met	8	low	low
	B	Mars orange	—	nil	3	—	—
	B	Mars orange	A	Trp	0	—	—
	AB	violet orange	B	Met	1	low	low
	B	bright orange	B	Met	1	low	low
	AB	reddish orange	—	nil	31	medium	high[b]
	B or AB	orange—reddish orange		various	3	medium	medium
	—	other colours	—	various or nil	6	low	low
2	A	violet red	—	nil	6	low	low
	A	violet red	A	Trp	0	—	—
	B	dark purple	—	nil	32	low	low
	B	dark purple	B	His	4	low	low
	B	dark purple	B?	His or choline	9	low	low
	A?	violet red	B	His	3	low	low
	AB	reddish purple	A	Trp	0	—	—
	AB	reddish purple	B	His	0	—	—
	AB	reddish purple	B	His or choline	5	low	medium
	A	violet red	—	various	7	low	low
	AB	reddish purple	—	various	6	low	low

Cross	Cross (genotypes)	Class	Colour	Size	Growth		Colour[a]	Colour		Deficient isolates	No.		
3	DII rpu his × AII vrd trp	3	violet red	small	inter	A	violet red	—	— [c]		14	inter	high
5	CII whi nic × AII vrd trp	2	violet red	small	inter	A	violet red	—	— [c]		8	inter	inter
9	CI pvb nic × BII mog met	1	white-pale mauve	large	medium	C	white with a mauve tinge	C	Nic or Ade		2	low	low
						C	white with a mauve tinge	—	nil		2	inter	inter
						C?	pale violet to purple	—			8	medium	inter
16	B dpu his × A22 org ade/nic	2	orange	very small	very high	A	orange	—	nil		52	very high	very high
						—	various colours	—	nil		3	intermediate	high
		3	orange	very small	very high	A	orange	—			108	very high	very high

[a] Names of colours, with initial capital letters, are from Ref.[13].
[b] 1 only.
[c] No deficient isolates.

from the mutation, the majority were orange. No tryptophan-requiring recombinants were obtained even when the culture medium was specially supplemented with trypto-phan. Of the orange isolates ten required methionine and resembled the orange parent B. Six other methionine requirers resembled the red-violet parent A; three of these were examined and found to sporulate. A number of apparent mutants were obtained having either complex growth requirements or markedly changed colours (purple, white). The best gibberellic acid production (medium yields) came from iso-lates which had no growth requirements and in colour resembled the orange shades of the diploids, intermediate between the parents. One-celled spores from nine defi-cient mutants were 7—11 µm long, with one at 14.4 µm.

Cross 2. The red-violet mutant used in Cross 1 was crossed with a purple defi-cient: requirements were respectively tryptophan and histidine. A purple, submerged, dendroid growth spread out from the original heterokaryotic colonies and became covered with a flat, velvety, light pink or mauve surface growth which later became wrinkled. This was again subcultured on an agar plate and prototrophs recovered. Seventy-two isolates from mutations of diploids were tested. No tryptophan-requiring strains were recovered. Both red-violet and purple isolates requiring hisitidine were obtained, suggesting that crossing-over has occurred, but owing to the similarity of the colours of the parents this could not be regarded as proven. An unexpected feature was the production of numerous isolates requiring either histidine or choline or biotin or other growth factors. The colours of the isolates were in the expected red-violet to purple range. As regards gibberellic acid production only low yields were given by the best of the isolates. Titres were lower than in Cross 1, perhaps because one of the defi-cient parents gave only trace production. Six prototrophic mutants had spores 14—16 µm long, while out of five deficients three were 9—11 µm long and two 13.4 and 14.6 µm.

Crosses 3 and 5. These two crosses involved the same red-violet parent with two different deficients (D11 *rpu his* and C11 *whi nic lys*, respectively). In both cases the ori-ginal piled-up heterokaryotic colony produced a mass of pink aerial hyphae, while fairly extensive dark-red submerged dendroid growth developed. The latter became covered with a pinky-white velvety overgrowth from which (on subculturing) proto-trophic spores were recovered. In both cases also the product was a new red-violet culture with small spores and a relatively high productivity (intermediate to high). The hybridisation appears to have proceeded directly to a special type of recombinant without any stable diploid phase. Such a process has not so far been recognised in fungi. It has not been reported in *Aspergillus nidulans*, but it has been observed by us with *P. chrysogenum*, and in *P. patulum* a cross of this type formed part of a successful strain-improvement programme; it is probably known to other workers.

In the present case the new strains closely resemble one of the parents, the red-violet A11 culture, but differ in the occurrence of sporulation. Many at-tempts have been made in the past to produce sporulation in 917 and its descendents such as 419, the parental type A. The spores of the recombinants were smaller in size than those of the sporulating C parent and may resemble those of the original A type of culture. They showed a low viability, usually less than 1%.

Cross 9. This cross involved the pale-blue or white C1 culture, requiring nicotinic acid (C1 *pvb& whi nic*) with a mutant of B1, orange, requiring methionine (B11 *mog met*). Both strains produced conidia. Flat spreading surface growth took place from the heterokaryotic colony. The diploids were white in colour with a tinge of mauve generally resembling the C parent. Only a few isolates from the mutated diploids were examined. No orange isolates were obtained (*i.e.* from B), most of the colonies being white or nearly white. Two deficients were obtained, one requiring nicotinic acid or adenine, the other adenine. In addition, a number of dark to pale violet isolates occurred, which may have been due to a mutational instability on the part of the C nucleus.

An interesting feature of this cross was the increase in productivity which occurred:

original parents:	medium
deficients:	low medium
best diploids:	intermediate
best recombinant:	intermediate, better than diploids
number of recombinants tested:	12

Cross 16. This was between a mutant of a high-yielding orange A parent and B21 *dpu his*. The heterokaryon showed dendroid growth on plate cultures. On transfer to fresh MM, corrugated dark-orange colonies developed with bright orange aerial growth. In two cases there was very extensive dark-red, dendroid, submerged growth. On subculture of marginal growth flat orange colonies were obtained with white over-growth. In some cases, prototrophs were isolated from the orange areas. They were orange in colour and resembled the A parent. No normal sporulation occurred but small particles (1—2 μm diameter) were observed in filtered culture suspensions. Micro-block cultures were made in which there were observed small mycelial side-branches, with swollen ends. The latter apparently consisted of a droplet of mucus containing numerous tiny particles, *i.e.* a process resembling the formation of unexpect-edly tiny microspores. These small particles were used as material for mutation work. Viability was very low (0.1% or less) and it is possible that viability was due to the presence of fragments of mycelium. However, in the course of mutation work many thousands of particles were inspected during counting and contamination of this type was never observed.

No deficients were obtained among the mutants isolated, and only a few coloured isolates. As regards gibberellic acid production, some of the prototrophs gave higher yields than the best parent and a further advance occurred among the mutants.

DISCUSSION OF RESULTS

The first crosses described involve at least two kinds of behaviour. In Crosses 1, 2 and 9 a fairly normal kind of diploidisation took place. Heterokaryons formed slowly but later grew more quickly. This was followed by diploidisation. Evidence for this was the production of prototrophs with large spores and the recovery of parental type recombinants after mutation. In one case crossing-over was observed, in another there was evidence for this, although it was not quite so convincing. Formation of the parental biochemical deficients (*cf.* Tables V and VI) involved a loss in gibberellic acid production. This was not recovered in the recombinants and isolates from Crosses 1 and 2 except in an individual case. In Cross 9, however, diploids and mutants were obtained giving higher titres than any of the parents.

In the other crosses (3, 5, 16) prototrophs arose from the heterokaryon without any apparent diploid stage. It is possible that temporary diploidisation occurred but there was no evidence for this. All the prototrophs isolated from the three crosses gave small or very small spores with only a low degree of viability. There is no explanation for this behaviour, although it has been observed in other moulds. A complication is the fact that one of the parents (type A) sporulates either not at all or with extreme difficulty. However, in other cases with this parent satisfactory diploids were obtained with spores of the expected large size.

Another feature of these unusual crosses (3, 5, 16) was the production of high-yielding prototrophs and mutants, especially the latter. In Cross 16 the best mutants (only one mutation step from the prototroph) were appreciably more productive than either of the parents.

The nature of the very small particles produced from the Cross 16 prototrophs is unknown. They appear to be very small defective microspores but no evidence has been obtained as to their composition.

GENERAL CONCLUSIONS AND DISCUSSION

Summarising the results obtained, it was found that with *P. patulum* it was very difficult or impossible to obtain crosses by the parasexual process. Even in the best cases the alleged diploids had spore sizes about 50% larger than the parents (rather than double the size) and recombinants were difficult to obtain. A number of other prototrophs were obtained in which the spore size was the same as that of the parents. These usually were extremely stable. It is possible that these were revertants, but they also arose in Experiment 3 in which doubly deficient strains were used and particular care was taken to check the stability of the parents. It is felt that these in fact arose from some kind of hybridisation process. The parental strains used throughout this work were production cultures which had undergone a long series of mutation and selection steps.

The crosses with *F. moniliforme* involved both wild strains and mutants as parents. Several different types of behaviour were observed. In some cases diploidisation appeared to occur, marked by increased spore size, followed by recombination. Strains of the A type, resembling the nonsporing parent, were obtained which showed sporulation. In other cases prototrophs were obtained which were very stable and possessed small or very small spores. Some of these gave increased yields of gibberellic acid, compared with the already high-yielding parents. Increased yields of gibberellic acid were also observed in a cross between two sporing strains (Cross 9).

In such a highly developed and precise subject as microbial genetics it is perhaps dangerous to speculate on the significance of these experiments; however, with due reservations, a few comments will be made.

1) The experiments show that with fungi, along with the normal effects produced by the parasexual process, other types of hybridisation seem to be possible, capable of producing only limited changes. In this process temporary diploidisation may occur but there is no evidence that this actually happens. Crosses in which this seemed to occur in the present work arose with *P. patulum* and also in Crosses 3, 5 and 16 with *F. moniliforme*.

2) The formation of apparent diploids, but with spores smaller than expected, and with a greater degree of stability, was observed with *P. patulum*. In unpublished experiments with *P. chrysogenum* we have noticed similar effects on a number of occasions, though in this case normal diploids were also obtained.

3) The production of what seemed to be conidiophores giving rise to tiny particles instead of normal microspores in *F. moniliforme* probably reflects a damaged system for spore production. It was surprising that such small particles retained a small but significant degree of viability.

4) It is suggested that the results observed reflect the behaviour of defective strains of the fungi involved. These defects arise either naturally, *e.g.* the loss of sporulation in wild gibberellic acid-producing strains, or by mutation in the industrial strains of *P. patulum*. It seems likely that as a result of this the possibilities for nuclear fusions in heterokaryons become extensively limited due to damage or distortions to the chromosomes. As a result of this only partial fusions of nuclear material occur. The strains produced have nuclei of the usual or only partly increased size and this is reflected in the spore sizes which are observed. In this connexion may be mentioned the realisation, in recent years, that mutations cause much more extensive damage than may appear at first sight. The results obtained probably reflect the choice of strains basically unsuited for crossing, a limitation that has to be accepted in an industrial situation but which is not favourable for progress in genetic research.

A consideration of the results reported in our paper and data presented at the Symposium[14] gives rise to the following tentative suggestion. This is that in the parasexual process nuclear fusions, in the process of diplodisation, are variable in the degree of completion that is achieved. This view is based on the considerations (1) that the controlling conditions present in sexual fusions are absent (*i.e.* the special apparatus associated with conjugation),   (2) the varying types of diploids produced in crossing, even when homozygous diploids are expected to be produced, and   (3) the production of prototrophs, following crossing, with spore sizes smaller than expected (this has ben observed in *P. patulum* and *F. moniliforme*, and also by us in *P. chrysogenum*). Where well formed diploids are obtained, as is the case in carefully planned genetic work, fusion is usually complete and results are obtained which parallel sexual hybridisation. Where industrial strains are used, which have been submitted to extensive mutation and selection, conditions are more difficult and hybridisation leads to a variety of prototrophic forms, some of which achieve a high degree of completion in diploidisation; others achieve only a limited degree of success. The lack of completion in diploidisation could account for the difficulty which is sometimes observed in the production of recombinants.

REFERENCES

1. PONTECORVO G., ROPER J. A., SERMONTI G.: Brit. Pat. 719,313 (1951); U. S. Pat. 2,820,742 (1951).
2. CALAM C. T.: Progr. Ind. Microbiol. **5**, 1 (1964).
3. CALAM C. T., p. 435 in *Methods in Microbiology*, Vol. IIIa. Academic Press, London 1970.
4. MACDONALD K. D.: Paper at 1st Internat. Symp. Genetics of Industrial Microorganisms, Prague 1970. Abstract Book, p. 176.
5. BALL C.: Paper at 1st Internat. Symp. Genetics of Industrial Microorganisms, Prague 1970. Abstract Book, p. 178.
6. ELANDER R. P., ESPENSHADE M. A., PATHAK S. G., PAN C. H.: Paper at 1st Internat. Symp. Genetics of Industrial Microorganisms, Prague 1970. Abstract Book, p. 180.
7. MACDONALD K. D., HUTCHINSON J. M., GILLETT W. A.: J. Gen. Microbiol. **33**, 365, 375, 385 (1963).
8. SPECTOR C., PHINNEY B. O.: Ann. N. Y. Acad. Sci. **144**, 204 (1967).
9. SOKOLOVA E. V., EROKHINA L. I.: Conf. Experimental Mutagenesis in Animals, Plants and Microorganism, Moscow 1965. Theses, p. 122.
10. BUXTON E. W., p. 183 in C. S. Holton et al. (Eds.): *Plant Pathology. Problems and Progress, 1908—1958*. Madison 1959.
11. TUVESON R. W., GARBER E. D.: Bot. Gaz. **121**, 74 (1959).
12. WOLLENWEBER H. W., REINKING O. A.: *Die Fusarien*. Verlag P. Parry, Berlin 1935.
13. WILSON: *The Wilson Horticultural Colour Chart*. British Colour Council, London 1938.
14. CALAM C. T., DAGLISH L. B., GAITSKELL W. S.: Paper at 1st Internat. Symp. Genetics of Industrial Microorganisms, Prague 1970. Abstract Book, p. 175.

The Biosynthesis of Microbial Metabolites from the Point of View of an Organic Chemist

CH. TAMM

Institute of Organic Chemistry, University of Basle, Switzerland

In elucidating the biosynthesis of natural products a number of problems arise for the organic chemist. In the first stage he has to establish the nature of the basic building blocks of the secondary metabolite. Several methods are available. The presumed precursor can be labelled by a radioactive atom such as ^{14}C or tritium and the incorporation of the precursor tested by feeding experiments. When radioactivity is incorporated into the natural product, degradations must be carried out to locate the radioactive atoms. An exact balance has to be established by applying unequivocal methods for the isolation of, if possible, all the sites of radioactivity incorporated, by chemical transformation and degradation. The methods have to be carefully selected and, if necessary, new degradations have to be devised. The use of doubly labelled precursors is advantageous in the resolution of specific stereochemical problems. For instance, if a hydrogen shift is involved as demonstrated by the use of tritium, ^{14}C serves as an internal standard. — A second possibility is offered by the use of precursors labelled with other isotopes such as deuterium, ^{15}N or [18]O. Since these species are devoid of radioactivity, their incorporation has to be determined by mass spectrometry. But since these measurements require a certain concentration of the isotope, a relatively high level of incorporation into the metabolite is needed. Furthermore, the experimental conditions used must exclude an exchange of the isotopes with the medium. — A third method operates with ^{13}C-labelled compounds. For the localization of the label no chemical degradation has to be carried out, since nuclear magnetic resonance spectroscopy allows the identification of specific carbon atoms in complex organic molecules due to the large differences in carbon chemical shifts. The use of ^{13}C-labelled substrates simplifies biosynthetic studies since the ^{13}C-isotopic excess incorporated at different sites can be located and identified by characteristic chemical shifts and an increased signal intensity over the natural abundance peaks. At present, the need of sufficient concentration of ^{13}C in the substrate for the detection of the chemical shifts in the spectra and the special instrumentation are the only limiting factors of this technique. Otherwise, the method offers many advantages over conventional tracer methods. A recent example is the investigation of the biosynthesis of radicinin with sodium 1-[13]C- and 2-^{13}C-acetate[1].

Having recognized the basic building blocks, testing of later intermediates follows by feeding corresponding labelled compounds. These generally have to be synthesized. In this connection, the isolation of minor metabolites of an organism is most important since they can represent biogenetic intermediates of the major metabolite. Also the technique of adding unnatural substrates to block certain steps in the biosynthetic pathway can be useful as it can permit a build-up of intermediates. Results of such feeding experiments and isolations give the sequence of the biosynthetic transformations and a study of the mechanisms of the single steps can follow. Such mechanisms must take into consideration the conditions (temperature, pressure, solvent pH *etc.*) which are limited for biological systems, and the stereochemistry if centres of chirality are involved. Some of the problems outlined are illustrated by our biosynthetic studies on the two macrolide antibiotics *brefeldin A* and *phomin*.

Brefeldin A, $C_{16}H_{24}O_4$, is a metabolite of *Penicillium brefeldianum* Dodge[2,3,3a]. Containing 16 C atoms, this metabolite is one of the smallest macrolide antibiotics. In contrast to the compounds isolated from actinomycetes it contains no sugar moieties and is devoid of nitrogen. Its 13-membered lactone ring is substituted with

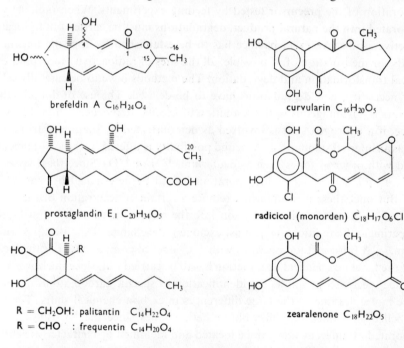

brefeldin A $C_{16}H_{24}O_4$

curvularin $C_{16}H_{20}O_5$

prostaglandin E_1 $C_{20}H_{34}O_5$

radicicol (monorden) $C_{18}H_{17}O_6Cl$

R = CH$_2$OH: palitantin $C_{14}H_{22}O_4$
R = CHO   : frequentin $C_{14}H_{20}O_4$

zearalenone $C_{18}H_{22}O_5$

Fig. 1. Brefeldin A and related natural products.

a single methyl group. The even number of carbon atoms and the relatively simple structure indicate acetate and malonate as precursors, respectively. Curvularin[4] and radicicol[5] are the only known macrolides isolated from higher fungi which are built up only from acetic acid. Fig. 1 shows the structure of brefeldin A and related natural products.

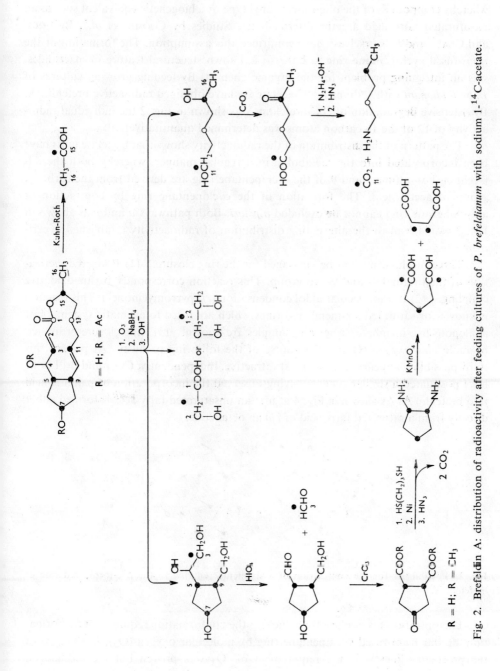

Fig. 2. Brefeldin A: distribution of radioactivity after feeding cultures of *P. brefeldianum* with sodium 1-^{14}C-acetate.

It was reasonable to assume that also brefeldin A is made from acetic acid, whereby compounds of the oligo-β-carbonyl type or a biogenetic equivalent such as an unsaturated fatty acid are the intermediates. Studies by COOMBE et al.[6], BU'LOCK and CLAY[7] and by ourselves[8] have confirmed this assumption. The formation of the trans-fused cyclopentane ring, a hitherto unknown structural feature in macrolides, was an intriguing problem for the organic chemist. By feeding growing cultures of P. brefeldianum with sodium 1-^{14}C-acetate we have obtained radioactive brefeldin A. By extensive degradation of ^{14}C-brefeldin A as shown in Fig. 2 the individual radio-activity of 12 of the 16 carbon atoms was determined quantitatively.

The pattern of the distribution of the radioactivity shows that 8 acetate units have been incorporated into the metabolite in a regular manner, whereby both the ad-jacent carbon atoms 5 and 9 of the cyclopentane ring are derived from the carboxyl group of acetic acid. The formation of the cyclopentane ring by contraction of a cyclohexane ring cannot be excluded a priori. Both pathways a and c as shown in Fig. 3 would lead to the alternating distribution of radioactivity established experi-mentally.

Three mechanisms can be envisaged for the ring closure: (1) Radical dimeriza-tion of the original 4- and 9-oxo group. This reaction corresponds to an oxidative coupling. (2) Intramolecular aldol condensation (cf. pyrromycinone[9]). This reaction is known for saturated six-membered rings which are fused to aromatic systems. For analogous five-membered rings no examples are known. (3) Ring closure promoted by molecular oxygen (O_2) in the course of the following pathway. At present this third possibility appears to us as most attractive. Independently, CLAY and BU'LOCK[7] have postulated a similar scheme having observed the incorporation of palmitic acid into brefeldin A. As shown in Fig. 4 at first an unsaturated fatty acid is formed either directly from a saturated fatty acid or via an octaketide.

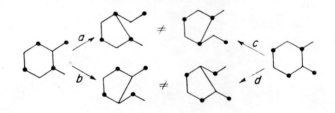

Fig. 3. Possibilities for the contraction of a six-membered ring to the five-membered ring of brefeldin A.

Lactonization takes place for achieving the conformation required for the forma-tion of the trans-fused cyclopentane ring by molecular oxygen (O_2). In the case of the prostaglandines[10-13] the requirement of O_2 was proven but the mechanism postulated for the ring formation cannot be adopted for brefeldin A because of the different distribution of the oxygen functions. The mechanism proposed for brefeldin A

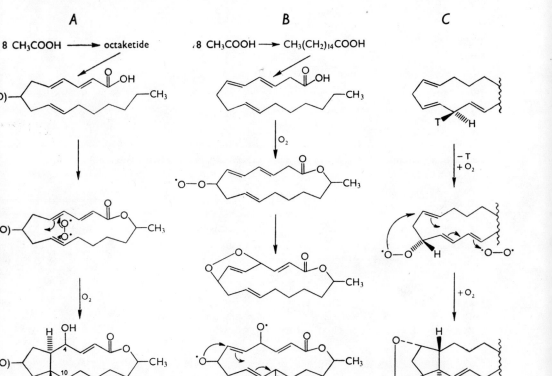

Fig. 4. Hypothetical pathways for the biosynthesis of brefeldin A. *A* Handschin, Sigg and Tamm[8], *B* Bu'Lock and Clay[7] (incorporation of 9-^{14}C-palmitate), *C* biosynthesis of prostaglandin E_1 (Hamberg and Samuelsson[13]).

explains the presence of a hydroxyl group at carbon 4 which originates from the methyl group of acetic acid. The 10-hydroxyl formed during the ring closure is eliminated to form the Δ^{10}-unsaturation. The 7-hydroxyl group must be introduced by a separate oxygenation in the course of the formation of the metabolite. Bu'Lock and Clay[7] have postulated an alternative mechanism for the O_2-induced ring formation (Fig. 4). It has to be pointed out that both pathways are still hypothetical.

Phomin, $C_{29}H_{37}NO_5$, and 5-dehydrophomin, $C_{29}H_{35}NO_5$, are cytostatically active metabolites which we have isolated from cultures of a *Phoma* sp. (strain S

TABLE I

Phomin and Related Metabolites

Product	Formula	Producer	Refs
Phomin = cytochalasin B	$C_{29}H_{37}NO_5$	*Phoma* sp. S 298, *Helminthosporium dematioideum*	14 16, 17
5-Dehydrophomin = cytochalasin A	$C_{29}H_{35}NO_5$	*Phoma* sp. S 298, *Helminthosporium dematioideum*	14 16, 17
Cytochalasin C	$C_{30}H_{37}NO_6$	*Metarrhizium anisopliae*	16, 17, 21, 22
Cytochalasin D = zygosporin A ·	$C_{30}H_{37}NO_6$	*Metarrhizium anisopliae* *Zygosporium masonii*	16, 17, 21, 22 18
Zygosporin D	$C_{28}H_{35}NO_5$		
Zygosporin E	$C_{30}H_{37}NO_5$	*Zygosporium masonii*	20
Zygosporin F	$C_{32}H_{39}NO_7$		
Zygosporin G	$C_{30}H_{37}NO_5$		

298)[14,15]. Independently, ALDRIDGE *et al.*[16,17] have isolated from cultures of *Helminthosporium dematioideum* cytochalasin A and B which were identical with 5-dehydrophomin and phomin. The cytochalasins C and D are metabolites of *Metarrhizium anisopliae*[16,17]. From *Zygosporium masonii* Japanese workers[18-20] have isolated zygosporin A and isozygosporin A. The former metabolite proved to be identical with cytochalasin D[21,22]. The zygosporins D, E, F and G belong to the same class of compounds[23] (*cf.* Tab. I).

For the elucidation of the structure of phomin and 5-dehydrophomin, the degradation of di-O-acetylphomin by ozone with a $NaBH_4$ reduction of the intermediary ozonide proved to be the key reaction[14,15]. The British group has carried out an independent structure determination using different arguments[16,17]. The interrelation with (+)-pulegone established the absolute configuration of the metabolites[14,15]. The essential steps are summarized in Fig. 5. The assignment of configuration was confirmed by the X-ray analysis of the phomin-$AgBF_4$ complex[24] (Fig. 6).

Phomin and 5-dehydrophomin represent a novel type of macrolide antibiotic, the large ring being fused to a highly substituted octahydroisoindole system. The 14-membered lactone ring of phomin corresponds to that of *e.g.* erythromycin, the aglycone of which is built from acetate and propionate units. The γ-lactam moiety of phomin points to a relationship with secondary metabolites derived from α-amino

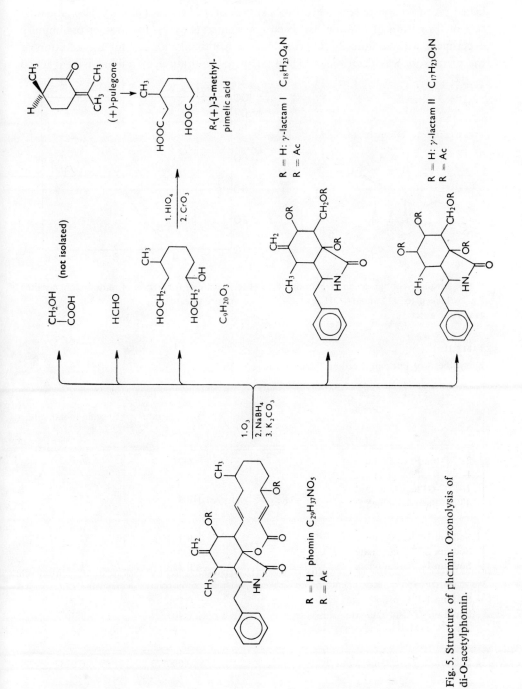

Fig. 5. Structure of phomin. Ozonolysis of di-O-acetylphomin.

acids. In order to evaluate these possibilities a number of potential ^{14}C- and ^{3}H-labelled precursors were fed to growing cultures of *Phoma* sp. (strain S 298), after the rate of formation of phomin had been determined (Fig. 7). By another preliminary experiment with sodium 2-^{14}C-propionate the optimal moment for the addition of the precursors was determined. The rate of incorporation of radioactivity reached

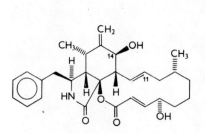

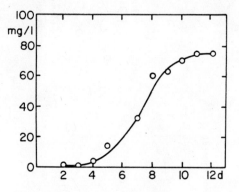

Fig. 6. Absolute configuration of phomin; Ag^{+} attached to Δ^{11} and 14-OH, 1 H$_2$O as crystal water.

Fig. 7. Rate of formation of phomin (mg per liter of culture filtrate).

TABLE II

Biosynthesis of Phomin: Feeding Experiments

Precursor	Incorporation	
	Absolute rate %	Specific molar rate % $\times 10^3$
2-^{14}C-D,L-Phenylalanine	6.7	60
1-^{14}C-D,L-Phenylalanine	4.9	15
U-^{14}C-L-Phenylalanine^a		
4'-T-L-Phenylalanine^a	10.4	55
Sodium 1-^{14}C-acetate	0.123	52
Sodium 2-^{14}C-malonate	1.04	12
Sodium 1-^{14}C-malonate	0.333	1.78
Sodium 1-^{14}C-propionate	0.296	1.26
Sodium 2-^{14}C-propionate	0.405	5.36
Me-^{14}C-L-Methionine	1.96	9.47
Sodium 2-^{14}C-D,L-mevalonate	0.000773	0.0142

^{a} Ratio 1 : 10.

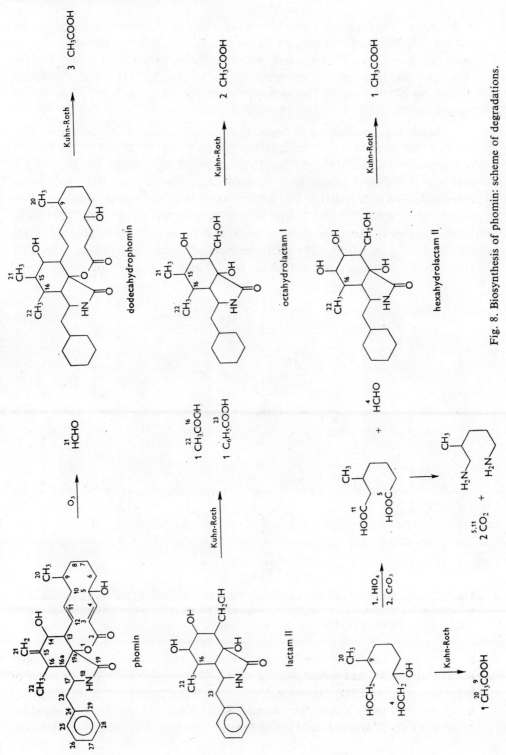

Fig. 8. Biosynthesis of phomin: scheme of degradations.

a maximum when the precursor is fed on the sixth day of incubation. Since this rate differed only by a factor of two, all the precursors were added to the cultures at the most convenient time, the beginning of the incubation before sterilization. The precursors fed and their levels of incorporation are listed in Table II. For the determination of the distribution of radioactivity the degradations outlined in Fig. 8 were carried out.

Phomin was transformed to dodecahydrophomin. Kuhn-Roth oxidation of the latter yielded three moles of acetic acid (C atoms 9 and 20, 15 and 21, 16 and 22). Ozonolysis of di-O-acetylphomin gave C-21 as formaldehyde on the one hand (reduction with H_2-Pd) and γ-lactam I, γ-lactam II and 3-methyloctane-1,7,8-triol on the other (reduction with $NaBH_4$). Kuhn-Roth oxidation of γ-lactam II yielded one mole of each acetic acid (C atoms 22 and 16) and benzoic acid (C-atoms 23—29). Octahydro-γ-lactam II gave two moles of acetic acid (C atoms 15 and 21, 16 and 22), whereas hexahydro-γ-lactam II yielded one mole of acetic acid (C atoms 16 and 22). Kuhn-Roth degradation of 3-methyloctane-1,7,8-triol led to one mole of acetic acid

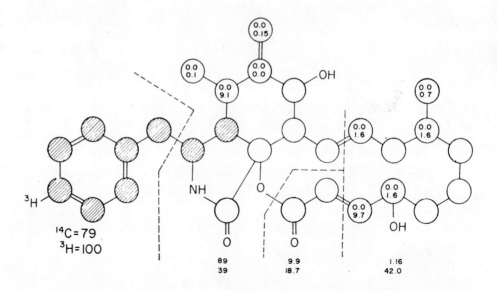

Fig. 9. Distribution of radioactivity in phomin: Part I (per cent of total activity). *Upper numbers:* 1-^{14}C-propionate; *lower numbers:* 2-^{14}C-propionate; *hatched circles:* U-^{14}C-L-phenylalanine: : 4'-^{3}H-L-phenylalanine (1 : 10).

(C atoms 9 and 20). In each case, the acetic acid was degraded further by the Schmidt method with NaN_3 and polyphosphoric acid to methylamine and CO_2. Cleavage of the octanetriol with HIO_4 gave formaldehyde (C-4) and 5-methyl-7-hydroxyheptanal which was oxidized to 3-methylpimelic acid. Schmidt degradation of the latter yielded 2 moles of CO_2 (C atoms 5 and 11) and 2-methyl-1,5-diaminopentane (α-methylcadaverine). In this manner the radioactivity of the carbon atoms 4, 9, 16, 20, 21, and 22, the pair C-5/C-11, by difference C-15 and C-1/C-2 and finally 23—29 were determined.

Some of the results are summarized in Fig. 9 and 10 in which the activities of the individual C atoms are given as percentage of total activity of phomin.

Phenylalanine shows a very high rate of incorporation. The feeding experiment with U-^{14}C-4'-^{3}H-L-phenylalanine (^{14}C/^{3}H = 1/10) demonstrates conclusively that

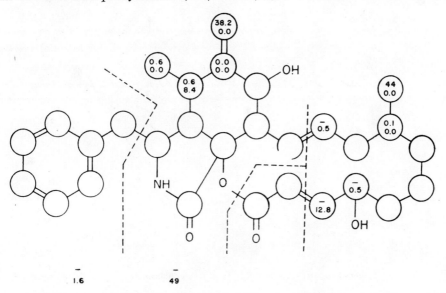

Fig. 10. Distribution of radioactivity in phomin: Part II (per cent of total activity). *Upper numbers: Me*-^{14}C-methionine; *lower numbers:* 1-^{14}C-acetate.

no loss of the carboxyl group has occurred during the incorporation, the benzoic acid isolated possessing 79% or 7/9 of the ^{14}C-activity and 100% of the ^{3}H-activity of phomin. After feeding 1-^{14}C- and 2-^{14}C-acetate an alternating distribution of radioactivity is found. Nine acetate units are incorporated. The same alternating distribution is found for 1-^{14}C- and 2-^{14}C-malonate, whereby the activity of carbon atoms 16 and 22 is lower, indicating that these C atoms are the origin of the chain. Thus acetate acts as the chain starter and malonate as the chain propagator. 1-^{14}C-Propionate did not exhibit the expected labelling at C-14 and C-8, the activity being concentrated in the γ-lactam moiety of phomin. 2-^{14}C-Propionate was incorporated, showing a distribution pattern like acetate and malonate. It behaves as a C_2-unit and not as a C_3-unit, carbon 2 corresponding to the carboxyl group of acetic acid. Such a transformation is known. *Me*-^{14}C-Methionine showed a very high rate of incorporation. 82% of the phomin activity was concentrated in carbon atoms 20 and 22. The remaining 18% of the phomin activity has not yet been localized.

These results demonstrate clearly that the building blocks of phomin are 1 unit of phenylalanine, 9 units of acetate or malonate and 2 units of methionine (Fig. 11).

Phomin differs from the known branched-chain macrolide antibiotics such as erythronolide A (aglycone of erythromycin) which is a heptapropionate. The C_2-units
are possibly transformed to a partially reduced octa- or nonaketide or to an oxygenated unsaturated C_{16}- or C_{18}-fatty acid.

A possible clue to one of the late steps in the biosynthesis of phomin comes from
the discovery of the cytochalasins C and D which in principle show the same struc-

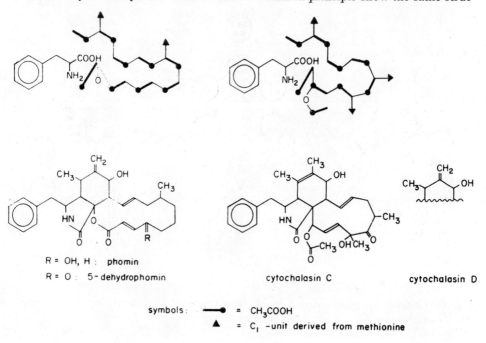

R = OH, H : phomin
R = O : 5-dehydrophomin

cytochalasin C   cytochalasin D

symbols :   ——●   = CH_3COOH

              ▲   = C_1 –unit derived from methionine

Fig. 11. The building blocks of phomin, 5-dehydrophomin and the predicted building blocks of
cytochalasin C and D.

tural patterns as phomin and 5-dehydrophomin with the exception that the lactone
group is replaced by a carbon—carbon bond. Therefore, similar patterns of the building blocks can be predicted for cytochalasin C and D (Fig. 11). It is very likely that an
analogous intermediate possessing such a C—C bond occurs in phomin biosynthesis,
the lactone group formed by a Baeyer-Villiger type oxidation. For the combination of
these elements leading to the intermediates and finally to phomin and to the cytochalasins the reaction mechanisms have to be elucidated. These reactions involve complicated enzyme-induced stereoelectronic processes. Investigation of the individual
steps in cell-free systems and with enriched or pure enzymes should follow. Such
studies are very difficult but represent a most challenging task for the growing field of
bio-organic chemistry. — Since in the case of phomin and the cytochalasins no such
experimental data are available it is too early to propose a scheme of reactions leading
to the metabolites that would not be speculative.

REFERENCES

1. TANABE M., SETO H., JOHNSON L,: J. Am. Chem. Soc. 92, 2157 (1970).
2. HÄRRI E., LOEFFLER W., SIGG H. P., STÄHELIN H., TAMM C.: Helv. Chim. Acta 46, 1235 (1963).
3. SIGG H. P.: Helv. Chim. Acta 47, 1401 (1964).
3a. WEBER W. P., HAUSER D., SIGG H. P.: Helv. Chim. Acta 54, 2763 (1971).
4. BIRCH A. J., MUSGRAVE O. C., RICKARDS R. W., SMITH H.: J. Chem. Soc. 3146 (1959).
5. MIRRINGTON R. N., RITCHIE E., SHOPPEE C. W., TAYLOR W. C., STERNHELL S.: Tetrahedron Letters 365 (1964).
6. COOMBE R. G., FOSS P. S., WATSON R. T.: Chem. Comm. 1229 (1967).
7. BU'LOCK J. D., CLAY P. T.: Chem. Comm. 237 (1969).
8. HANDSCHIN U., SIGG H. P., TAMM C.: Helv. Chim. Acta 51, 1943 (1968).
9. OLLIS W. D., SUTHERLAND I. O., CODNER R. C., GORDON J. J., MILLER G. A.: Proc. Chem. Soc. 347 (1960).
10. BERGSTRÖM S., DANIELSSON H., KLEINBERG D., SAMUELSSON B.: J. Biol. Chem. 239, PC 4006 (1964).
11. SAMUELSSON B.: J. Amer. Chem. Soc. 87, 3011 (1965).
12. HAMBERG M., SAMUELSSON B.: J. Amer. Chem. Soc. 88, 2349 (1966).
13. HAMBERG M., SAMUELSSON B.: J. Biol. Chem. 242, 5336 (1967).
14. ROTHWEILER W., TAMM C.: Experientia 22, 750 (1966).
15. ROTHWEILER W., TAMM C.: Helv. Chim. Acta 53, 696 (1970).
16. ALDRIDGE D. C., ARMSTRONG J. J., SPEAKE R. N., TURNER W. B.: Chem. Comm. 26 (1967).
17. ALDRIDGE D. C., ARMSTRONG J. J., SPEAKE R. N., TURNER W. B.: J. Chem. Soc. (C) 1667 (1967).
18. HAYAKAWA S., MATSUSHIMA T., KIMURA T., MINATO H., KATAGIRI K.: J. Antibiotics 21, 523 (1968).
19. TSAKUDA Y., MATSUMOTO M., MINATO J., KOYAMA K.: Chem. Comm. 41 (1969).
20. MINATO H., MATSUMOTO M.: J. Chem. Soc. (C) 38 (1970).
21. ALDRIDGE D. C., TURNER W. B.: Chem. Comm. 923 (1969).
22. ALDRIDGE D. C., TURNER W. B.: J. Antibiotics 22, 170 (1969).
23. MINATO H., KATAYAMA T.: J. Chem. Soc. (C) 45 (1970).
24. McLAUGHLIN G. M., SIM G. A., KIECHEL J. R., TAMM C.: Chem. Comm. 1398 (1970).

Using Mutants to Elucidate the Pathways of Biosynthesis of Secondary Metabolites of Fungi

C.H. HASSALL

Department of Chemistry, University College of Swansea, Singleton Park, Swansea, Great Britain

The distinction between primary and secondary metabolites of microorganisms is generally understood. The former includes products, such as amino acids and coenzymes, which are essential for continued viability. Secondary metabolites, on the other hand, are not essential for the growth of the organism in the normal environment. Nevertheless, they may be produced in substantial yield; they include products of commercial significance such as antibiotics, growth factors, toxins and other biologically active compounds. Knowledge of the biosynthesis of these compounds is desirable not only to enhance our general understanding of the nature and scope of enzyme-catalysed processes but also to facilitate commercial production of particular compounds. This may result from improved fermentation procedures or, as in the use of 6-aminopenicillanic acid for synthesizing various penicillins[1], from an increase in the range of biologically active compounds. There is another application of knowledge of biosynthesis that deserves mention. The natural pathway may serve as a model for synthesis, *in vitro*, of complex organic compounds. This will be illustrated in examples that follow.

In the case of primary metabolites of microorganisms, the manner of their biosynthesis has been elucidated very extensively through investigations using blocked mutants. However, these procedures, which are well illustrated by the elegant investigations on phenylalanine and related shikimic acid-derived amino acids[2], cannot be applied in the same way to secondary metabolites. The convenient techniques for the selection of blocked mutants in the former case depend on the fact that the pathway leads to an "essential" compound. Some mutants relating to secondary pathways have arisen from what appeared to be spontaneous degeneration of the wild type. Subculturing resulted in the isolation of a blocked mutant. Comparison of the structures of metabolites isolated from cultures of these mutants with those of the original wild type, has led to suggestions concerning the sequence of biosynthesis. For example, the use of mutants of *Penicillium islandicum*[3] and *Aspergillus terreus*[4] has contributed to our understanding of the biosynthesis of anthraquinones and of citrinin, respectively.

Evidently, it is possible to produce mutants of microorganisms with blocks in a pathway leading to a particular secondary metabolite, by irradiation with UV-light

or the use of chemical mutagens. Thus HOLKER[5,6] has produced blocked mutants of
Aspergillus versicolor by UV-irradiation and, through this, isolated new metabolites
I, II and **III** which were related in structure to averufin (**IV**). This led to a revision of
the original formulation of averufin. Similarly MARTIN and his co-workers[7] have obtained blocked mutants of *Streptomyces erythreus* by UV-irradiation or treatment
with the chemical mutagen, ethyleneimine. They have established that metabolites **V**,

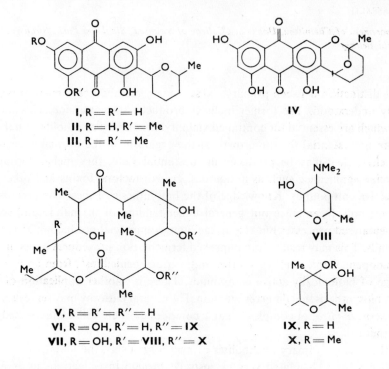

I, R = R' = H
II, R = H, R' = Me
III, R = R' = Me

IV

V, R = R' = R'' = H
VI, R = OH, R' = H, R'' = IX
VII, R = OH, R' = VIII, R'' = X

VIII

IX, R = H
X, R = Me

VI accumulated by these mutants are intermediates in the biosynthesis of erythromycin B (**VII**). These, and the later examples which will be discussed, certainly establish that such studies with mutants are fruitful. However, they have found relatively
limited use. What is missing, for secondary metabolites, is a general simple procedure
for selecting from the variety of strains produced in a mutation experiment, those that
have blocks relating to the pathway of biosynthesis which is under investigation. In
our studies of complex phenolic metabolites, the selection has been facilitated in two
ways. Simple, sensitive colour tests and paper or thin-layer-chromatography procedures
for separating phenols are readily available. Secondly, we have found that if, after
UV-irradiation or treatment with mutagens, attention is concentrated on the remaining viable strains with differences in gross morphology, the proportion of interesting
blocked mutants is significantly enhanced. Our experience of mutation techniques has
been largely with *Penicillium* and *Aspergillus* species. In some cases as when *P. citri-*

num was employed for the study of the pathway of citrinin biosynthesis[8], the organism did not produce a great variety of phenolic compounds and these were closely related to the relatively short pathway to citrinin. Consequently, it was probably to be expected that over a thousand cultures were examined for each selection of a blocked mutant of this strain. In other cases, and *Aspergillus rugulosus* was an example of this, the wild type produced a great variety of complex phenolic compounds and various distinct pathways were involved. In these circumstances, mutation experiments gave relatively high yields of mutant strains with different capacities for phenol biosynthesis, but great caution had to be exercised in making proposals concerning the relationships of metabolites produced by these different mutants.

The following examples have been selected to illustrate the use of mutation techniques in investigations of the biosynthesis of secondary metabolites. It will be apparent that these studies may produce valuable evidence which supplements that of ^{14}C-labelling. This can be expected to lead to proposals for biosynthetic pathways that may be investigated in a definitive way using enzymes.

CITRININ

The antibiotic citrinin (**XVI**), a metabolite of several *Penicillium* and *Aspergillus* species as well as the Australian plant *Crotolaria crispata*[9], was first isolated by HETHERINGTON and RAISTRICK[10]. Extensive degradative[11] and synthetic studies[12,13] have defined the structure and configuration[14]. It was established by three independent ^{14}C-labelling experiments[15-17] that all but three carbon atoms were derived from acetate; these three were provided by methionine but the stage at which incorporation occurred was undefined. Thus at the outset of our investigation there was evidence to the effect that the ^{14}C-labelling pattern of citrinin was as shown. There had been suggestions[17,18] for the mode of biosynthesis of citrinin based on this labelling pattern but all differed in detail from that indicated below. The scheme we have proposed was the result of identifying metabolites with structures related to citrinin and produced by the wild-type, or blocked mutant strains of *P. citrinum* and *A. terreus* that were selected after UV-irradiation or treatment with chemical mutagens. The structures of the new metabolites marked (+) were defined. The postulated pathway of biosynthesis accounts for the co-occurrence of the metabolites. It is noteworthy, in this case, that, as already remarked, very large numbers of strains of *P. citrinum* had to be examined in order to obtain mutants with blocks in the pathway to citrinin. Thus, in one series of experiments using both UV-irradiation and chemical mutagens, only six of 1172 cultures which were examined had lost the capacity to synthesize citrinin and only two of these produced a metabolite **XIV** which was related in structure to citrinin.

A partial *in vitro* synthesis of citrinin from **XIV** based on the stages proposed for the biosynthesis, has been achieved using diboran to give dihydrocitrinin (**XV**) and

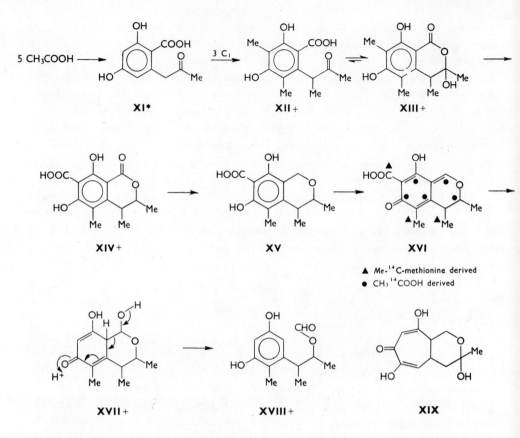

manganese dioxide to convert this to citrinin. The stage **XIV** to **XV** is similar to one proposed[20] for the formation of sepidonin (**XIX**) from the corresponding lactone. The compound **XVIII** presumably derived from decarboxycitrinin (**XVII**) by the reverse aldol transformation which is shown, was identified as a metabolite of a particular strain of *P. citrinum*.

GEODOXIN

The evidence for the sequence sulochrin (**XX**) to geodoxin (**XXI**) has been reviewed in previous publications[21,22]. Blocked mutants of *A. terreus* relating to both the chlorinated and the unchlorinated series of compounds have been isolated. Moreover, it has been established that similar transformations may be achieved *in vitro* in good yield[23]. Contrary to an earlier proposal, it appears that sulochrin is derived from the cor-

---

* Although C-acetylorsellinic acid (**XI**) has not been isolated from cultures containing citrinin it is known as a fungal metabolite[19].

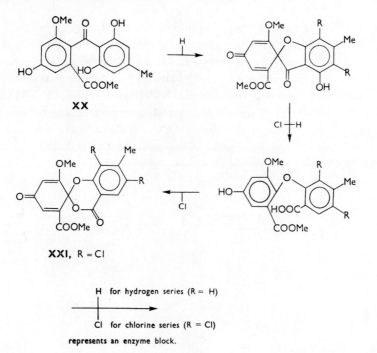

H for hydrogen series (R = H)

Cl for chlorine series (R = Cl)

represents an enzyme block.

responding anthraquinone questin (**XLI**). This is indicated both by an investigation using a cell-free enzyme system[24] and by an incorporation study utilizing double-labelled questin in cultures of *A. terreus*[25].

PHENOLIC METABOLITES OF *A. rugulosus*

Exploratory studies using this organism showed that it produced a rich variety of novel phenolic metabolites. Moreover, the species appeared to be well suited to genetical studies as it was closely related to *A. nidulans* which has been the subject of extensive investigations[26] using both the sexual and parasexual techniques. Mutation experiments using UV-irradiation or chemical mutagens have resulted in the isolation of a large number of blocked mutants with differing capacities for phenol biosynthesis. As mentioned earlier, the selection of these mutants with different capacities for phenol biosynthesis was facilitated by concentrating attention on strains that differed in pigmentation and morphology from the wild type. This study is still incomplete but the evidence which has been obtained makes it possible to propose, already, the course of biosynthesis of particular novel phenolic metabolites and to recognize, in a more general way, some relationships of others.

The monobenzenoid derivatives which have been distinguished are related in structure to *o*-orsellinic acid (**XXII**), the well-known acetate−malonate-derived[27,28]

metabolite of many species of *Penicillia* and *Aspergilli*. The compounds **XXIV**—
XXVII[29,30] which are novel are probably interrelated by the transformations represented in the scheme. Particular mutants with blocks between these stages have been
identified. There is good evidence from ^{14}C-labelling studies by PETTERSSON[31] that
o-orsellinic acid is incorporated into flavipin (**XXVIII**) and fumigatin (**XXIX**), metabolites of *A. fumigatus*. It is of interest that metabolites with related structures have
been isolated from other fungi: quadrilineatin (**XXX**)[32]; gladiolic and dihydrogladiolic
acids (**XXXI, XXXIII**)[33]; cyclopolic and cyclopaldic acids (**XXXII, XXXIV**)[34]; fomecin
A, B (**XXXV, XXXVI**)[35].

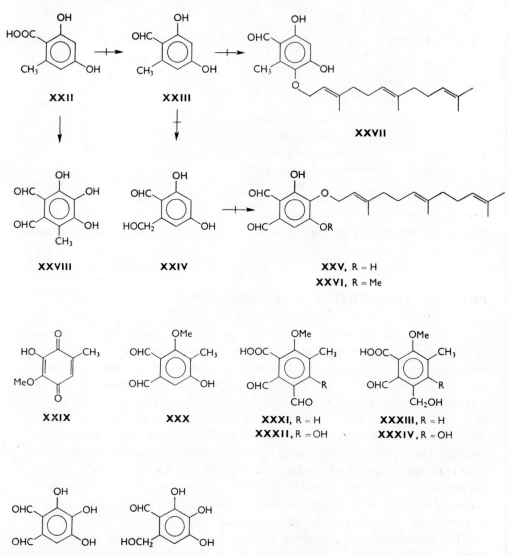

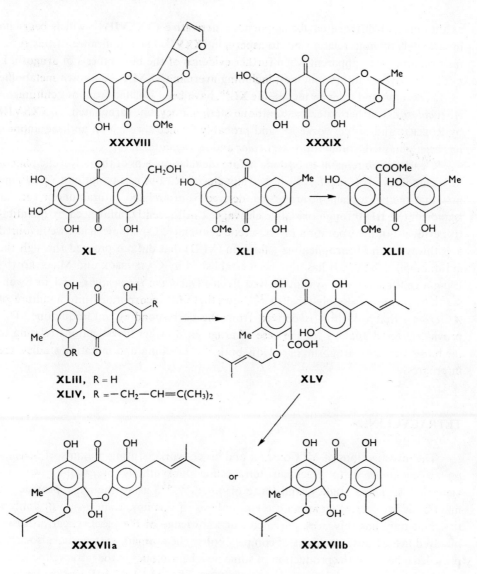

XXXVIII   XXXIX

XL   XLI   XLII

XLIII, R = H
XLIV, R = —CH₂—CH=C(CH₃)₂

XLV

or

XXXVIIa   XXXVIIb

In one of our earliest experiments, a mutant strain was isolated that differed from the wild type in having fluffy morphology and producing the novel metabolite, asperugin. The major phenolic metabolite of the wild type, arugosin, could not be found in the mixture of compounds produced by the "fluffy" mutant. When strains bearing the marker "fluffy" were crossed with others carrying the wild-type allele a 1 : 1 segregation pattern was obtained and the "fluffy" character could be recovered in all potential recombinant forms. Evidently the character "fluffy" was under genetic control in the strict sense[36]. The structure of the metabolite arugosin has been defined as the unusual di*iso*pentenylbenzophenone derivative (**XXXVII**)[37], which bears no immediately obvious relationship to asperugin (**XXVI**). The significance of this result may become more apparent when further evidence of the biosynthesis of arugosin is available. Several anthraquinones, including averufin (**XXXIX**)[5], a known metabolite of *A.versicolor*, and the new metabolite **XL**[38], have been isolated from our cultures of *A. rugulosus*. Furthermore, the xanthone sterigmatocystin, formulated as **XXXVIII** by ROBERTS and his co-workers[39] and probably formed through an anthraquinone[6], has been obtained from both *A. versicolor* and *A. rugulosus*.

Clearly much remains to be done before the relationships of the polycyclic phenolic metabolites of *A. rugulosus* will be fully elucidated but it is tempting to postulate now that the benzophenone arugosin is derived from a related anthraquinone such as islandicin (**XLIII**) through oxidative cleavage. Until recently, influenced by quantitative evidence of incorporation of ^{14}C-acetic acid and ^{14}C-malonic acids, we favoured a pathway to the benzophenone sulochrin (**XLII**) that did not proceed through the anthraquinone **XLI**[40]. It has now been established by GATENBECK and MALSTROM[24], using a cell-free enzyme system derived from *Penicillium frequentans*, and in Swansea[25], by means of doubly-labelled ^{14}C-questin (**XLI**) incorporated into a culture of *A. terreus*, that sulochrin is derived from the corresponding anthraquinone. This provides a good analogy for the case of arugosin. Further investigations relating to the biosynthesis of arugosin, based on both ^{14}C-labelling and mutation studies, are in progress.

TETRACYCLINES

The investigations of McCORMICK and his co-workers using mutants of *Streptomyces aureofaciens* to elucidate tetracycline biosynthesis are well-known (reviews[41-43]). These studies, and those of others[44-46], using mutants, provide the main basis for current views on the biosynthesis of this important group of antibiotics. Not only has this work enriched our knowledge of the biochemical processes involved in the transformations of complex polycyclic aromatic systems but also it has provided a basis for the production of some novel antibiotics in the tetracycline series.

It is now well established[41] that the pretetramids (**XLVI, XLVII**) are biosynthetic precursors of the tetracyclines. It is suggested that they are converted through the compounds **XLVIII** *etc.* to **LIII**. Our knowledge of the stages in the biosynthesis of 6-methylpretetramid itself is still limited. It appears from ^{14}C-labelling studies that the tetracyclic system is acetate—malonate derived and that the 6-methyl group of **XLVII** is incorporated from L-methionine[47,48] but the stages involved are still undefined. Calculations of TURLEY and SNELL[49] based on their evidence or that of GATENBECK[50] for incorporation of 2-[14]C- and 1-^{14}C-acetic acids into tetracyclines has led them to suggest that different portions of the tetracyclic structure of oxytetracyclines are labelled by acetate units at different rates. In particular, ring A is markedly diffe-

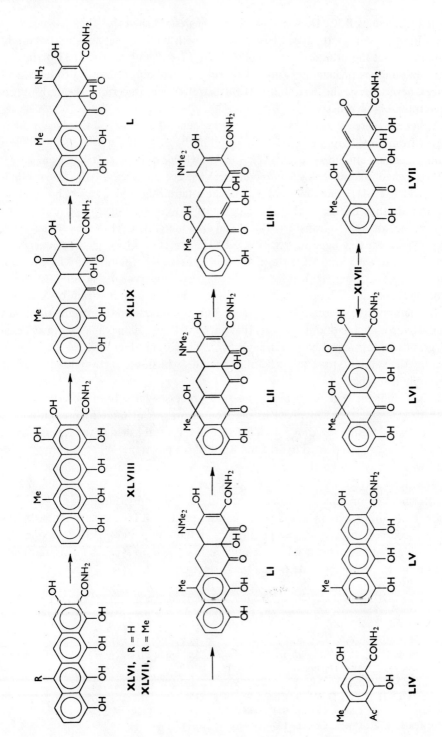

rentiated from rings B, C, D. A recent study[51] has established that the incorporation into positions 6, 6a, 7—10, 10a, 11 is in accord with uniform head-to-tail linking of acetate units, but the situation regarding rings A, B is still ambiguous. The isolation of the anthraquinones protetrone[42] and ekatetrone[52], probably "shunt" products, emphasizes, nevertheless, the desirability of further evidence concerning the early stages of pretetramid biosynthesis.

Although there is good evidence for the general outline of the biosynthetic sequence from 6-methylpretetramid to tetracycline, based on the studies using blocked mutants, there are still important gaps in our knowledge of details. It has been shown that the blocked mutant strain *S. aureofaciens* V655 accumulates 4-hydroxy-6-methyl-pretetramid (**XLVIII**) but does not synthesize a tetracycline. However, both strain V655 and another ED1369, which is capable of converting 6-methylpretetramid to a tetracycline, are able to transform the 12a-hydroxyderivative **XLIX** to a tetracycline. This provides convincing evidence that **XLVIII** precedes **XLIX** in the sequence[42]. However, the description of the stages by which the transformation of **XLIX** to **LIII** occurs is, as yet, incomplete. It is interesting to remark that recently a mono-O-gluco-side of this series was isolated from the culture fluid of a mutant strain of *S. aureofaciens*, the aglycone of this glucoside gave 6-methylpretetramid on reduction with phenol—hydroiodic acid; this suggests that carbohydrate residues may serve as "carrier" groups in the sequence to tetracycline[52]. They could facilitate the process by increasing the solubility of tetracyclic intermediates which, unfunctionalized, may be remarkably insoluble.

It is significant that *in vitro* oxygenation processes have been observed which appear to resemble those involved in the transformation of 6-methylpretetramid *in vivo*. Studies[53,54] using the ring A model **LIV** and **LV**, established that oxygenation in 0.1N sodium hydroxide led to hydroxylation in the 4-position. When 6-methylprete-tramid was treated under these conditions, the oxidation involved both the 4- and the 6-positions to give **LVI**[55]. However, when the solvent was changed to alkali—di-methylformamide or aqueous magnesium chloride—dimethylsulphoxide, there was a remarkable difference in the hydroxylation pattern; 12a-, rather than 4-hydro-xylation occurred to give an excellent yield of the compound **LVII**[56]. Evidently, it is possible that such stages might be utilized in an *in vitro* synthesis of tetracycline based on the biosynthetic route from 6-methylpretetramid.

REFERENCES

1. DEMAIN A. L., p. 29 in *Biosynthesis of Antibiotics*. Academic Press, New York 1966.
2. SPRINSON D. B.: Adv. Carbohydrate Chem. **15**, 235 (1960).
3. KIKUCHI M., NAKAHARA M.: Bot. Mag. Tokyo **74**, 463 (1964).
4. HASSALL C. H., JONES D. W.: J. Chem. Soc. 4189 (1962).
5. HOLKER J. S. E.: Chem. Comm. 911 (1966).
6. HOLKER J. S. E.: Chem. Comm. 1574 (1968).

7. MARTIN J. R., ROSENBROOK W.: Biochemistry **6**, 435 (1967); MARTIN J. R., PERUN T. J., GIROLAMI R. L.: Biochemistry **5**, 2852 (1966).
8. CURTIS R. F., HASSALL C. H., NAZAR M.: J. Chem. Soc. (C) 85 (1968).
9. EVERT A. J.: Ann. Bot. Lond. **47**, 913 (1933).
10. HETHERINGTON C. A., RAISTRICK H.: Phil. Trans. **220** B, 269 (1931).
11. BROWN J. P., ROBERTSON A., WHALLEY W. B., CARTWRIGHT N. J.: J. Chem. Soc. 859, 867 (1949); CARTWRIGHT N. J., ROBERTSON A., WHALLEY W. B.: J. Chem. Soc. 1563 (1949).
12. JOHNSON D. H., ROBERTSON A., WHALLEY W. B.: J. Chem. Soc. 2971 (1950).
13. WARREN H. H., DOUGHERTY G., WALLIS E. S.: J. Amer. Chem. Soc. **71**, 3422 (1949).
14. HILL R. K., GARDELLA L. A.: J. Org. Chem. **29**, 766 (1964).
15. BIRCH A. J., FITTON P., PRIDE E., RYAN A. J., SMITH H., WHALLEY W. B.: J. Chem. Soc. 4567 (1958).
16. SCHWENK E., ALEXANDER G. J., GOLD A. M., STEVENS D. F.: J. Biol. Chem. **233**, 1211 (1958).
17. RODIG O. R., ELLIS L. C., GLOVER I. T.: Biochemistry **5**, 2451, 2458 (1966).
18. WHALLEY W. B., p. 49 in *Recent Developments in the Chemistry of Natural Phenolic Compounds*. Pergamon Press, London 1961.
19. OXFORD A. E., RAISTRICK H.: Biochem. J. **27**, 634, 1473 (1933).
20. DIVEKAR P. V., RAISTRICK H., DOBSON T. A., VINING L. C.: Canad. J. Chem. **43**, 1835 (1965).
21. HASSALL C. H., p. 51 in *Biogenesis of Antibiotic Substances*. Publ. House Czech. Acad. Sci., Prague 1965.
22. CURTIS R. F., HARRIES P. C., HASSALL C. H., LEVI J. D.: Biochem. J. **90**, 43 (1965).
23. HASSALL C. H., LEWIS J. R.: J. Chem. Soc. 2312 (1961).
24. GATENBECK S., MALSTROM L.: Acta Chem. Scand. **23**, 3493 (1969).
25. CURTIS R. F., HASSALL C. H., PARRY D. H., *in press*.
26. PONTECORVO G., ROPER J. A., HEMMONS L. M., MACDONALD K. D., BUFTON A. W. J.: Adv. Genet. **5**, 141 (1953).
27. MOSBACH K.: Acta Chem. Scand. **14**, 457 (1960); Naturwiss. **48**, 525 (1961).
28. GATENBECK S., MOSBACH K.: Acta Chem. Scand. **13**, 1561 (1959).
29. BALLANTINE J. A., HASSALL C. H., JONES G.: J. Chem. Soc. 4672 (1965).
30. BALLANTINE J. A., HASSALL C. H., JONES B. D., JONES G.: Phytochemistry **6**, 1157 (1967).
31. PETTERSSON G.: Acta Chem. Scand. **17**, 1323, 1771 (1963).
32. BIRKINSHAW J. H., CHAPLEN R., LAHOZ-OLIVER R.: Biochem. J. **67**, 155 (1957).
33. GROVE J. F.: Biochem. J. **50**, 648 (1952).
34. BIRKINSHAW J. H., RAISTRICK H., ROSS D. J., STICKINGS C. E.: Biochem. J. **50**, 610 (1952).
35. McMORRIS T. C., ANCHEL M.: Canad. J. Chem. **42**, 1595 (1964).
36. HASSALL C. H., LAWRENCE K.: J. Gen. Microbiol. **35**, 483 (1964).
37. BALLANTINE J. A., FRANCIS D. J., HASSALL C. H., WRIGHT J. L. C.: J. Chem. Soc. (C) 1175 (1970).
38. BALLANTINE J. A., *unpublished results*.
39. BULLOCK E., ROBERTS J. C., UNDERWOOD J. G.: J. Chem. Soc. 1479 (1962).
40. CURTIS R. F., HARRIES P. C., HASSALL C. H., LEVI J. D., PHILLIPS D. M.: J. Chem. Soc. (C) 169 (1966).
41. McCORMICK J. R. D., p. 73 in *Biogenesis of Antibiotic Substances*. Publ. House Czech. Acad. Sci., Prague 1965.
42. McCORMICK J. R. D., p. 113 in *Antibiotics*, Vol. II, *Biosynthesis*. Springer-Verlag, Berlin 1967.
43. McCORMICK J. R. D., p. 163 in *Genetics and Breeding of Streptomyces*. Yug. Acad. Sci. Arts, Zagreb 1969.
44. BLUMAUEROVÁ M., MRAČEK M., VONDRÁČKOVÁ J., PODOJIL M., HOŠŤÁLEK Z., VANĚK Z.: Fol. Microbiol. **14**, 215 (1969).

45. BLUMAUEROVÁ M., HOŠŤÁLEK Z., MRAČEK M., PODOJIL M., VANĚK Z.: Fol. Microbiol. **14**, 226 (1969).
46. CATLIN E. R., HASSALL C. H., PRATT B. C.: Biochim. Biophys. Acta **156**, 109 (1968).
47. BIRCH A. J., SNELL J. F., THOMPSON P. J.: J. Chem. Soc. 425 (1962).
48. MILLER P. A., McCORMICK J. R. D., DOERSHUK A. P.: Science **123**, 1030 (1956).
49. TURLEY R. H., SNELL J. F., p. 95 in *Biosynthesis of Antibiotics*. Academic Press, New York 1966.
50. GATENBECK S.: Biochem. Biophys. Res. Comm. **6**, 422 (1961).
51. CATLIN E. R., HASSALL C. H., PARRY D. H.: J. Chem. Soc. 1363 (1969).
52. PODOJIL M., VANĚK Z., VOKOUN J., CUDLÍN J., BLUMAUEROVÁ M., VONDRÁČEK M., HASSALL C.H.: Paper at 1st Internat. Symp. Genetics of Industrial Microorganisms, Prague 1970. Abstract Book, p. 106.
53. HASSALL C. H., WINTERS T. E.: J. Chem. Soc. (C) 912 (1967).
54. HASSALL C. H., WOOTTON G.: J. Chem. Soc. (C) 2805 (1969).
55. HASSALL C. H., WINTERS T. E.: J. Chem. Soc. (C) 1558 (1968).
56. HASSALL C. H., THOMAS G. J.: Chem. Comm. 1053 (1970).

The Biosynthesis of Cephalosporin C

J. Nüesch, H.J. Treichler and M. Liersch

Research Laboratories of the Pharmaceutical Division of CIBA Ltd., Basle, Switzerland

INTRODUCTION

The true penicillins and cephalosporin C form a group of antibiotics characterized by many common features with regard to their chemical structure, as well as to their biosynthesis[1,2]. Nevertheless, these antibiotics can be divided into two groups on the basis of certain aspects of their biosynthesis (Table I).

1. Classical penicillins which are all N-acyl derivatives of 6-aminopenicillanic acid (APA). Their acyl side chains are generally of a nonpolar aliphatic or aromatic type (*e.g.* phenylacetic acid). The main and industrially most important producers belong to the fungal species *Penicillium chrysogenum*. Many other fungi belonging to the genera *Penicillium, Aspergillus* and other related ascomycetes are known to produce these compounds[3]. Their formation depends qualitatively and quantitatively on the side-chain precursors in the culture medium[4]. The nucleus, APA, can be understood as a condensation product of cysteine and valine[1,2]. *P. chrysogenum* is able to grow and produce in very simple synthetic media, thus building-up all the precursors from glucose, ammonia and sulfate. Sulfate is the optimal sulfur donor for APA[5].

2. The second group contains the β-lactam heterocyclic antibiotics with D-α-aminoadipic acid (AAA) as an acyl side chain and APA (penicillin N) or 7-aminocephalosporanic acid (ACA) (cephalosporin C) as nuclei. The principal producers of penicillin N and the only known fungi producing cephalosporin C are from the genera *Emericellopsis* and *Cephalosporium*[6,7]. Contrary to the formation of the classical penicillins the addition of side-chain precursors to the medium has no influence on the antibiotics formed. They are invariably cephalosporin C or penicillin N[1,2]. On the other hand, methionine enhances the production of these two compounds to a remarkable degree. This amino acid is the key substance in the biosynthesis of both antibiotics[8,9,10]. The formation of APA and ACA however, is in analogy to the first group based on cysteine and valine[1].

Judging from the literature the amount of investigations dealing with aspects of biosynthesis of secondary metabolites and antibiotics in particular is quite impressive. Nevertheless, our knowledge of the enzymatic and genetic background and the control

TABLE I

Biosynthetic Relationship of Penicillins and Cephalosporin C mechanisms involved is still rather restricted. *E.g.*, we do not yet know how primary and secondary metabolism are connected and how the enzymes involved are genetically determined and regulated. Therefore, the aim of our contribution is not a review of the biosynthesis of cephalosporin C, especially as such already exist[1,2], but rather an attempt to investigate the pathway of formation of cephalosporin C in its biological entity. The investigator of secondary metabolism is faced with certain specific difficulties. The end products are not essential for the producers, this rendering difficult the interpretation of results with certain mutants such as production-minus strains. On the other hand, the microbiologist working in this field can take advantage of all the knowledge accumulated in the last two decades with regard to biosynthesis of primary metabolites, genetics, regulation and synthesis of enzymes.

Our approach to the biosynthesis of cephalosporin C is based on the first genetic investigation of the producing organism *C. acremonium* with regard to induction and selection of specific mutants and aspects of cytology and somatic recombination. In the second step, special strains have been used for the study of the biosynthesis of cephalosporin C; especially the metabolism of the sulfur-containing amino acids and its connection to the formation of antibiotics.

MUTAGENESIS IN *C. acremonium*

MATERIALS AND METHODS

Organism and preparation of conidial suspensions for mutagenic treatment. *C. acremonium* C.M.I. 49137 mutant 8650 was used for mutagenic treatment. To prepare conidia for subsequent treatment, mycelium from 20 ml agar slants was suspended in 10 ml of 0.2 M phosphate buffer (pH = 7.2) and inoculated at a ratio of 1% into shake flasks containing a seed medium (I) of the following composition: 23.5 g corn-steep dry, 20.0 g sucrose, 4.5 g ammonium acetate and tap water to 1000 ml. The medium (100 ml) was dispensed into shake flasks, stoppered with cotton plugs and autoclaved (120 °C, 20 min). The pH after sterilisation was between 6.8 and 7.0. The flasks were incubated at 25 °C on a rotary shaker at 250 r.p.m. The cultures were harvested after 4 days when maximal production of conidia was reached. Pure suspensions of conidia have been obtained by glass-wool filtration of the culture broth. Suspensions of washed conidia were adjusted to a titer of 10^7 per ml in 0.2 M phosphate buffer (pH = 7.2).

Mutagenic treatment. Preliminary trials with different chemical and physical mutagens showed that N-methyl-N'-nitro-N-nitrosoguanidine (NG, Aldrich Chem. Co.) was an ideal agent for the induction of mutants in *C. acremonium*. Therefore, all subsequent mutagenic treatments have been carried out with NG. 36 ml of the conidial suspension were mixed with 4 ml of a 1% NG solution. The solvent was dimethylformamide—water 1 : 10. The reaction mixture was agitated in a water bath at 27 °C. For the survival curve, as well as for the determination of the rate of mutant induction, 2 ml aliquots were withdrawn at appropriate intervals. The samples were immediately filtered through a sterile Millipore filter (HAWP 04700, pore size 0.45 μm), the conidia were resuspended from the filter in phosphate buffer and after appropriate dilution, used for plating on a complete medium V for the estimation of the survival rate.

Composition of medium V (in g):

Difco yeast extract	4.0
glucose	10.0
NaCl	1.0
$FeSO_4.7H_2O$	0.1
$MgSO_4.7H_2O$	0.1
DL-methionine	0.1
DL-cysteine	1.0
L-lysine	0.5
DL-α-aminoadipic acid (dissolved in $NaHCO_3$)	0.1
DL-valine	1.0
casamino acids	1.0
Difco agar	25.0
tap water	to 1000 ml

The agar was sterilized separately in 800 ml of water in the autoclave (120 °C, 20 min), whereas the other ingredients were dissolved in 200 ml of water and sterilized by filtration. The two solutions were mixed in a water bath at 50 °C.

Isolation of mutants. After 7 days of incubation on complete medium V at 26 °C all colonies were replica-plated onto minimal medium II.

Composition of medium II (in g):

sucrose	15.0	
glucose	2.5	
$NaNO_3$	5.0	
K_2HPO_4	3.0	
$MgSO_4.7H_2O$	0.2	
$FeSO_4.7H_2O$	0.05	
KCl	0.5	
trace-element solution (containing, in g:	2 mi	
$FeSO_4.7H_2O$		1.0
$CuSO_4.5H_2O$		0.15
$ZnSO_4.7H_2O$		1.0
$MnSO_4.4H_2O$		0.1
$K_2MoO_4.2H_2O$		0.1
distilled water		to 1000 ml)
Bacto agar	25.0	
distilled water	to 1000 ml	
pH before sterilisation	7.3	
after sterilisation	7.0	

The medium was sterilized in the autoclave at 120 °C for 20 min. The colonies which failed to grow on the minimal medium were transferred from the master plate to nutrient agar slants and retested on minimal medium. The conidial clones from the isolated auxotrophic mycelium were tested for their requirements by auxanography.

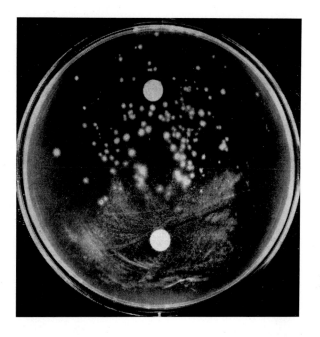

Fig. 1. "Inositol-less death" of growing cells and preservation of mutant cells of the inositol-requiring strain 8650 *inl* on inositol-less minimal medium.

Mutant enrichment techniques. The filtration enrichment technique of WOODWARD, DE ZEEUW and SRB[11] and the "inositol-less-death" method of LESTER and GROSS[12] have been successfully applied for the enrichment of mutants. For filtration enrichment, cultures containing 100 ml of conidial suspensions in a minimal medium III with a titer of 10^6 per ml were incubated on a rotary shaker in 500 ml shake flasks at 25 °C and 250 r.p.m. until the formation of short hyphae took place.

Composition of medium III (in g):

$(NH_4)_2SO_4$	2.5
KNO_3	5.0
$MgSO_4.7H_2O$	0.2
KH_2PO_4	0.2
$CaCO_3$	5.0
trace-element solution	10 ml
maltose	40.0
methyl oleate	7.0
inositol	2.0
distilled water	to 1000 ml
pH before sterilisation	7.3
after sterilisation	7.0

The medium was sterilized in the autoclave at 120 °C for 20 min. The incubation lasted usually about 48 to 60 h. The mycelial fragments were removed from the nongerminating conidia by subsequent filtration, the latter were washed three times by alternate centrifugation and reinoculated into minimal medium III for repeated enrichment.

To provide optimal conditions for "inositol-less death", conidia of the inositol-auxotrophic strain 8650 *inl* were adjusted to 10^7 per ml in 0.2 M phosphate buffer and preincubated on a rotary shaker for 12 h at 25 °C to deplete them of endogenous growth factors. Subsequently the conidia were spread on minimal agar II in a concentration of 10^6 per plate. A filter disc with 1% inositol was added as a control. The expected growth zone around this disc could already be observed after 72 h of incubation at 25 °C, while the conidia outside this zone were exposed to conditions of inositol deprivation. After further 240 h of incubation, another disc containing yeast extract, casamino acids and inositol (each component in a concentration of 3%) was added opposite to the control disc. The plates were again incubated for 96 h. The colonies which appeared subsequently around this disc were isolated and analyzed in order to detect auxotrophic mutants (Fig. 1).

RESULTS

THE RELATIONSHIP BETWEEN SURVIVAL RATE AND MUTANT INDUCTION

The survival curve of *C. acremonium* by 0.1% NG treatment and its relation to the mutation rate is shown in Fig. 2. The survival rate was computed from colony counts on complete-medium plates and the auxotrophic mutant count from the corresponding minimal-medium replica plates. The highest mutation rate occurred after 20 min of exposure to the mutagenic agent. After this time, 98% of the original colony-forming conidia have been killed and the mutation frequency was 0.5025%, taking into consideration the total of 105 auxotrophic mutants isolated from 2×10^4 survivors. In all experiments, very reproducible mutant yields were observed.

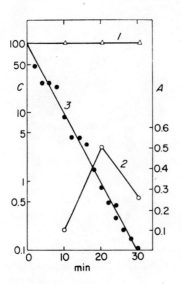

Fig. 2. The lethal and mutagenic effect of NG on *C. acremonium*. *C* colony-forming conidia, %, *A* auxotrophic mutants, %; *1* control survival in buffer, *2* mutant induction, *3* inactivation by NG.

THE EFFECT OF ENRICHMENT TECHNIQUES

The filtration enrichment technique applied resulted in a 6-fold increase with regard to the amount of isolated auxotrophic mutants. With this method, the spectrum of mutants and the frequency distribution was similar to that without enrichment. On the other hand, the "inositol-less death" method increased the amount of mutants by a factor of 20. However, this method led to the selection of predominantly vitamin-requiring strains. Furthermore, a number of mutants were isolated which could not be recovered by standard procedures. It was possible to detect mutants which were unable to use glucose, fructose and sucrose as carbon sources. They grow on gluconate or glycerol. A number of mutants requiring malonate, acetoacetate, lactate, pyruvate and butyrate could also be isolated. The conidia from the inositol-deficient strain 8650 *inl* rapidly lose their viability when allowed to germinate on an inositol-less medium. They grow quite abnormally and form tight clumps (Fig. 3 *upper left*) compared with the filamentous form of the wild type (Fig. 3 *upper right*). The loss of viability of the conidia is sharply reduced by imposing another growth-factor requirement on the inositol-deficient strain. Conidia of a double mutant 8650 *inl met* were spread on an inositol-less plate and inositol was added in two filter discs (1 and 0.1%) whereas methionine was added in a third disc (1%). A sickle-shaped growth zone was observed after 96 h of incubation outlining the area with optimal inositol and methionine concentration (Fig. 4). The conidia inside the growth zone toward the methionine disc could not be subcultured and were obviously damaged by inositol deprivation (Fig. 3 *upper left*) whereas the conidia outside the sickle-shaped growth zone looked normal (Fig. 3 *lower*) and were still viable. For the routine isolation the two disc methods (Fig. 1) as described under *Material and Methods* proved to be helpful in that they permitted a control of the starting strain and an easy isolation of colonies appearing after inositol deprivation.

THE FREQUENCY OF DISTRIBUTION OF SINGLE-REQUIREMENT MUTANTS

Table II shows the frequency distribution of 164 single auxotrophic mutants isolated after the optimal NG treatment without enrichment procedures. Mutants of

TABLE II

Frequency Distribution of NG-Induced Mutans in *C. acremonium*

N°	Requirement for	Number of mutants
1	Nicotinic acid	18
2	Pyridoxine (vitamin B_6)	16
3	Tryptophan or indole	13
4	Thiamine (vitamin B_1)	12
5	Arginine	11
6	Phenylalanine or tyrosine	10
7	Folic acid	9
8	Serine or glycine	8
9	Lysine	8
10	Inositol	7
11	Biotin	6
12	Choline	6
13	Leucine	5
14	Glutamic acid	5
15	Adenine, guanine	5
16	Uracil	4
17	Pantothenic acid	4
18	Alanine	4
19	Proline	3
20	Valine, isoleucine	2
21	Methionine, cysteine, homocysteine, cystathionine	3
22	Methionine, homocysteine	1
23	Cysteine, threonine	1
24	Threonine	1
25	α-Lipoic acid	1
26	Riboflavin	1

24 different requirements were found, including 13 amino acid, 3 nucleic acid base and 9 vitamin-deficient strains. The unequal frequency distribution of the mutants may indicate that they did not occur randomly but rather according to a specific pattern of distribution where certain defects were produced with high frequency while others were relatively rare. *E.g.* vitamin-deficient strains are frequently isolated whereas the isolation of a methionine-less, cysteine- or valine-less strain was a rather rare event. A high percentage of mutants gave no growth response to any of the supplements tested. It seems that many mutants (approx. 25%) have either multiple, complex or unusual growth-factor requirements.

CYTOLOGY

Cytological preparations were made in order to observe the structure of the chromatin of the nuclei and to determine the number of nuclei present in mature hyphal cells and in conidia. Stained preparations of the original haploid strain of *C. acremonium* CMI 49137, mutant 8650, were compared with preparations of a newly synthesized somatic diploid strain, which was obtained from a parasexual cross between two haploid monoauxotrophic strains of the above-mentioned mutant 8650. Hyphae in anastomosis were also stained before and after nuclear fusion had occurred. Anastomosis has been observed between hyphae of the haploid and the somatic diploid strain.

STAINING METHOD

The HCl-Giemsa staining procedure according to C. F. ROBINOW*, was used[13]. A loopful of a mycelial or conidial suspension was spread on a cover-slip, which was previously coated with albumin (5 ml of fresh egg-white added to 95 ml of distilled water, shaken and filtered immediately before use). The dried films were fixed in a modified Helly solution ($K_2Cr_2O_7$ 3%, $HgCl_2$ 5% in distilled water; immediately before use 0.6 ml of formalin, stored over powdered chalk, was added to 10 ml of this solution). The fixation was carried out in "Columbia staining dishes" for cover-slips** for 10 to 15 min. After fixation, the films were rinsed with tap water and stored for 10 min in 70% alcohol and finally washed thoroughly in tap water. By changing the dishes the cells were treated with saline (0.9%) for 90 min at 55 °C, rinsed again with tap water and hydrolysed for 8 min in 1 M-HCl at 60 °C. After hydrolysis they were rinsed with tap water and distilled water. The Giemsa solution was prepared by adding 0.5 ml of Giemsa stock solution*** to 10 ml of 0.1 M phosphate buffer (pH = 7.0). The films were left for 3 h in the staining solution. To keep the cover-slips upright, the "Columbia jars" were also used for the staining procedure. After a final examination of the stained films in buffer the cover-slip was mounted over a drop of tap water and blotted with blotting paper. Considerable pressure was applied with the thumb in a rotary direction. To prevent drying, the preparations were sealed with "Glyceel", a quickly hardening sealer for wet mounts†.

---

* Technique for the study of the chromatin of the nuclei of *Aspergillus* and *Penicillium* in fixed preparations, *private communication* (1967).
** A. H. Thomas Co., Philadelphia, Pa 19105, P. O. B. 779
*** Merck Co., Darmstadt, W. Germany.
† ESBE Co., Toronto 19, Canada.

RESULTS

The hyphal cells of the haploid strain of *C. acremonium* were found to be uninucleate without exception (Fig. 5). The hyphae of the newly synthesized somatic diploid strain also proved to be in a strict uninucleate condition (Fig. 6). Furthermore, the diploid strain appeared to be larger in nuclear size and cell volume. The conidia from the haploid strain were also found to be strictly uninucleate (Fig. 7). Conidia with more than one nucleus could not be detected so far. The few conidia which have been observed from the diploid strain also proved to be uninucleate. The pictures of hyphal anastomosis were taken after the nucleus of one of the participant cells had migrated into the other cell (Fig. 8). Anastomosis was preferentially observed between hyphae of the haploid strain.

HETEROKARYOSIS AND MITOTIC SEGREGATION

The techniques employed here have been described by several investigators[14-17]. Briefly, the process consists of mixing and plating auxotrophs of suitable genotypes on defined media, and analyzing any resulting prototrophs to determine whether they may be heterokaryons, heterozygous diploids, somatic recombinants or revertants.

THE SYNTHESIS OF HETEROKARYONS

Preliminary experiments designed to synthesize heterokaryons in *C. acremonium* showed that the heterokaryotic condition is not easily achieved in this organism. Occasional heterokaryosis was obtained, however, by employing germinating conidia on defined media. All of these experiments were carried out with monoauxotrophic strains of different colors (cross A and B, shown in Table III). Filtered and washed conidia from the two monoauxotrophs of cross A (M26 R, *his grn* × M26 D118, *leu whi*)* were first plated on unsupplemented and supplemented minimal medium (II with DL-histidine and DL-leucine respectively) in order to determine the frequency of back mutation and the viable count. Viable counts approximated to 10^6 conidia per ml; in none of these experiments were revertants recovered. In addition each combination of auxotrophs was checked for cross feeding by parallel streaking on unsupplemented minimal agar plates (II). For the synthesis of heterokaryons the two auxotrophic conidial suspensions (10^6 per ml each) were mixed thoroughly and added to the surface of minimal medium (II) supplemented with the growth factors required by the parent strains. The mycelial mat which resulted after 7—10 days of incubation at 26 °C was removed and washed in four changes of distilled water. The shredded fragments were added to the surface of unsupplemented minimal agar II. Cross B (M26 H302, *val ylo* × M26 D118, *leu whi*)* was carried out in the same way. Tiny stunted colonies

---

* *grn* green, *whi* white, *ylo* yellow.

Table III

Heterokaryons Obtained from Crosses A and B

Cross		Number of presumptive heterokaryotic colonies		
		without proto-trophic fans	with proto-trophic fans	with proto-trophic fans[a]
A	his grn × leu whi	253	21	4
B	val ylo × leu whi	312	25	0

[a] Identified as somatic diploids.

appeared at the periphery of the fragments after 2 weeks of incubation at 23°C. Material from those colonies was tranferred to fresh unsupplemented minimal agar plates and their prototrophic behaviour was validated by continued growth after transfer. The transfer was successively repeated three times with incubation periods of 10 days at 23 °C. It was only possible to maintain the prototrophy continuously if larger mycelial units of about 20 to 50 cells were picked up for transfer from those colonies; hyphal tips and single cells, on the other hand, proved to have regularly lost the prototrophic behaviour after transfer. The colonies are further characterized by an extremely slow growth and by their inability to form conidia. Cytologically only a low proportion of about 10 per cent of the hyphal cells were found to contain two nuclei in the same cytoplasmic unit; the overwhelming majority of cells still proved to be uninucleate. It appeared that most of the multinucleate cells observed so far in stained preparations were carrying imperfect or leaky cross walls joining two adjacent cells to a single cytoplasmic unit. This would suggest that the prototrophy was accomplished primarily by unstable and unbalanced heterokaryons and most probably also by cross-feeding mediated by anastomosis between the uninucleate component strains. Single cells obtained from hyphal tip isolations (conidia were not available) from the partially heterokaryotic colonies were plated on minimal medium (II supplemented with DL-histidine or DL-leucine) resulted in the recovery of the parental genotypes, which had been used in synthesizing the heterokaryon.

DIPLOIDY AND INDUCTION OF MITOTIC SEGREGATION

The above studies with heterokaryosis had actually been designed for the primary purpose of obtaining heterozygous diploids and subsequent recovery of mono- and diauxotrophic parental and nonparental segregants after haploidization. Prototrophic mycelia emerging from the above-described colonies were regularly observed in about 8 per cent of the colonies after an incubation period of three weeks at 23 °C on minimal medium II. They also showed a restored ability for conidia formation. The myce-lial fans themselves and their conidia were found to be prototrophic and, based on cytological stainings of the nuclei and the size of the hyphae, most probably haploid. The isolated mycelia do not undergo any segregation by use of p-fluorophenylalanine (FPA) which has been proved to be an effective haploidization agent in fungi[18,19] as well as in yeasts[20,21]. Therefore, the high frequency of sectoring suggests that the prototrophic fans may represent products of mitotic segregation and subsequent haploidization. This would be possible if mitotic segregation had occurred early in the development of an unstable diploid clone within the heterokaryotic colony giving rise to a fast growing haploid prototrophic fan. Any attempt to explain the proto-trophy in terms of back mutations would have to imply an unusually high reversion rate compared with the control plating of both parental genotypes. This would be very unlikely in view of a back-mutation frequency of less than 10^6, particularly when compared to the high frequency of sectoring (8%). Two combinations (cross A and B) of auxotrophs were tried in an attempt to synthesize diploids. What appeared to be a heterozygous diploid was finally obtained from cross A.

The presumptive diploid strain was isolated as a prototrophic mycelial fan in the same way as the haploid prototrophs. The hyphae of this strain were easily distin-guishable from the haploid ones by their unusual large size and thickness (Fig. 6). Staining of the nuclei of the diploid mycelium revealed again strictly uninucleate hyphal cells with nuclei about twice as large in diameter as the haploid ones (Fig. 6). The formation of conidia was observed in the diploid strain, but it was extremely poor compared with any of the haploid strains. The diploid strain appeared to be very stable, for spontaneous haploidization has not been so far observed. Several attempts were made to induce spontaneous haploidization. Successful mitotic segregation was finally achieved by the use of FPA as follows.

INDUCTION OF MITOTIC SEGREGATION

Mycelial fragments were spread on gradient plates, containing 0.45% FPA in the top layer. The plates were prepared by the method described by W. BRAUN[22]. The complete medium V was used for both layers. Each plate was inoculated with about 10^3 single mycelial fragments, because conidia were not available in an appropriate concentration from the respective diploid strain. At the end of the incubation time of 7 days at 30 °C, a strong inhibition of growth was observed, starting in the middle of the plates and increasing with FPA concentration. Twenty-eight colonies growing in inhibited zones (about 0.05 to 0.3% FPA) were picked and small hyphal fragments were streaked on complete medium (V). Fifteen of the colonies taken from the FPA plates were found to be still diploid. Evidence for diploidy was given morphologically (hyphal and nuclear size, observed in stained preparations) as well as by the poor formation of conidia or inability to do so, which is a characteristic peculiarity of the

TABLE IV

Effect of FPA Treatment on a Diploid Strain of *C. acremonium*

Total of colonies isolated[a]	Diploid or partially disomic colonies[a]	Genotype of the haploid conidia[b]				
		number of mixed colonies analysed	parental segregation		nonparental segregation	
			genotype	number	genotype	number
28	15	13	+ leu whi	18	+ + whi	17
			+ leu grn	28	+ + grn	25
			his + grn	22	his leu grn	1
			his + whi	17	his leu whi	2

[a] After FPA treatment.

[b] After FPA induced mitotic segregation of the somatic diploid $\dfrac{his}{+} : \dfrac{+}{leu}$ grn whi

(white phenotype).

diploid strains observed so far. The remaining thirteen colonies exhibiting a satisfactory formation of conidia were examined cytologically. They were found to contain both haploid and diploid hyphae. The respective dilution streaks of hyphal fragments on complete medium V resulted in colonies of the original diploid strain and newly induced haploid mitotic segregants. They were obtained in unequal proportions. Only a minority (varying between 5 and 20%) of the colonies of the respective dilution streak were identified as haploid segregants. On the other hand, nearly all the conidia produced by the fifteen original mixed colonies proved to be haploid; they were most probably formed by the haploid hyphal minority component. Therefore, the analysis of those conidia was a useful tool for the recovery of the genotypes of the mitotic segregants. Each of the thirteen colonies was resuspended in 5 ml of phosphate buffer and the conidia were separated from the hyphae by glass-wool filtration. The conidia were plated on complete medium (V) and single colonies derived from each were checked for their requirements. The genotypes of ten different conidia taken from each of the thirteen mixed colonies were analyzed, the results being listed in Table IV. Analysis of the conidia indicates that the segregation of parental and nonparental genotypes with respect to the auxotrophic markers is equally likely for only three of the four combinations, which are expected for a random assortment. The exceptionally low frequency of the double auxotrophic segregants might be due to the low germination rate, which has been found in reconstruction experiments with the conidia of three double auxotrophic segregants, which could be isolated so far. It might be possible for this reason that haploid segregants are produced, in which all four combinations are indeed equally likely but the recovery of the latter type might be rendered more difficult. The markers are for the present assumed to be independent and,

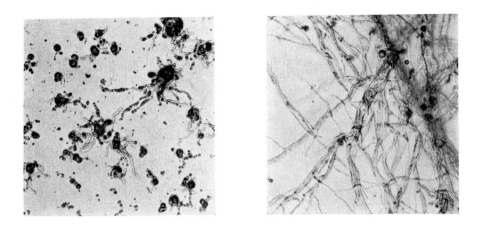

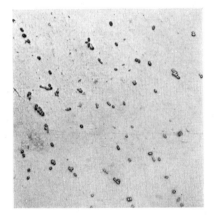

Fig. 3. *Upper left:* Germinated conidia of the strain 8650 *inl met* (inositol- and methionine-requiring) on methionine-containing minimal agar without inositol;

upper right: Germinated conidia of the wild strain 8650 on minimal agar without inositol;

lower: Resting conidia of the strain 8650 *inl met* on inositol-containing minimal agar without methionine.

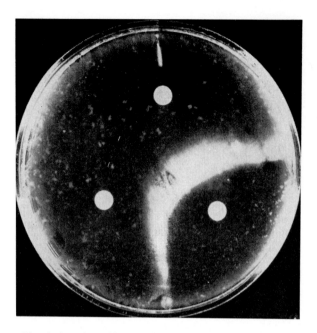

Fig. 4. The effect of inositol and methionine deprivation on germinating conidia of the double mutant 8650 *inl met.*

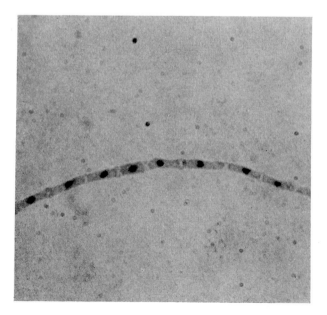

Fig. 5. Uninucleate hyphae of the haploid strain of *C. acremonium,* stained with Giemsa after hydrolysis. × 1300.

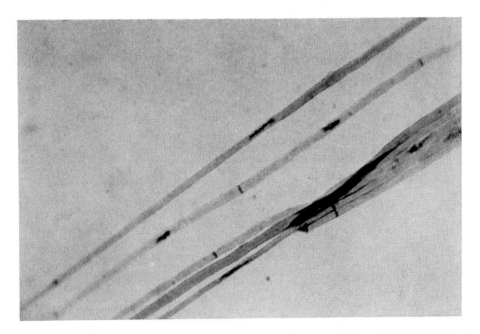

Fig. 6. Uninucleate hyphae of the somatic diploid strain of *C. acremonium*, stained with Giemsa after hydrolysis. × 1300.

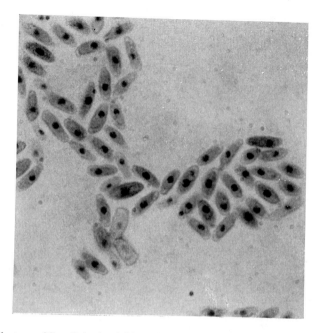

Fig. 7. Uninucleate conidia of the haploid strain of *C. acremonium* stained in the same way as the hyphae. × 1300.

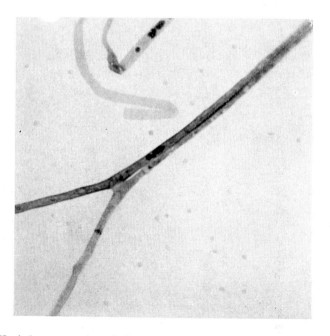

Fig. 8. Hyphal anastomosis and plasmogamy, stained as mentioned above. × 1800.

therefore, on different linkage groups. Whether the colors, which exhibit a phenotypic random assortment, are also controlled by single loci needs to be established in further experiments, namely by sectoring of heterozygous mitotic recombinant diploids.

Our results indicate that the haploid strain of *C. acremonium* is able to undergo plasmogamy and karyogamy leading to a relatively stable diploid strain, which can be induced to mitotic segregation. These investigations suggest the possibility of carrying out genetic analyses in *C. acremonium*, a fungus lacking a well-defined sexual cycle.

METABOLIC PATTERNS OF CEPHALOSPORIN C FERMENTATION

The relationship between growth and some physiological factors such as sugar and methionine consumption and cephalosporin C formation is represented in Fig. 9. The production of the antibiotic occurs mainly in a phase of slow cell-mass increase, rapid sugar and methionine consumption between 72 and 150 h. Usually under all conditions tested the net synthesis of cephalosporin C ceases 12—36 h after methionine and sugar have been exhausted. At this moment lytic effects are evident, *e.g.* the

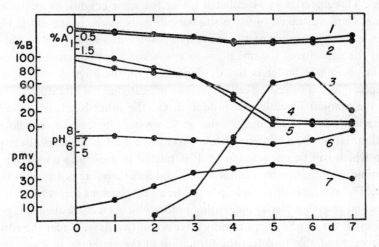

Fig. 9. General metabolic pattern of cephalosporin C fermentation by *C. acremonium. 1* CO_2,% A, *2* O_2, % A, *3* cephalosporin C, relative units, *4* total sugar, % B, *5* methionine, % B, *6* pH, *7* growth, pmv (packed mycelial volume, %).

ammonia level as well as the pH rise and consequently the cephalosporin C titer and the cell mass tend to decrease. The phase-bound feature of cephalosporin C biosynthesis has been further demonstrated in the following way. From fermentations in tanks samples of culture broths have been withdrawn at various intervals of the growth phases (trophophase) and the phase of accumulation of the antibiotic (idiophase). The mycelium has been separated from the broth, washed several times and

TABLE V

Cephalosporin C Synthesis (μg/ml) with Cell Suspensions and Disrupted Cells of Different Ages

Age of culture h	Incubation, h					
	Cell suspension			Disrupted cells		
	0	24	48	0	24	48
48	—	trace	trace	—	—	—
72	—	83	228	—	—	—
96	—	10	61	—	—	trace
120	61	85	72	62	81	62

depleted of endogenous cephalosporin C by incubation in phosphate buffer at pH 7 for 3 h. Thereafter, a part of the mycelial mass has been dispensed in a ratio of 30 g per litre (wet weight) in 0.2 M phosphate buffer of pH 7 and incubated in shake flasks on a rotary shaker under standard conditions for cephalosporin C formation. From the same mycelial mass, disrupted-cell suspensions have been prepared with the aid of an X-press. This material was incubated under the same conditions as the mycelial suspensions. Care was taken to use disrupted-cell suspensions which did not contain more than 100 subculturable units per ml. The results are summarized in Table V. It seems that the capacity of the mycelium to synthesize a certain amount of cephalosporin C in phosphate buffer is strictly dependent on the physiological age of the culture. Special media or resting cell systems as indicated in the literature[10,22] had no effect on the amount of antibiotic produced. On the other hand, trials where DL-methionine was added to the buffer solution showed an increase in cephalosporin C fermentation. However, methionine proved to have a dual effect on cephalosporin C formation which will be discussed later. It is related to growth as a nitrogen source and to antibiotic formation as a sulfur donor and was therefore not used in these experiments. The synthesis of cephalosporin C in all suspensions seems to be based on precursors and reserve material preexisting in the cells. As a consequence it proceeds at a relatively low rate and drops rapidly after about two days. Under the conditions chosen we were not able to detect the formation of the antibiotic in a suspension of disrupted cells. Corresponding data from the literature are somewhat contradictory[22,23]. At the end of the production phase, after 120 h, the mycelium contains a certain amount of cephalosporin C which is slowly released into the medium. Therefore, this preparation of disrupted cells shows antibiotic activity which, however, does not increase with time.

In order to evaluate further the growth dependence of cephalosporin C synthesis we undertook a series of experiments where various inhibitors like griseofulvin, amphothericin B and a mixture of KCN, iodo- and fluoroacetate were added to a culture in a synthetic production medium after 48 and 72 h, respectively. The results in Fig. 10

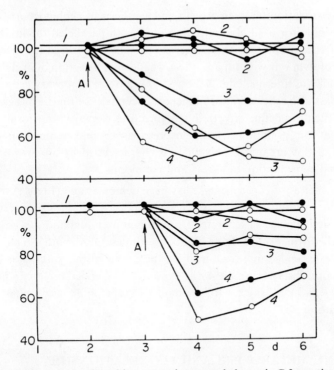

Fig. 10. The inhibition of growth and its connection to cephalosporin C formation in *C. acremo-nium*. *Closed symbols:* formation of cephalosporin C, %; *open symbols:* growth, %; *1* control, *2* griscofulvin, 200 µg/ml, *3* KCN—fluoroacetate—iodoacetate, 50 µg/ml, *4* amphotericin B, 50 µg/ml; A addition of inhibitor.

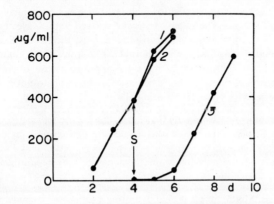

Fig. 11. The importance of media composition with regard to cephalosporin C synthesis (µg/ml) in *C. acremonium*. *1* control, *2* resuspension in culture liquid, *3* mycelium suspended in fresh medium; S separation of mycelium from culture liquid.

show a clear correlation between growth and synthesis of cephalosporin C. Independently of the inhibitor used, the inhibition of growth paralleled the inhibition of the formation of the antibiotic. Amphothericin B had in general the strongest effect whereas griseofulvin was without any effect whatever. The action of cycloheximide, on the other hand, was intermediate. It showed only a weak temporary growth-inhibiting effect which was also reflected in its influence on cephalosporin C synthesis. Although it is rather difficult to define growth in a filamentous organism our results lead to the conclusion that the formation of cephalosporin C is bound to a certain phase in the growth cycle of this organism. This phase is characterized by a slow increase in cell mass and rapid consumption of sugars and methionine.

In the last series of experiments it has been demonstrated (Fig. 11) that the composition of the medium influences the cephalosporin C-synthesizing system. If mycelium in the active phase is separated from its culture liquid and subsequently resuspended in it the formation of the antibiotic continues. On the other hand, if the same mycelium is resuspended in a fresh medium the course of biosynthesis is interrupted. This interruption lasts for about 48 h. It seems that under these conditions a new growth phase is induced which is followed once more after 48 h by a phase of renewed antibiotic synthesis.

METHIONINE METABOLISM AND ITS RELATIONSHIP TO CEPHALOSPORIN C BIOSYNTHESIS

MATERIALS AND METHODS

Microbiological assays of cephalosporin C and penicillin N. Cephalosporin C was quantitatively determined using an agar diffusion test with *Alcaligenes faecalis* ATCC 8750[24] as the test organism. The strain was specially selected for high resistance against penicillin N and the P-type cephalosporins. Penicillin N, on the other hand, was assayed with a selected strain of *Sarcina lutea* ATCC 9341. This strain shows a high resistance against the steroid-like antibiotics of the cephalosporin P-type and also against cephalosporin C. Therefore, it was not necessary to destroy any of the antibiotics in the culture broths when the appropriate test organism was used. The determination of cephalosporin C included always deacetylcephalosporin C which possesses about one-third of the activity of cephalosporin C itself.

Paper chromatography of cephalosporin C and penicillin N. The separation of cephalosporin C, deacetylcephalosporin C, penicillin N and the cephalosporin P group was done by descending chromatography using Whatman N° 1 paper with butanol—acetic acid—water (11 : 3 : 11) as solvent[25]. The presence of the antibiotics was demonstrated by bioautograms with *Neisseria catharralis* or the other above-mentioned test organisms.

Methionine analysis. Methionine was analyzed using a thin-layer chromatography technique. 1 μl of culture broth was dropped on a thin-layer plate (SL 254, silica gel with fluorescent additive, layer thickness 0.1 mm), dried and developed with a mixture of *n*-butanol—acetic acid—0.1M-NaCl (20 : 5 : 20 $v/v/v$) as solvent. After drying, the plates were sprayed with a ninhydrin solution. A semiquantitative evaluation was carried out by comparing the spots of the samples with those of standard solutions of DL-methionine in various concentrations. The determination of methionine according to the method of LA RUE[26] could not be employed because of its relative nonspecificity; *e.g.* it responds to various mercapto-compounds.

Determination of 2-oxo- and 2-hydroxy-4-methylthiobutyric acid. 1 µl of culture broth is dropped onto a cellulose thin-layer plate (layer 0.1 mm, E. Merck, Darmstadt) and developed for 4 h with a mixture of 2-propanol—formic acid—H_2O (77 : 4 : 19 $v/v/v$) as solvent. The presence of the compounds is shown by spraying with a 1 : 1 (v/v) mixture of 1.7% $FeCl_3.6H_2O$ and 2% $K_3[Fe(CN)_6]$. The substances appear as deep-blue spots with R_F values of 0.77 (oxoacid) and 0.87 (hydroxyacid).

Tracer experiments. *Me*-^{14}C-L-Methionine, ^{35}S-L-methionine and 3-^{14}C-L-serine (Radiochemical Centre, Amersham, England) have been used for studies of incorporation in cephalosporin C. The separation of the radioactive antibiotics was done by paper chromatography (*see above*). The procedure of detection of the labeled compounds and measurement of radioactivity followed the method of CALTRIDER and NISS[10].

Culture conditions for production of cephalosporin C. Normally, growth from a slant containing complete medium V was suspended in 5 ml of 0.2 M phosphate buffer pH 7. With the mycelial suspension thus obtained 500 ml shake flasks with 100 ml of seed medium I were inoculated at a ratio of 1%. The flasks were incubated on a rotary shaker with 250 r.p.m. at 25 °C for 72 h. These seed cultures were used to inoculate the synthetic production medium III which is based on media studies of OTT *et al.*[8]. This medium served as a basis for all production and incorporation studies. Usually 500 ml baffled shake flasks with 100 ml of medium III were inoculated with 5% of seed cultures or washed mycelial suspensions from seed cultures. The incubation was carried out on a rotary shaker with 250 r.p.m. at 25 °C. Maximal antibiotic activity was reached after 120 to 144 h.

RESULTS AND DISCUSSION

It is generally accepted that methionine, especially D-methionine, has a unique stimulatory effect on cephalosporin C as well as penicillin N biosynthesis in fungi of the genus *Cephalosporium*[8,27]. Nevertheless, there is still uncertainty with regard to the role played by this compound in the biosynthesis. On the basis of a certain structural analogy it was first believed that D-methionine might be a precursor of the AAA side chain[28]. However, it was found that AAA did not replace methionine and that practically no radioactivity from 1-^{14}C-methionine was incorporated into penicillin N[29]. Today two different hypotheses still exist with regard to the role played by methionine in the biosynthesis of the cephalosporins. DEMAIN[2,30] assumes on the basis of experiments in a chemically defined medium, where norleucine, a structural analogue of methionine, could completely replace methionine, that the latter was not a precursor of cephalosporin C but rather an inhibitor of a cysteine-degrading enzyme. It is generally accepted that cysteine is a direct precursor of ACA as well as APA[1,31]. The second hypothesis by CALTRIDER and NISS[10] is based on experiments in a complete medium where the incorporation of ^{35}S-L-methionine and 3-^{14}C-DL-serine into cephalosporins was studied. Their data suggest that methionine plays an essential role as a sulfur donor for the cephalosporin nucleus. The transfer of sulfur from methionine to cysteine seems to follow the pathway of transsulfuration[32] with cystathionine as intermediate.

It is evident that the methionine metabolism offers an excellent possibility for studying the relation between primary and secondary metabolism in the particular case of cephalosporin C biosynthesis. Hence we investigated the biosynthesis of sulfur-containing amino acids in *C. acremonium*, its connection to the biosynthesis of cephalosporin C and the question of the high methionine concentration needed for optimal formation of the antibiotic.

The study of methionine metabolism is mainly based on investigations with auxotrophic mutants of *C. acremonium*. The auxanographic pattern of mutant 8650 *slp* in Table VI shows that this strain needs a sulfur-containing amino acid for growth. Although final biochemical confirmation has still to be obtained one might conclude that this mutant must be blocked in its sulfate utilization. Therefore, it is unable to produce cysteine from serine and sulfate. Such mutants have been isolated from *Neurospora* and other fungi[33,34]. The interchangeability of all the sulfur-containing amino acids as growth factors for mutant 8650 *slp* makes it very probable that *C. acremonium* possesses, as in *Neurospora* and other fungi[32,35], the cystathionine-mediated pathway of reverse transsulfuration. Thanks to this mechanism the fungus can synthesize methionine from cysteine or cysteine from methionine. Further evidence for this as-

TABLE VI

Growth Requirements of *C. acremonium* 8650 *slp* in Minimal Medium II

Addition (1 g/l)	Growth
None	—
DL-Methionine	+
L-Methionine	+
D-Methionine	+
DL-Methionine-DL-sulfoxide	+
S-Methyl-L-cysteine	+
DL-Homocysteine	+
DL-(+)-*allo*-Cystathionine	+
L-Cysteine	+
DL-Norleucine	—
DL-Ethionine	—
DL-Serine	—
DL-Homoserine	—
Na_2SO_4	—
Na_2SO_4 + DL-serine	—
Na_2SO_4 + DL-homoserine	—

TABLE VII

Growth Requirements of Two Mutants Derived from Strain 8650 *slp* in Minimal Medium II

Addition (1 g/l)	Growth of mutant	
	8650 *slp met* 68/3	8650 *slp cys* 28/1
DL-Methionine	+	−
DL-Homocysteine	+	−
DL-(+)-*allo*-Cystathionine	+	−
DL-Cysteine	−	+

sumption has been obtained by mutants derived from strain 8650 *slp* (*see* Table VII). Whereas mutant 8650 *slp met* 68/3 depends on methionine or homocysteine or cysta- thionine, strain 8650 *slp cys* 28/1, on the other hand, is strictly cysteine-deficient. In view of the reverse transsulfuration, this pattern of growth requirements leads to the assumption that the first strain may be deficient in its cystathionine γ-synthase where- as the second strain seems to lack γ-cystathionase activity[32,35,36]. For final proof the corresponding biochemical tests are actually carried out.

The influence of methionine on growth and cephalosporin C formation of the sulfate non-assimilating mutant was compared with the wild strain in the synthetic medium III. The mutant strain showed a net increase in the production of the anti- biotic (*see* Table VIII) in comparison with the wild strain. That there is a connection between the introduction of a mutation in the sulfate-assimilating pathway and cephalosporin C formation is further supported by the isolation of revertants to the sulfate-assimilating wild type. Such strains produce the same amount of cephalo-

TABLE VIII

Influence of DL-Methionine on Growth and Cephalosporin C Formation of Mutant 8650 *slp* and its Parent Strain in Synthetic Medium III

Addition	g/l	Strain			
		parent		mutant 8650 *slp*	
		growth pmv %	cephalosporin C μg/ml	growth pmv %	cephalosporin C μg/ml
None	—	20	105	—	—
(NH₄)₂SO₄	5	20	43	—	—
	1	33	145	21	250
	2	27	189	34	534
DL-Methionine	4	25	123	26	810
	8	21	91	24	850

TABLE IX

Comparison of Cephalosporin C Formation of a Sulfur Non-Assimilating Mutant with its Parent Strain and its Revertant

Medium	Strain	Cephalosporin C µg/ml
IIIa	8650 *slp* (SO_4^{2-}—✗→ cysteine)	760
	8650 *slp*+ (parent strain)	350
	8650 *slp*+ (revertant)	292
IIIa without methionine	8650 *slp*	—
	8650 *slp*+ (parent strain)	210
	8650 *slp*+ (revertant)	191

sporin C as the parent strain (*see* Table IX). If one assumes that the assimilation of sulfate occurs mainly through the synthesis of cysteine from serine and sulfate, then a mutation which leads to an interruption of this pathway would necessarily be related to the mechanism of formation of cephalosporin C. BENKO *et al.*[33] have demonstrated that mutants of *Aspergillus nidulans* blocked in the formation of sulfide from sulfate no longer respond to inhibitors of L-methionine permease. Their results suggest strongly that various fungi possess a sulfide-regulated L-methionine permease. With regard to mutant 8650 *slp* we may therefore assume that the defect in the sulfate-depending cysteine formation parallels a deinhibited state of L-methionine permease which allows a continuous flow of methionine into the cells. DENNEN and CARVER[37] found a relation between sulfur sources, sulfatase activity and formation of cephalosporins in *C. acremonium*. Whereas sulfate represses the sulfatase, organic sulfur sources such as methionine, cysteine and others, derepress the enzyme to various degrees. Although there was a relation between increasing ability to synthesize the sulfur-containing antibiotics and sulfatase activity, high sulfatase activity was not essential for the formation of the cephalosporins. The authors assume that high-yielding mutants may have a decreased ability to degrade methionine for metabolic processes other than antibiotic synthesis. GOULDEN and CHATTAWAY[38] found in a high-penicillin-yielding mutant of *P. chrysogenum* that the increase of penicillin formation is paralleled by an increase in activity and decrease in end-product inhibition of acetohydroxyacid synthase by valine, an enzyme initiating valine synthesis, one of the precursors of the penicillin nucleus. This situation resembles the one encountered in the sulfate non-assimilating mutant, where, in contrary to the regulation of the formation, the transport of an important precursor seems to be deregulated.

In contrary to the situation in *P. chrysogenum* where inorganic sulfur is the optimal sulfur source[5] for penicillin formation, the synthesis of the cephalosporins depends strictly on methionine. Although it was shown that cysteine had a sparing effect

TABLE X

Influence of DL-Methionine, L-Cysteine and DL-(+)-*allo*-Cystathionine on Growth and Cephalosporin C Formation of Mutant 8650 *slp*

Addition	g/l	Growth pmv %	Cephalosporin C μg/ml
DL-Methionine (control)	4	22	822
L-Cysteine	1	29	225
	2	28	303
	4	52	450
DL-(+)-*allo*-Cystathionine	0.4	30	—
	0.8	27	—
	4	26	241

on methionine consumption[10,37] no organic sulfur source was found which could fully replace methionine with the exception of a few methionine derivatives. Methionine has not only a stimulatory effect on the biosynthesis of cephalosporins but also affects growth and morphology of the mycelium of *C. acremonium*[10]. In our experiment, a sulfate-assimilating wild strain (*see* Table VIII) was inhibited in its growth and antibiotic formation with high DL-methionine concentrations whereas a sulfate non-assimilating mutant was not influenced even at higher concentrations of DL-methionine. Contrary to methionine, cysteine and cystathionine (*see* Table X) as well as homocysteine induce a higher growth rate of a filamentous type in contrast to the swollen irregularily septated and fragmented hyphae in methionine-containing media. In addition, much lower yields of cephalosporin C are obtained. As CALTRIDER and NISS[10], we were not able to demonstrate a stimulatory effect of DL-norleucine on cephalosporin C formation as stated by DEMAIN *et al.*[30].

In spite of the rather clear evidence of the role of methionine as a sulfur donor in the cephalosporin C biosynthesis[10] we made further incorporation studies with mutants deficient in sulfate assimilation and serine biosynthesis. Under the condition that the formation of the antibiotic under study is not impaired by the introduction of a mutation to auxotrophy, strains deficient in the synthesis of presumable precursors have great advantages over wild strains in that the fate of the compound can be observed much more easily. In our case we did not find a negative influence of auxotrophy on antibiotic formation, provided that the nutrient was added in the proper concentration which is in general much in excess of concentrations needed for growth (*see* Table XI). Table XII demonstrates that the incorporation of ^{35}S from L-methionine into cephalosporin C occurred without any molar dilution of radioactivity.

These results with a sulfate non-assimilating strain correspond exactly to the results of CALTRIDER and NISS[10] obtained with a wild strain. On the other hand,

TABLE XI

Cephalosporin C Formation in Basal Medium III with Mutant 8650 *slp* and 8650 *slp ser* A 314

Additions	g/l	Cephalosporin C μg/ml	
		Mutant 8650 *slp*	Mutant 8650 *slp ser* A 314
—	—	—	—
DL-Methionine	4	810	—
DL-Methionine	4	—	822
L-Serine	4		
DL-Methionine	8	928	—
DL-Methionine	8	—	870
L-Serine	8		
L-Cysteine	4	488	—
L-Cysteine	4	—	259
L-Serine	4		

[14]C-methyl from L-methionine could not be detected in cephalosporin C[1,10]. 3-^{14}C-L-Serine also showed a very high incorporation into the antibiotic. There was no molar dilution of radioactivity in a serine-deficient strain and a rather low dilution in the corresponding wild strain (*see* Table XI). These results support very strongly the hypothesis of cystathionine-mediated cysteine formation from methionine and serine by means of the pathway of reverse transsulfuration. The question why extracellular cysteine is an inefficient precursor of the cephalosporin nucleus is still unanswered. However, it seems that only cysteine generated from cystathionine by the γ-cystathio-

TABLE XII

The Incorporation of ^{35}S-L-Methionine, *Me*-^{14}C-L-Methionine and 3-^{14}C-L-Serine into Cephalosporin C by Mutants of *C. acremonium*[a]

Strain	Substance	Incorporation % of total added	Dilution of radioactivity
8650 *slp*	^{35}S-L-methionine	7.43	1.05
	Me-^{14}C-L-methionine	—	—
	3-^{14}C-L-serine	1.57	2.65
8650 *slp ser* A 314	3-^{14}C-L-Serine	3.01	1.06

[a] Basal medium III with 4 g DL-methionine or L-serine per liter. 3-^{14}C-L-Serine and ^{35}S-L-methionine (0.1—0.2 μCi/ml) added after inoculation.

nase is optimally incorporated. In fact, mutant 8650 *slp met* 68/3 which is, as discussed before, probably deficient in its cystathionine γ-synthase and, therefore, no longer able to form methionine from cysteine is still as active as the wild strain with regard to cephalosporin C production. On the other hand, mutant 8650 *slp cys* 28/1 has a presumable defect in its γ-cystathionase and consequently needs cysteine for growth. This strain shows the same low cephalosporin C yield with cysteine as the wild strain.

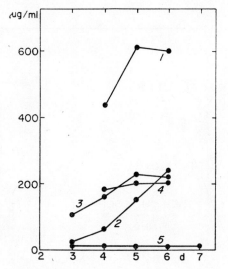

Fig. 12. Cephalosporin C formation (µg/ml) by mutant 8650 *slp cys* 28/1 in basal medium III. *1* control wild strain 8650 *slp*, medium III with 4g/l DL-methionine, *2* 8650 *slp cys* 28/1, medium III with 4 g/l L-cysteine, *3* 8650 *slp cys* 28/1, medium III with 2 g/l L-cysteine, *4* control wild strain 8650 *slp*, medium III with 4 g/l L-cysteine, *5* 8650 *slp cys* 28/1, medium III with 4 g/l DL-methionine.

Methionine has no influence on antibiotic synthesis in this mutant (*see* Fig. 12). Studies in this direction are still in progress. Other mutants, especially β-cystathionase-less strains have to be isolated and tested and biochemical investigations are needed for further clarity. Nevertheless, one may assume that a linkage exists between the γ-cystathionase and the cephalosporin-synthesizing system (*see* Fig. 13).

Another question which is still open with regard to methionine lies in the fact that for optimal cephalosporin C synthesis methionine has to be added far in excess of the theoretical amount needed for incorporation of sulfur into the nucleus of the antibiotic. This is in contrary to the growth requirements which can be satisfied by concentrations much lower than the ones needed for the synthesis of the cephalosporins. *C. acremonium* is able to use both stereoisomers for growth whereas only the D-form is of optimal efficiency in cephalosporin C biosynthesis. Therefore, it was supposed that methionine may play a secondary role in the biosynthesis of cephalosporin C. CALTRIDER *et al.*[39] presumed that methionine-grown cells differed in many respects from cultures grown in a sulfate-supplemented medium, *e. g.* they had an increased rate of respiration, higher rate of katabolism of acetate and glucose and higher specific activities of certain respiratory enzymes. We showed that the sulfate non-assimilating mutant, as well as the wild type are able to grow on methionine as the sole source of nitrogen. With the aid of *Me*-^{14}C-methionine we were able to demonstrate that up to

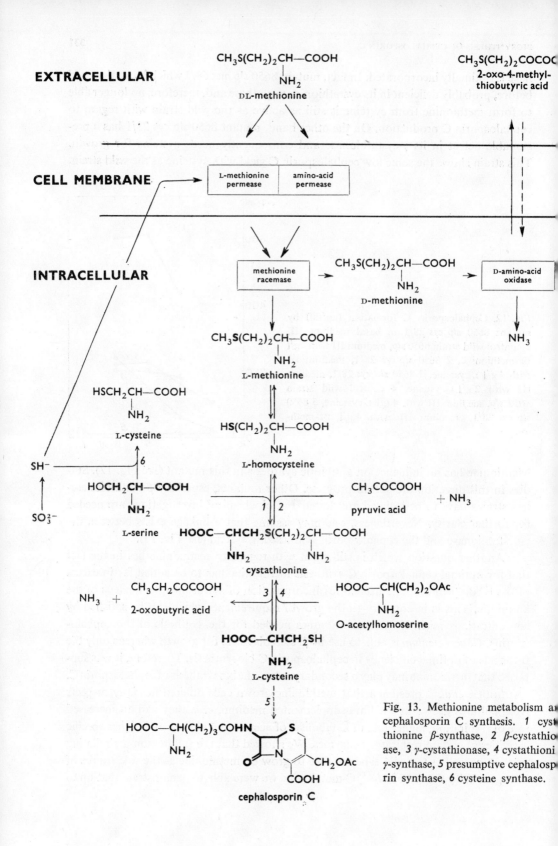

Fig. 13. Methionine metabolism and cephalosporin C synthesis. 1 cystathionine β-synthase, 2 β-cystathionase, 3 γ-cystathionase, 4 cystathionine γ-synthase, 5 presumptive cephalosporin synthase, 6 cysteine synthase.

70% of the radioactivity could be detected in the culture filtrate. This radioactive material was shown by paper chromatography (*see* Materials and Methods) to have an R_F value of 0.84, whereas the R_F value of methionine was 0.54. The compound has later been isolated and identified as 2-oxo-4-methylthiobutyric acid and its hydroxy form. Both compounds represent deaminated methionine. We were able to isolate from various strains of *C. acremonium* a soluble D-amino-acid oxidase with properties similar to analogous enzymes from *Neurospora*[34,40]. Although the isolated enzyme is strictly D-methionine-specific, whole cells, on the other hand, deaminate both stereoisomers of methionine. As we were not able to detect L-amino-acid oxidase activity in *C. acremonium* we presume that L-methionine is reversed into the D-form in the cells and subsequently deaminated. The deaminated compound is excreted into the medium and is only slowly re-used with time. This reaction explains why *C. acremonium* can grow and produce cephalosporin C in a medium containing methionine as the sole nitrogen source. It is understandable that high levels of methionine have to be added to the culture medium, because methionine is being continuously deaminated. The results discussed above are summarized in Fig. 13.

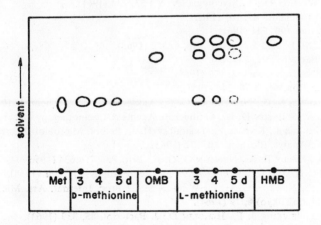

Fig. 14. The difference between D- and L-methionine consumption in *C. acremonium*. Met DL-methionine, OMB 2-oxo-4-methylthiobutyric acid, HMB 2-hydroxy-4-methylthiobutyric acid.

Although it has been clearly demonstrated in our experiments and others[10] that sulfur from L-methionine is incorporated into cephalosporin C, D-methionine proved to be more efficient than the L-form. In an experiment in which the uptake of D-methionine was compared with the uptake of L-methionine it was seen that the latter is much more rapidly taken up and subsequently deaminated (*see* Fig. 14). Thus D-methionine, which probably enters the cells by means of a nonspecific general amino-acid permease[33,41], is less rapidly metabolized and, therefore, ensures a long lasting pool of free methionine available for cephalosporin C synthesis.

REFERENCES

1. ABRAHAM E. P., NEWTON G. G. F., p. 1 in D. Gottlieb, P. D. Shaw (Eds.): *Antibiotics*, Vol. II. Springer-Verlag, Berlin—Heidelberg—New York 1967.
2. DEMAIN A. L., p. 29 in J. Snell (Ed.): *Biosynthesis of Antibiotics*, Vol. I. Academic Press, New York 1966.
3. COLE M.: Appl. Microbiol. **14**, 98 (1966).
4. THORN J. A., JOHNSON M. J.: J. Am. Chem. Soc. **72**, 2052 (1950).
5. DOYLE F. P., NAYLER J. H. C.: Adv. Drug Res. **1**, 1 (1964).
6. ELANDER R. P., STAUFFER J. F., BACKUS M. P.: Antimicrob. Agents, Ann. Meet., 91 (1961).
7. STAUFFER J. F., SCHWARTZ L. J., BRADY C. W.: Developments Ind. Microbiol. **7**, 104 (1966).
8. OTT J. L., GODZESKI C. W., PAVEY D., FARAN J. D., HORTON D. R.: Appl. Microbiol. **10**, 515 (1962).
9. DEMAIN A. L.: Trans. N. Y. Acad. Sci. **25**, 731 (1963).
10. CALTRIDER P. G., NISS H. F.: Appl. Microbiol. **14**, 746 (1966).
11. WOODWARD V. W., DE ZEEUW J. R., SRB A. M.: Proc. Nat. Acad. Sci. U.S.A. **40**, 192 (1954).
12. LESTER H. E., GROSS S. R.: Science **129**, 572 (1959).
13. ROBINOW C. F., MARAK J.: J. Cell Biol. **29**, 129 (1966).
14. PONTECORVO G., ROPER J. A., FORBES E.: J. Gen. Microbiol. **8**, 198 (1953).
15. ROPER J. A., PRITCHARD R. H.: Nature **175**, 639 (1955).
16. TUVESON R. W., COY D. O.: Mycologia (N. Y.) **53**, 244 (1961).
17. FANTINI A. A.: Genetics **47**, 161 (1962).
18. LHOAS P.: Nature **190**, 744 (1961).
19. MCCULLY K. S., FORBES E.: Genet. Res. **6**, 352 (1965).
20. GUTZ H.: J. Bacteriol. **92**, 5 (1966).
21. BRAUN W.: Bacteriol. Genet. **2**, 120 (1965).
22. DEMAIN A. L.: Clin. Med. **70**, 245 (1963).
23. SMITH B., WARREN S. C., NEWTON G. G. F., ABRAHAM E. P.: Biochem. J. **103**, 877 (1967).
24. CLARIDGE C. A., JOHNSON D. L.: Antimicrob. Agents & Chemotherapy 682 (1962).
25. NÜESCH J., GRUNER J., KNÜSEL F., TREICHLER H. J.: Pathol. Microbiol. **30**, 880 (1967).
26. LA RUE F. A.: Analyt. Biochem. **10**, 172 (1965).
27. MILLER G. A., KELLY B. K., NEWTON G. G. F.: Brit. Pat. 759,624 (1956).
28. KAVANAGH F., TUNIN D., WILD G.: Arch. Biochem. Biophys. **77**, 268 (1958).
29. ABRAHAM E. P., NEWTON G. G. F., WARREN S. C.: Proc. Internat. Ass. Microbiol. Symp. Appl. Microbiol. (Tokyo) **6**, 79 (1964).
30. DEMAIN A. L., NEWKIRK J. F., HENDLIN D.: J. Bacteriol. **85**, 339 (1963).
31. TROWN P. W., SHARP M., ABRAHAM E. P.: Biochem. J. **86**, 284 (1963).
32. KERR D. S., FLAVIN M.: Biochem. Biophys. Res. Comm. **31**, 124 (1968).
33. BENKO P. V., WOOD C. T., SEGEL I. H.: Arch. Biochem. Biophys. **122**, 783 (1967).
34. OHNISHI E., MACLEOD H., HOROWITZ N. H.: J. Biol. Chem. **237**, 138 (1962).
35. KERR D. S., FLAVIN M.: Biochem. Biophys. Res. Comm. **31**, 124 (1968).
36. FISCHER G. A.: Biochim. Biophys. Acta **25**, 50 (1957).
37. DENNEN D. W., CARVER D. D.: Can. J. Microbiol. **15** (2), 175 (1969).
38. GOULDEN S. A., CHATTAWAY F. W.: J. Gen. Microbiol. **59**, 111 (1969).
39. CALTRIDER P. G., HUBER F. M., DAY L. E.: Appl. Microbiol. **16**, 1913 (1968).
40. BENKO P. V., WOOD F. C., SEGEL I. H.: Arch. Biochem. Biophys. **129**, 498 (1969).
41. BENZ F., LIERSCH M., NÜESCH J., TREICHLER H. J.: European J. Biochem. **20**, 81 (1971).

Determination of the Biosynthetic Pathway for Pyrrolopyrimidine Nucleosides

ROBERT J. SUHADOLNIK

Research Laboratory, Albert Einstein Medical Center, Philadelphia, Pennsylvania, U.S.A.

Recent studies have established that purines, purine nucleosides or purine nucleo-tides serve as precursors in the biosynthesis of eight of the naturally occurring nucleo-side antibiotics[1]. While a purine ring serves as the carbon and nitrogen source for the biosynthesis of the purine ring of psicofuranine, decoyinine, 3'-deoxyadenosine, 3'-amino-3'-deoxyadenosine and 3'-acetamido-3'-deoxyadenosine, only the pyrimi-dine ring and possibly nitrogen 9 of a purine is involved in the biosynthesis of the pyrrolopyrimidine ring of toyocamycin, sangivamycin, and tubercidin[2,3] (Fig. 1). Carbon 8 of 8-^{14}C-adenine does not become carbon 6 of tubercidin or toyocamycin in the conversion of a purine to the pyrrolopyrimidine nucleosides[2,3]. The biosynthetic origin of pyrrole carbons 6, 5 and the cyano group of toyocamycin has been shown

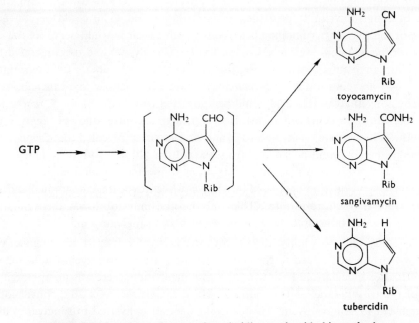

Fig. 1. Initial pathway for pyrrolopyrimidine-nucleoside biosynthesis.

335

to arise from ribose. In addition, a second ribose is covalently bound to the nitrogen attached to carbon 4 of the pyrimidine ring following the opening of the imidazole ring of a purine. This second ribose is added before pyrrole ring formation occurs and carbons 1', 2' and 3' of these two ribose units contribute equally to the formation of pyrrole carbons 6, 5 and the cyano carbon of toyocamycin, respectively[4].

Although toyocamycin, sangivamycin, and tubercidin are structurally related nucleoside antibiotics, little is known concerning their genetic relationship. This communication shows that   (1) a preformed purine is the precursor for toyocamycin, sangivamycin, and tubercidin;   (2) a *Streptomyces* strain has been found that produces toyocamycin and sangivamycin; and   (3) biochemical studies (*in vivo* and *in vitro*) show that there is a genetic relationship between toyocamycin and sangivamycin. The evidence available to date indicates that the biosynthesis of these two nucleoside antibiotics occurs on one branch during the conversion of a purine ribonucleoside or ribonucleotide to the pyrrolopyrimidine nucleosides, while tubercidin biosynthesis occurs on another branch.

Although the three pyrrolopyrimidine-nucleoside antibiotics had been isolated from numerous *Streptomyces* strains in thirteen independent laboratories, there had not been any reports on the simultaneous production of either tubercidin, toyocamycin, or sangivamycin by any one pyrrolopyrimidine-nucleoside-producing *Streptomyces*. Therefore, we had initially assumed that carbon 2 of the pyrimidine ring of 2-^{14}C-adenine was incorporated into carbon 2 of the pyrimidine ring of tubercidin, toyocamycin, or sangivamycin by separate pathways (Fig. 1). This reasoning required the conversion of the preformed purine ribonucleotide, GTP to toyocamycin, tubercidin, or sangivamycin. It precluded the possibility that GTP could be converted to one of the pyrrolopyrimidine nucleosides which in turn would serve as the precursor for the remaining two nucleosides. Experimentally, we had demonstrated that tubercidin, sangivamycic acid or sangivamycin were not precursors for toyocamycin by the toyocamycin-producing *Streptomyces*. Similarly, neither toyocamycin, sangivamycin nor sangivamycic acid could be converted to tubercidin by *S. tubercidicus* (the tubercidin-producing organism). However, a time course study of sangivamycin biosynthesis by the sangivamycin-producing *Streptomyces* revealed that toyocamycin and sangivamycin could be isolated from the culture filtrates 30 h after inoculation. Toyocamycin production reached a maximum 42 h after inoculation. At this time, sangivamycin production increased markedly while toyocamycin gradually disappeared from the culture filtrates. This was the first indication that there might be a biosynthetic relationship between toyocamycin and sangivamycin.

To study the possibility that toyocamycin was converted to sangivamycin, ^{3}H-toyocamycin was added to the culture filtrates of a sangivamycin-producing culture 60 h after inoculation. Following a 60-min incubation, there was no apparent uptake of the tritium-labelled toyocamycin. However, there was a 20% conversion of toyocamycin to sangivamycin. Toyocamycin was not converted to sangivamycin by the culture filtrate (free of mycelium). When cell-free extracts of the *Streptomyces* were prepared, there was a rapid conversion to sangivamycin. The enzyme responsible for this conversion has been partially purified. It is stable when stored in 50% glycerol at $-20°C$ and does not lose activity following dialysis; no cofactors appear necessary for the conversion of the nitrile to the carboxamide group. The pH optimum is 6.5. Enzyme activity, with crude extracts, is rapidly lost when incubations are performed at 40°C. Increasing concentrations of toyocamycin increased the formation of sangivamycin. The K_m for toyocamycin nitrile hydratase is 5×10^{-4}M. p-Hydroxybenzonitrile is a competitive inhibitor. The K_i is 5.7×10^{-3}M. The cations, Hg^{2+}, Cu^{2+}, Zn^{2+}, Fe^{3+}, and Fe^{2+} are inhibitors of this enzyme reaction. Hydroxylamine inhibits the hydratase. Pyridoxal phosphate is not a cofactor for this enzyme. Purification of the enzyme on DEAE results in a rapid loss of activity. These data strongly suggest that toyocamycin is the precursor for the biosynthesis of sangivamycin.

In an attempt to elucidate the manner in which water is added to the nitrile bond by toyocamycin nitrile hydratase, the carboxamidoxime of toyocamycin was added

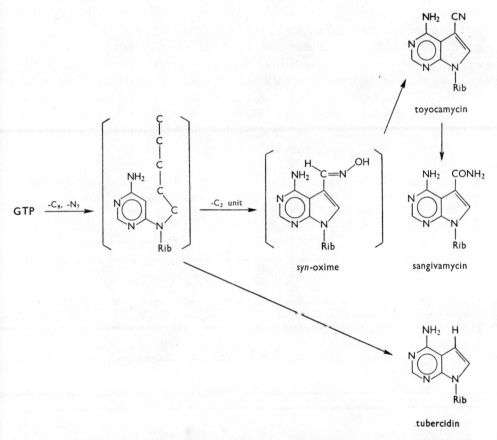

Fig. 2. Biosynthetic relationship of toyocamycin and sangivamycin compared with tubercidin.

to nitrile hydratase at a concentration of 10^{-2}M. Toyocamycin carboxamidoxime was not a substrate.

When cell-free extracts of the toyocamycin-producing organism were made, there was no nitrile hydratase activity. Similarly, the tubercidin-producing organism did not have this enzyme. Evidently, the accumulation of toyocamycin in the toyocamycin-producing organism is the result of a mutational block in the gene for this enzyme. Another possible explanation for the above observation is that the evolution of the *Streptomyces* that never produced this enzyme.

The absence of the nitrile hydratase in cell-free extracts of the tubercidin-producing organism strongly suggests that the hydratase is not present in the organism. This experimental observation, coupled with the *in vivo* and *in vitro* studies in which it was observed that toyocamycin, sangivamycin, or sangivamycic acid were not precursors for tubercidin biosynthesis, strongly suggest that the pathway for tubercidin biosynthesis does not involve the utilization of toyocamycin or sangivamycin as precursors.

These above data strongly suggest the following biochemical and genetic relationship in pyrrolopyrimidine-nucleoside biosynthesis: (1) the pyrimidine ring of the pyrrolopyrimidine-nucleosides arises from the pyrimidine ring of GTP;   (2) nitrogen 7 and carbon 8 of this imidazole ring of GTP is lost;   (3) the pyrrole carbons and the cyano group of toyocamycin arise from D-ribose;   (4) the sequence in the biosynthesis of the pyrrolopyrimidine nucleosides involves the conversion of the purine ribotide, GTP, to toyocamycin which is then converted to sangivamycin;   (5) tubercidin biosynthesis also involves the utilization of a preformed purine ribotide but the biosynthesis of this nucleoside must involve a branch point before pyrrole ring formation occurs (the evidence presented is summarized in Fig. 2).

This investigation was supported by Research Grant NIH AI-08932-10 from the United States Public Health Service and by Grant GB 17441 from the National Science Foundation.

Dr. Robert J. Suhadolnik is a Research Career Development Awardee 5-K3-GM-7100-10 of the United States Public Health Service.

REFERENCES

1. SUHADOLNIK R. J., p. 298 in *Nucleoside Antibiotics*. John Wiley & Sons, New York 1970.
2. SMULSON M. E., SUHADOLNIK R. J.: J. Biol. Chem. **242**, 2872 (1967).
3. UEMATSU T., SUHADOLNIK R. J.: Biochemistry **17**, 1260 (1970).
4. SUHADOLNIK R. J., UEMATSU T.: J. Biol. Chem. **245**, 4365 (1970).

Genetic Considerations and Erythromycin Production by *Streptomyces erythreus*

JOHN W. CORCORAN

Department of Biochemistry, Northwestern University Medical School, Chicago, Illinois, U.S.A.

Genetics is the discipline which explains metabolism in all its aspects. It is a language essential to our eventual understanding of the origin and maintenance of life. The formation and regulation of production of secondary metabolites can be explained by the geneticist, but *he* in turn must rely on the biochemist (chemist) for data leading to a full understanding of genetics. Thus a symbiosis between these (and other) disciplines must continue to exist until both are fully developed; a very remote condition.

One way in which the biochemist can help the geneticist is to delve deeper into the chemistry of living systems. He must study even more vigorously than at present all of the enzymes present in living systems. He must learn their chemical structure and he must learn how they act — even to the most minute detail. Only then will it be possible to study the individual enzymes in association and in the context of partial and complete cellular systems. These physiological studies are required for any understanding of the interrelationships between genetics and chemistry.

A biochemist as well as a geneticist working at this still relatively primitive time must make a decision as to what small part of the whole he can profitably study. Workers in all the biological disciplines are making these decisions, either intentionally or by default. One useful approach to this problem is to develop a model experimental system in which the ubiquitous biochemical pathways can be studied in great detail, but in addition to choose a system which possesses some unique chemical properties. The model system must become understood in fine detail, starting with isolated enzymatic processes and then later with its physiological behavior. Only after this can we ever do much more than speculate as to the relationship between genetics and the chemical expression of cellular development and function.

As a specific experimental system in which to study biochemical events in their relationship to the heredity of certain microorganisms the author and his associates have been concerned with the so-called macrolide antibiotics which are produced primarily by actinomycetes. The erythromycin family of macrolides has been studied from several points of view and the structures (Fig. 1) are now known to be formed from intermediates formed by at least two different biochemical pathways. The erythromycins are glycosides and the two deoxysugar moieties are presumably formed from glucose *via* nucleoside diphosphate derivatives. The aglycone of the erythromycins is a macrocyclic lactone (14-membered) and abundant evidence has been obtained to show that, although the structures are completely formed from propionate, they are formed by biogenetic reactions analogous to those involved in the formation of oligo-ketide-derived natural products[1]. The erythromycins are formed by *Streptomyces*

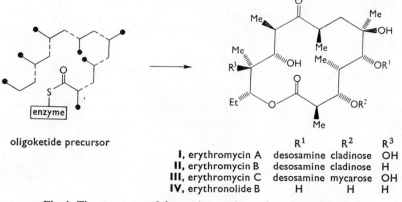

oligoketide precursor

	R^1	R^2	R^3
I, erythromycin A	desosamine	cladinose	OH
II, erythromycin B	desosamine	cladinose	H
III, erythromycin C	desosamine	mycarose	OH
IV, erythronolide B	H	H	H

Fig. 1. The structures of the erythromycins and erythronolide B.

erythreus and a large number of other macrolide structures are formed by other strep-tomycetes and at least one by a *Micromonospora* species (megalomicin[2]).

All macrolide lactones studied apparently are synthesized wholly or in part by similar mechanisms and thus we see in the streptomycetes and closely related micro-organisms a clear-cut case where a unique group of natural products appears. The same organisms produce palmitic and other common fatty acids reflective of so-called primary metabolism. Hence, we have here an interesting model system to study; one in which the approaches of the biochemist are needed if we are to understand fully how a particular streptomycete can produce two quite different classes of product when each depends on the same type of enzymatic apparatus. We are very far from under-standing our model system and the present discussion represents a preliminary survey based on scanty experimental evidence. Before considering this system further it will be necessary to briefly review the enzymology of fatty acid synthesis and to con-sider what general knowledge we have of some other oligoketide-type syntheses.

$$HOCH_2-\underset{\underset{CH_3}{|}}{\overset{\overset{CH_3}{|}}{C}}-CHOH-CO-NH-CH_2-CH_2-CO-NH-CH_2-CH_2-SH$$

pantetheine **(V)**

The synthesis of palmitic acid and the other usual fatty acids found in most if not all living cells depends on acetic and malonic acids in suitably activated form. Activation occurs through ATP and the formation of thiol esters, either with coenzyme A (CoA) or with pantetheine (V) in which the latter prosthetic group is bound by a phosphodiester link to a small protein. Both sorts of activation are usually of concern to the metabolism of the cell, since acetylCoA, obtained from glucose *via* pyruvate or from the breakdown of fatty acids, is the substrate for a carboxylation reaction which generates malonylCoA. The latter in turn donates the malonyl group to the pantetheinyl-phosphoprotein. Palmitate synthesis involves one acetyl and seven malonyl groups. The synthesis (Fig. 2) involves the formation of β-oxothiol esters which

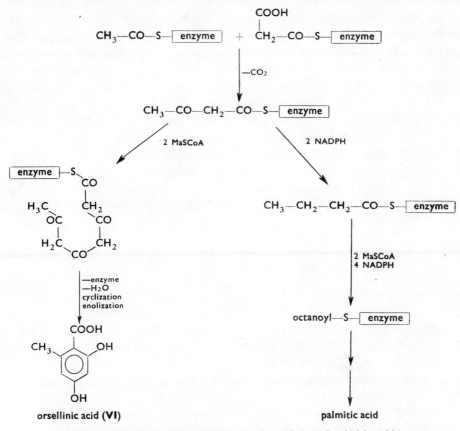

Fig. 2. A comparison of the biogenesis of orsellinic and palmitic acids.

remain attached to the pantetheinyl-protein and the β-oxo group is reduced to a methylene function before chain extension takes place.

This latter fact is of particular significance for our discussion since no free or enzyme-bound oligooxothiol ester ever exists. There is a variation in how different cells accomplish the synthesis of palmitic acid. Animal cells, yeasts, plants and possibly fungi have all the enzymes required for palmitate synthesis associated in a multienzyme complex, in which the small protein bearing the phosphopantetheinyl prosthetic group is also incorporated. Many bacteria, on the other hand, seem to have the same enzyme activities, but as separable entities. It is not yet clear if there is a sharp line of evolutionary differentiation between the two types of fatty acid synthase, and in fact some evidence that there may not be one is implicit in the reports from BLOCH's laboratory concerning *Mycobacterium phlei* which seems to have two synthases — one of each type[3].

The living world abounds in substances made from the same two substrates that are used for the formation of palmitic acid. A wide variety of both aromatic and non-aromatic molecules are produced by many cells from acetic and malonic acid units. Abundant evidence exists to suggest that the biogenetic mechanisms involved are very similar to those just described for palmitic acid. The details of the enzymatic processes are not nearly so well worked out but sufficient information is available (*cf.* discussion of 6-methylsalicylic acid, *below*) to link the two reaction sequences at the enzyme level as well as that of the substrates used. The principal difference appears to be that while there is no oligo-β-oxothiol ester (oligoketide) structure involved in palmitate biogenesis the incorporation of acetyl and malonyl groups into aromatic structures requires the formal existence of such oligooxo intermediates. In other words, the β-oxo function generated at each step in the elongation of the chain is not reduced to a β-methylene group (Fig. 2). Apparently the oligoketide biosynthetic enzymes are very similar indeed to those of the more ubiquitous palmitate synthase and yet they differ in at least one functional aspect. The problem posed above — how to use the tools of biochemistry as an aid in studying the genetic behavior and evolution of cells — is clearly involved in explaining the biogenetic relationship between oligoketide products and palmitic acid.

What are the possible explanations for the production both of common fatty acids and oligoketide-derived products by the same cell? There are two principal possibilities. One is that a single synthase is capable of producing both types of chemical product. The other is that two or more synthases are produced by the cell, one of which produces common acids with the other(s) being responsible for the oligoketide products. The author has speculated along the lines of the first possibility[4] as it would apply to the biogenesis of the erythromycin (non-aromatic, oligoketide-derived macrolide) lactones. So far, no evidence has been obtained to support this concept. It may still be correct, but in the face of proof that some cell populations which form oligoketide-derived products do possess two different fatty acid synthases the second alternative has become rather attractive. It is somewhat easier to visualize the regulation of production of the two different products, *e.g.* palmitate and 6-methylsalicylic acid (VII), if different synthases exist. Under these circumstances one could predict that the two enzyme systems might have different affinities (K_m's) for the acetyl and malonyl moieties required for both products. Other differences such as in cofactor requirements or in pH dependences or even in enzyme concentration as a function of the age of the cell can readily be invoked to explain how two competing enzyme systems can be integrated into the balanced biochemical activities of the producing organism.

Some of the same factors might explain regulation if there is but a single synthase in the cell; for instance a relative lack of NADPH during certain times in the cell growth cycle might favor the formation of oligoketides at the expense of acids like palmitic which require substantial reducing power for their synthesis [see discussion of alternariol (VIII) below]. As discussed before, the only way to provide convincing answers to questions such as these is to isolate and characterize the enzymes responsible for making these products. In addition to giving us answers as to how many synthases there are and how the cell regulates primary vs. secondary metabolism such characterization will be useful to the geneticist, for it should permit answers to questions which he will ask concerning the evolutionary development of the cells concerned. Chemical study of the purified synthase(s) may permit such a judgment in the same manner as has a careful comparative study of the amino-acid sequences in cytochrome c derived from different species[5].

Several experimental systems have been defined from which we have learned much about the enzymology responsible for producing oligoketide-type substances. Orsellinic (VI) and 6-methylsalicylic (VII) acids are two simple examples of products produced by the oligoketide pathways. Orsellinic acid synthesis is biochemically simpler than is the case for 6-methylsalicylic acid (MSA; Fig. 3). In the former example the hypothetical tetraketide progenitor has merely to cyclize, dehydrate and enolize in order to produce a thiol-bound form of orsellinic acid. With 6-methylsalicylic acid synthesis an additional reduction of one of the oxo functions is required to produce the structurally simpler product. From the enzymatic point of view more work has been done with 6-methylsalicylate synthase than with orsellinate synthase. BASSETT and TANENBAUM[6] showed that acetylCoA is utilized for MSA synthesis in a cell-free extract from Penicillium patulum. LYNEN and TADA[7] extended this finding by showing that both acetyl- and malonylCoA's are required as well as a reducing agent, NADPH. Other characteristics of the system likewise suggested a close similarity to a fatty acid synthase. A long time was required before these initial findings were extended, largely due to the instability of the enzyme preparation. LIGHT has recently reviewed this progress[8] in which work from LYNEN's laboratory[9] as well as by LIGHT and HAGER[10] has conclusively demonstrated that the MSA synthase is a multienzyme complex. DIMROTH et al.[9] have moreover shown that the MSA synthase is a multienzyme complex which is different from a second multienzyme complex in the same extract which produces palmitic acid. Thus for the first time there is a specific suggestion as to the relationship between the enzyme apparatus which makes a oligoketide and that responsible for a saturated fatty acid. They seem to be similar but are physically as well as functionally distinguishable. If this should turn out to be the general pattern, we can anticipate answering some of the questions raised above. Comparisons of the enzymes (amino acid sequence, substrate specificities, affinities for substrates and inhibitors, pH dependence, concentration as a function of the life cycle of the mycelium etc.), can be made. From such studies we may be able to evaluate the relative evolutionary development of the two synthases, learn how the mycelium regulates the utilization of the two identical substrates by two different synthases and especially learn how the synthase determines what the product of the synthesis will be. Most other oligoketide-derived products are made (formally) from much longer oligo-ketide chains and almost nothing is known about the factors determining their terminal length, stabilization of the chain during the biosynthesis, and the nature of the product when more than one is possible.

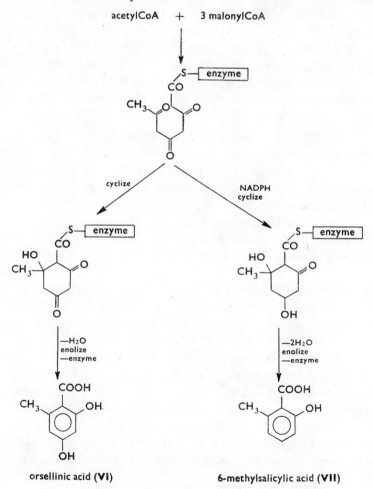

Fig. 3. Biogenesis of orsellinic and 6-methylsalicylic acids.

Another interesting oligoketide-derived secondary metabolite is alternariol (VIII). Like orsellinic acid (VI) alternariol can be made formally from an oligoketide precursor by cyclization, dehydration and enolization reactions. The progenitor is a heptaketide and thus alternariol more than orsellinic acid represents an acetate/malonate-derived product comparable to palmitic acid. Studies by GATENBECK's group in Sweden[11,12] on cell-free extracts of *Alternaria tenuis* have demonstrated that alternariol is formed, as predicted, from one acetyl and six malonyl entities. The coenzyme A esters are utilized as are some simpler thiol esters (*e.g.* with pantetheine). No enzyme-free inter-mediates were detected and replacement of the acetylCoA by propionylCoA caused the formation of a homologue of alternariol with a C-ethyl group instead of the C-methyl function. An attempt was made by GATENBECK and HERMODSSON[11] to inves-tigate possible regulatory factors involved in determining the fate of acetyl and malo-

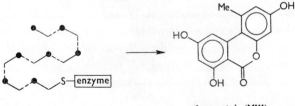

alternariol (**VIII**)

nyl donors in *A. tenuis*. In extracts supplied with NADPH the utilization of radio-active acetylCoA for "lipid" synthesis was large relative to the formation of alter-nariol. When the NADPH was not supplied the relative synthesis of alternariol was stimulated. Unfortunately, the system studied was unstable and a second variable was introduced into these experiments, namely a change in the ionic strength of the buffer. Thus it is not proven conclusively that the level of reducing agent can play a major role in regulating the synthesis of fatty acids with respect to the oligoketide, alternariol (**VIII**). Nonetheless, this finding is most interesting and the system deserves further study.

The model system developed and under study in the author's laboratory concerns the erythromycins (Fig. 1). The aglycone of the known erythromycins is a 21-carbon fatty acid made by the polymerization of seven propionate subunits. Numerous oxygen atoms are present in the structures and their placement is suggestive of an oligoketide origin. Few secondary metabolites are derived even in part from propionic acid and of these the largest number are noted in the macrocyclic lactones present in the macro-lide group of antibiotics. These in turn, as mentioned above, are mostly produced by a fairly restrictive group of organisms, mostly streptomycetes. The example of the erythronolides is unique in that only propionate or its biological equivalent 2-methyl-malonic acid is involved in the biogenesis (*cf.* review[1]). A variety of circumstantial evidence has been accumulated, all of which indicates that the biogenesis of the eryth-ronolides is mechanistically analogous to the formation of orsellinic and 6-methyl-salicylic acids, alternariol and other oligoketide products. Thus, even though only 3- and 4-carbon substrates (thiol esters of propionate and 2-methylmalonate) are utilized for erythronolide biosynthesis the process and the enzymic apparatus used must be compared to the formation of palmitic acid from acetate and malonate by the produc-ing organism (*S. erythreus*). The comparison is all the more interesting in that *S. ery-threus* seems to make no other propionate-derived products other than the erythronol-ides; all other lipids are typical of the usual bacterial distributions.

Preliminary studies of the comparative enzymology of erythronolide and palmitate biogenesis by *S. erythreus* were done by CORCORAN *et al.*[13]. A mycelium-free
system was partially purified by fractional precipitation with ammonium sulfate. The
protein precipitating at ammonium sulfate concentrations between 60 and 90 per cent
of saturation had the ability to form fatty acids from acetyl- and malonylCoA and the
synthesis depended on both (Table I), indicating that acetylCoA carboxylase activity
was not sufficient to permit the synthesis of the malonylCoA from acetylCoA. The
product of the synthesis was identified by DARBY and CORCORAN[14] as a mixture in
which palmitic acid predominated. If propionylCoA was substituted for acetylCoA
fatty acids were also formed, presumably with odd-numbered chain lengths. The activity of the partially purified enzyme mixture was about as high with the propionyl
substrate as it had been with acetylCoA (Table I). The fatty acid synthase of *S. erythreus* has been studied further in the author's laboratory, but its properties have made
progress most difficult. The maximum purification of the synthase activity has been
about 400-fold, using a combination of chromatographic steps (DEAE- and TEAE-
cellulose and Sephadex). The synthase is unstable and the procedures used give erratic
results. The biosynthetic activity is maximal at pH 7.8 and it is at least 4-fold lower
at either pH 6.7 or 8.5[15]. No physical examination has been possible so that we do not
know by a direct assay if the synthase is a multienzyme complex. However, an indirect indication that it may be comes from a preliminary investigation of the distribution of radioactive β-alanine in the constituents of *S. erythreus* after the organism
has grown in the presence of this substrate. Assuming that the utilization of the
β-alanine is solely for the biosynthesis of pantetheinyl moieties (coenzyme A, pante-
theinyl-proteins of fatty acid or oligoketide synthases, *etc.*) we would predict that the
presence of radioactivity in protein fractions indicates the presence of phosphopante-

TABLE I

Fatty Acid Biosynthesis by Extracts of *S. erythreus*[a]

Additions				d.p.m.	
1-^{14}C-AcetylCoA	1-^{14}C-PropionylCoA	MalonylCoA	2-Methyl-malonylCoA	Expt. 1[c]	Expt. 2[d]
+	−	+	−	5780[b]	1290
+	−	−	−	165	34
−	+	+	−	266	700
−	+	−	−	25	13
−	+	−	+	31	16

[a] Experimental details *see* Ref.[13].
[b] Indicates the addition of 1,3-^{14}C-malonylCoA as well as malonylCoA.
[c] Supernatant.
[d] $(NH_4)_2SO_4$ precipitate.

theinyl-prosthetic groups and hence suggests the presence of fatty acid synthase activity — even if biological activity cannot be detected. We have found[16] that radioactivity from β-alanine added to the growth medium of *S. erythreus* is present in cell-free extracts and is distributed into two fractions. One represents very small organic molecules of the size of coenzyme A (retarded by Sephadex G-15 chromatography) and the other includes very large molecules. The high-molecular-weight fraction is possibly a mixture but no component seems to have a molecular weight less than about 800,000. If the assumptions behind this experiment are valid, this result suggests that the fatty acid synthase(s) of *S. erythreus* is a multienzyme complex. The experiment also should suggest the same thing about the erythronolide synthase, although there is no evidence yet as to whether this activity is due to the same complex as that producing palmitate.

The partially purified protein fraction which was studied by CORCORAN et al.[13] also was able to form radioactive nonvolatile and apparently neutral molecules from propionylCoA alone or better from both propionyl- and 2-methylmalonylCoA (Table II). The radioactive products were examined by chromatographic methods, and several substances were found to be present. The most interesting was very similar in chromatographic mobility to 6-deoxyerythronolide B (IX), a known sugar-free precursor of the lactones present in the erythromycins (*see below*). Insufficient material was available to allow a final judgment but slight differences in mobility indicated that this unknown product is not identical to 6-deoxyerythronolide B. There are several as yet unavailable sugar-free lactones which may be intermediates in the biogenesis of erythronolide B (IV) which has been found by MARTIN[17] to be the acceptor of the first sugar moiety during erythromycin B formation. Chemical synthesis of some of these is being attempted, and accumulation of more of the enzymatic product may also permit a direct chemical identification of its structure. At this stage we are not completely certain that any of these products are even related to the erythronolides. However, if these are artifacts they are at least made solely from propionate and their physical behavior is very similar to known erythronolides on partition chromato-

TABLE II

Methyl*iso*butylketone-Extractable Material Produced by Extracts of *S. erythreus*[a]

Additions		d.p.m.	
1-^{14}C-PropionylCoA	2-MethylmalonylCoA	Expt. 1[b]	Expt. 2[c]
+	+	38000	337000
+	—	—	265000

[a] Experimental details *see* Ref.[13]
[b] Supernatant.
[c] $(NH_4)_2SO_4$ precipitate.

graphy in solvent systems which are sensitive to slight structural changes in these known compounds. Although most of our attention has been focused on purification of the palmitate synthase from *S. erythreus* we have recently repeated the experiments of CORCORAN *et al.*[13] and have again obtained radioactive neutral material, made from highly purified propionyl- and 2-methylmalonylCoA by a fresh extract of *S. erythreus*[18]. Until the products of this synthesis are identified and until the enzymes concerned are characterized our model system cannot help us to answer the questions discussed earlier.

The metabolic fate of the erythromycin lactone, once formed and released from the synthase which makes it, is also no longer directly reflective of the oligoketide/palmitate biogenetic problem whose understanding has been the main initial aim of our experimental model system. However, as stressed above it is only by knowing in detail

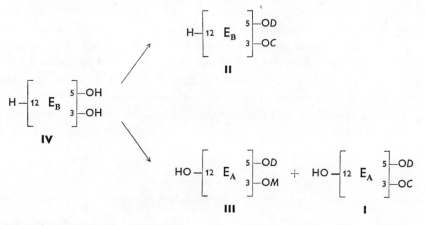

Fig. 4. Formation of the erythromycins from erythronolide B. D = D-desosaminyl, C = L-cladinosyl, M = L-mycarosyl.

the overall metabolism of *S. erythreus* that we will be able to explain the physiology of this organism. Studies of other enzyme activities in *S. erythreus* are underway. No significant results are available yet, save for some related to the metabolic fate of the erythromycin lactones. A variety of studies (reviewed by CORCORAN and CHICK[1]) have shown that the propionate-derived lactone is converted further into a bis-glycoside with two deoxysugars, which are made from D-glucose and also subjected to O-, N- and C-methylation at some stage before the formation of the biologically active erythromycins (*cf.* Fig. 5). Experiments with intact *S. erythreus* had suggested that the formation of erythromycin B and A may not be directly related, but that some sort of precursor—product relationship exists between erythromycins C and A (Fig. 4). Initially erythromycin A was reported to be the precursor of the C form (formal removal of the O-methyl group of the cladinosyl moiety) but more convincing data from MARTIN *et al.*[17,19-21] have suggested the reverse order (*see* Fig. 5). An elegant series of experiments utilizing mutants of *S. erythreus* has allowed MARTIN to propose a relatively detailed sequence of biogenetic steps in the biogenesis of the erythromycins (Fig. 5). MARTIN has detected a number of sugar-free lactones as metabolic products of his various mutant strains, but several of these seem to be shunt metabolites and not directly involved in erythromycin formation. As obligate intermediates MARTIN pro-

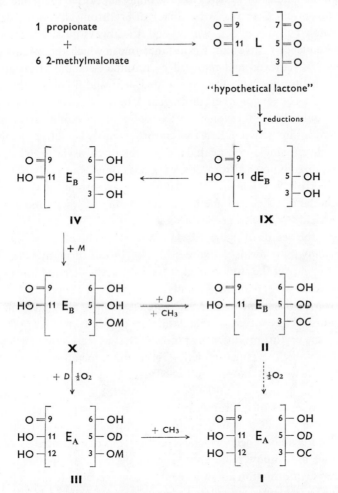

Fig. 5. Biogenetic pathways leading to the known erythromycins.

poses that 6-deoxyerythronolide B (**IX**) is formed before erythronolide B (**IV**), the acceptor of some glycosyl group (possibly a mycarosyl group itself) to yield the stable monoglycoside mycarosylerythronolide B (**X**). This intermediate is capable of accepting either the desosaminyl group directly or some as yet unidentified precursor which is fashioned into it after being joined to the lactone. In any event erythromycin B (**II**) can be formed. The exact sequence of steps leading to the oxygenated lactone present in erythromycins C (**III**) and A (**I**) is not clear. The first product isolated is erythro-mycin C, which has the oxygenated lactone in addition to the desosaminyl substitu-tion. The predicted single-step intermediate with the B form of the lactone and the sugars of erythromycin C, or its alternative the hydroxylated lactone bearing only the mycarosyl group has not been detected.

Although the studies of MARTIN have strongly supported the sequence described, in which erythromycin C serves as precursor of erythromycin A, the data were ob-tained with whole cultures of *S. erythreus* and it is always possible that a more com-plex reaction sequence is involved; for instance one in which the mycarosyl group of erythromycin C is removed and replaced by an intact cladinosyl group from some donor. Such substitutions have been speculated on[22] and if correct this would allow the possibility of a conversion of erythromycin A to the C form as well as the reverse. This, although unlikely, would explain the two differing viewpoints about the precursor − product relationship between these two erythromycins[1]. Evidence in support of the simpler possibility, a single-step methylation of the mycarosyl group in erythromycin C (**III**) leading directly to erythromycin A (**I**), has come from study of a crude myce-lium-free extract of *S. erythreus*. Nonradioactive erythromycin C when incubated with this extract and radioactive methyl-labeled S-adenosylmethionine is tranformed in reasonable yield ($\sim 5\%$) to erythromycin A[23].

Obviously, the details of the glycosylation reactions involved in the biogenesis of the erythromycins must await the successful conclusion of many careful enzymatic studies. Like the part of the study aimed at the oligoketide-derived portion of the ery-thromycins this aim when realized will still not permit even a superficial understanding of how *S. erythreus* has come to its present state. We have absolutely no insight into why this organism has evolved or maintained its ability to make apparently useless molecules like the erythromycins. It is probable that at least some of the constituent parts of the antibiotics (*e.g.* sugar moieties) reflect other metabolic pathways in *S. ery-threus* but of these we have no present knowledge. Thus it is clear beyond doubt that before we can take a close look at the genetic substance in *S. erythreus* and follow the metabolic consequences of its transcription and translation we must document and characterize these chemical activities. In this effort the geneticist can help us. In attempting to study an organism which is genetically as complicated as a fungus or an actinomycete we must know more about the life cycle of the organism and in parti-cular have precise information about its genetic homogeneity. It would be a great pity indeed to spend much effort in defining a model experimental system like that under study in our own laboratory and later find out that our strain of *S. erythreus*, al-though homogeneous by morphological and metabolic criteria, is genetically impure. By this it is meant to find that it possesses different DNA's and variable relative amounts of them after alterations in culture conditions, *etc*.

In summary, it is quite clear that even though we are a long way from understand-ing the precise relationship between the genetics of *S. erythreus* and its physiology we are ready for the active cooperation of the geneticist. Such collaboration should prove fruitful, both for the solution of the problems discussed in this chapter and for the better understanding of gene action and regulation.

˙ *The author gratefully acknowledges the following support for the work performed in his own laboratory: Abbott Laboratories, American Heart Association, Inc. (Nos. 65 G 126, 68—799); National Institutes of Health, Institutes of Allergy and Infectious Diseases (Nos. AI 06758, AI 09158) and General Medical Studies (GM 2545); National Science Foundation (No. GB 18157); Northwestern University Medical School (Dean Richard H. Young).*

REFERENCES

1. CORCORAN J. W., CHICK M., p. 159 in J. F. Snell (Ed.): *Biosynthesis of Antibiotics*. Academic Press, New York 1966.
2. MALLAMS A. K., JARET R. S., REIMANN H.: J. Am. Chem. Soc. **91**, 7506 (1969).
3. MATSUMURA S., BRINDLEY D. N., BLOCH K.: Biochem. Biophys. Res. Comm. **38**, 369 (1970).
4. CORCORAN J. W., p. 131 in Z. Vaněk, Z. Hošťálek (Eds.): *Biogenesis of Antibiotic Substances*. Academic Press, New York and London 1965.
5. NOLAN C., MARGOLIASH E.: Ann. Rev. Biochem. **37**, 727 (1968).
6. BASSETT E. W., TANENBAUM S. W.: Biochim. Biophys. Acta **40**, 535 (1960).
7. LYNEN F., TADA M.: Angew. Chem. **73**, 513 (1961).
8. LIGHT R. J.: J. Agric. Food Chem. **18**, 260 (1970).
9. DIMROTH P., WALTER H., LYNEN F.: European J. Biochem. **13**, 98 (1970).
10. LIGHT R. J., HAGER L. P.: Arch. Biochem. Biophys. **125**, 326 (1968).
11. GATENBECK S., HERMODSSON S.: Acta Chem. Scand. **19**, 65 (1965).
12. SJÖLAND S., GATENBECK S.: Acta Chem. Scand. **20**, 1053 (1966).
13. CORCORAN J. W., CHICK M., DARBY F. J., p. 35 in Proc. 5th Internat. Congr. Chemotherapy, Vienna 1967.
14. DARBY F. J., CORCORAN J. W., *unpublished results*.
15. DARBY F. J., BOROWSKI P. T., CORCORAN J. W., *unpublished results*.
16. YOUNG M. R., ROSSI A., CORCORAN J. W., *unpublished results*.
17. MARTIN J. R., PERUN T. J., GIROLAMI R. L.: Biochemistry **5**, 2852 (1966).
18. ARORA K. L., YOUNG M. R., CORCORAN J. W., *unpublished results*.
19. MARTIN J. R., ROSENBROOK W.: Biochemistry **6**, 435 (1967).
20. MARTIN J. R., PERUN T.: Biochemistry **7**, 1728 (1968).
21. MARTIN J. R., GOLDSTEIN A. W.: *Progress in Antimicrobial and Anticancer Chemotherapy*. Yurisha Ltd. 1969.
22. SPÍŽEK J., CHICK M., CORCORAN J. W.: Antimicrob. Agents & Chemotherapy 138 (1965).
23. MCALPINE T., CORCORAN J. W., *unpublished results*.

Molecular Basis of Polygenic Inheritance in the Biosynthesis of Chlortetracycline

Z. Hošťálek and Z. Vaněk

Department of Biogenesis of Natural Substances, Institute of Microbiology, Czechoslovak Academy of Sciences, Prague, Czechoslovakia

INTRODUCTION

Quantitative inheritance and the genetic basis of productivity are major problems in the improvement of industrial strains. So far, most microbial breeders have accepted the concepts of classical genetics which explain the continuous variability of quantitative characters (size, shape and other features mostly of physiological character) by the existence of a large number of genes, controlling the realization of the phenotype.

The so-called multiple gene inheritance was an accepted hypothesis during the first half of this century. It proceeded from the view that the phenotypic expression of a character is under the control of a series of genes, each of which contributes to its formation but in itself has a relatively small effect. MATHER[1] called the assumed determinants as polygenes. The fundamental postulate of polygenic inheritance is that the degree of expression of quantitative characters is determined by the action of a number of genes displaying a cumulative (additive or multiplicative) effect without complete dominance.

In spite of the fact that biometric methods indicated the chromosomal character of the so-called polygenic systems in many models, the hypothesis was difficult to confirm experimentally. The role of the individual genes could not be exactly demonstrated and it was difficult to determine whether the variability is a question of genotype as a whole or whether it is determined by the environmental influence. The role of the environment in the expression of the final phenotype is much greater in quantitative than in qualitative features.

Progress in molecular genetics developed the concept of multiple gene inheritance through the hypothesis of so-called modifiers. In this case several minor genes are invested with a modifying effect, capable of altering the degree of expression of other genes. It is assumed that the modifiers (diluting genes) alter the phenotypic effect of "major" genes responsible for qualitative characters. WAGNER and MITCHELL claim, albeit with a certain scepticism[2], that the interaction of factors responsible for quantitative inheritance will not be established until the physiological and chemical approach is successfully applied to the analysis of their effects. Genetic analysis alone, regardless of the biochemistry and physiology of the character investigated, cannot unravel the basis of gene interaction.

B.D. DAVIS believes the current state in the study of polygenic inheritance to be primitive[3], and one must certainly agree with the view of SERMONTI[4] that *"all these concepts are approximate and superficial, since on a molecular basis all lose their value"*.

In the present work we have attempted a detailed analysis of the factors responsible for the great variability in secondary metabolite production, using the biosynthesis of chlortetracycline (CTC), an oligoketide secondary metabolite produced by *Streptomyces aureofaciens*, as the model (Fig. 1).

It was the aim of this work to establish in what way the CTC biosynthesis is affected by the activity of other metabolic pathways involving acetylCoA and how the rate of enzyme reactions of such alternative pathways might be reflected in the

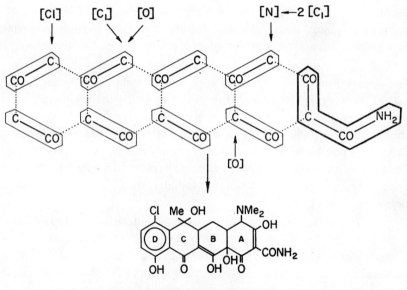

chlortetracycline

Fig. 1. Building units of chlortetracycline.

degree of antibiotic production. Two principal objectives were of interest: the role of environment on the final phenotype effect and the expression of hereditary changes in production activity at the biochemical (metabolic) level. From the results we have attempted to outline a concept for the existence of regulatory mechanisms controlling the biosynthesis of CTC as a quantitative feature.

ENVIRONMENTAL CONTROL OF CHLORTETRACYCLINE BIOSYNTHESIS

CULTIVATION CONDITIONS

The yield of CTC during submerged cultivation depends to a certain degree on the cultivation conditions, *i.e.* medium composition, aeration intensity, *etc.* Of the numerous factors affecting the biosynthesis of tetracyclines[5] the action of phosphates[6] is worthy of consideration. The concentration of orthophosphate in the medium fundamentally affects the phenotypic expression, since the antibiotic is produced within a narow range of phosphate concentrations and any increase in phosphate level results in intensified vegetative growth and decrease of antibiotic production.

The intimate inhibitory mechanism relating the orthophosphate level to the biosynthesis of CTC is not yet clearly understood. However, the results so far obtained indicate that the phosphate concentration affects the equilibrium between glycolysis

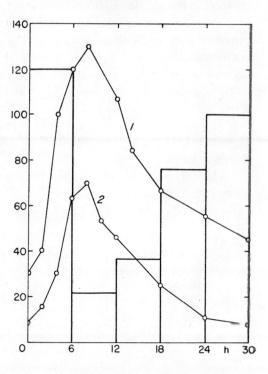

Fig. 2. Course of Q_{O_2} (*1*) and the level of pyruvic acid (µg/ml, *2*) in a submerged culture of *S. aureofaciens* Bg. The column graph shows the effect of aeration interruptions on the final production of CTC in % of control-culture yields. The aeration was stopped every hour for 10 min, *i.e.* in the first group between hour 0 and 6, in the second group between hour 6 and 12, *etc.*

and direct oxidation of sugars. BORETTI *et al.* demonstrated an inhibitory effect of phosphate on some enzyme reactions of the pentose cycle[7] and using radiorespirometry with specifically labelled glucoses, we were able to demonstrate that at higher phosphate concentrations the proportion of glycolysis in overall sugar oxidation is higher[8].

Biosynthesis of CTC is also markedly decreased by periodically stopping aeration for short intervals during the initial phases of fermentation[9]. The culture is most sensitive to the interruption of aeration between the 6th and 12th hour of growth (Fig. 2). A culture grown under such initial conditions is later characterized by low production of CTC and by a general increase in oxidative processes (Q_{O_2}, dehydrogenase activity) throughout the whole cultivation[10]. The proportion of glycolysis in the total glucose oxidation (determined with specifically labelled glucoses) is higher in these cultures than in controls[8].

The yield of CTC is increased substantially by the presence of benzyl thiocyanate in the medium (about 10 μM). The enhancement of the production is accompanied by the decrease in the rate of sugar consumption[11] and in the participation of glycolysis in glucose oxidation[12]. The formation of the individual enzymes of the citric-acid cycle is also decreased[13]. Benzyl thiocyanate stimulates only if added to the culture during the first 12 h of cultivation. A later addition has no effect.

Interruption of aeration thus affects the metabolism of *S. aureofaciens* in a way opposite to the addition of benzyl thiocyanate. In an interrupted culture the increase of glycolysis is accompanied by a drop of production, whilst in a culture containing benzyl thiocyanate, both the yield of CTC and the participation of the pentose cycle are increased. It was also shown that the addition of benzyl thiocyanate will cancel the unfavourable effect of interrupted aeration on CTC biosynthesis[8].

The positive effect of benzyl thiocyanate and the negative effect of interrupted aeration on the production of CTC are most pronouncedly manifested in low-production strains, close to the standard type. In production variants the effect is much less expressed (Table I). α-Naphthylacetic acid and other stimulators of CTC biosyntheis show similar effects[14,15].

This may serve as evidence for a number of genetic changes in the metabolism of high-production variants and of their saturation by various physiological mutations — environmental control does not play any significant role under such conditions.

TABLE I

Influence of Interrupted Aeration and Benzyl Thiocyanate on the Production of CTC in Different Strains of *S. aureofaciens*[a]

Strain	Interrupted culture[b]		Benzyl thiocyanate[c]		Control
	μg/ml	% of control	μg/ml	% of control	μg/ml
Bg	137	15	2005	223	897
111158	605	34	2780	157	1775
8425	1710	51	4450	133	3328

[a] Sucrose—corn-steep—soybean meal medium.
[b] The aeration was stopped every hour for 10 min between 6 and 12 h of cultivation.
[c] 12 μM.

METABOLIC PHASES AND CONTROL OF BIOSYNTHESIS

Benzyl thiocyanate and interrupted aeration have a pronounced effect on CTC production at a time when the biosynthesis of the tetracene nucleus of the antibiotic has not yet occurred, *i.e.* at the level of primary metabolism. The rate of biosynthesis of the antibiotic is thus a secondary consequence of the changes caused and the results point out the importance of primary metabolism, especially of the proportion of the individual pathways of carbohydrate metabolism for CTC biosynthesis.

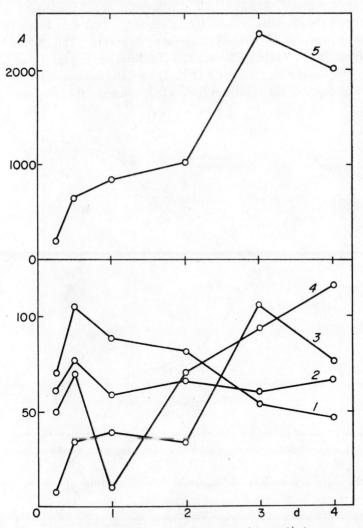

Fig. 3. Specific activity of tricarboxylic-acid-cycle enzymes (*A*, nmol/min per mg protein) in the mycelium of a submerged culture of *S. aureofaciens* RIA 57 during cultivation. *1* aconitate hydratase, *2* isocitrate dehydrogenase (NADP), *3* citrate synthase, *4* fumarate hydratase, *5* malate dehydrogenase.

The first phase of cultivation during which the phenotypic expression of the geno-type is determined is characterized by a high metabolic activity. The RNA content of the mycelium is highest during this period, attaining 30% weight[16]. The rate of protein synthesis, as followed by the incorporation of labelled leucine, reaches a maxi-mum at about 12 h of cultivation and subsequently drops[17]. Maximal lipid synthesis also occurs at this time[18]. At 12 h, a maximum of activity of enzymes participating in lipogenesis is reached; this includes malate dehydrogenase (decarboxylating)[19] and acetylCoA carboxylase[20]. A maximum is reached also by the activity of the pyruvate dehydrogenase complex[21]. The first 12 h of cultivation are characterized by a rapid rise of activity of the oxidative process, the values of Q_{O_2} and dehydrogenase activity reaching here a sharp peak[10]. The cytochrome content (first of all of cytochrome b) in the mycelium is highest between 10 and 12 h of cultivation[22]. A maximum is reached here also by the intracellular level of ATP[23]. The tricarboxylic-acid cycle serves here first of all as a source of intermediates for protein synthesis, the amount of enzymes

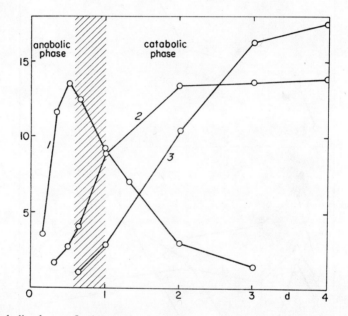

Fig. 4. Metabolic phases of submerged culture of *S. aureofaciens*. *1* incorporation of ^{14}C-leucine (c.p.m. $\times 10^{-4}$), *2* dry weight (mg/ml), *3* chlortetracycline (μg/ml $\times 10^{-2}$).

catalyzing the initial reactions of the cycle predominating at the beginning of cultivation over enzymes of the terminal reactions (Fig. 3).

After 12 h, the metabolic rate of the culture decelerates and the first traces of CTC begin to appear in the medium. This period is characterized by changes in the ribosomal pool and ribosomal subunits. Sedimentation analysis as well as electron microscopy showed[24] that at about 12 h, the cytoplasm contains large polysomes rich in nucleotides. After 24 h protein synthesis ceases and ribosomes aggregate, apparently as a result of initial CTC synthesis.

It is rather likely that the metabolic flow of biosynthetic intermediates after slowing down of the rate of protein synthesis is differently channelled. One of the pathways now in play may be the biosynthesis of the tetracene nonaketide. After the slowing down of protein synthesis, the increase in mycelial dry weight continues and the antibiotic begins to accumulate. When growth stops secondary metabolism becomes more active and CTC production accelerates (Fig. 4). During this phase the great variation breadth in the rate of antibiotic production by different strains is observed. In the low-production strain, katabolic processes accompanied by the formation of reserve substances, such as polyphosphate[13], predominate, while in the production variants, intense biosynthesis of CTC takes place.

The metabolic phases and their relation to secondary biosynthesis have been discussed many times[25]. Usually, symptomatic criteria were used for characterizing the phases, such as growth or production of secondary metabolites. The present results obtained by studying the biosynthesis of oligoketides indicate that an important role in metabolic differentiation is played by protein synthesis and its regulatory mechanisms. In both the biosynthesis of CTC[26] and bikaverins, the red anthraquinoid pigments produced by *Gibberella fujikuroi*[27], the cessation of protein synthesis and associated metabolic processes is the starting signal for the onset of secondary metabolism.

The initial cultivation period, characterized by intense protein synthesis, may be fittingly termed the anabolic phase. The second period, setting in after cessation of protein synthesis, is characterized by katabolic processes. The prevailing part of the metabolic activity is concentrated on the degradation of sugars and conservation of the energy thus produced. These clearly defined forms of metabolism are found only in *S. aureofaciens* strains close to the standard type. In production variants, katabolic processes are suppressed and secondary biosynthesis proceeds at their expense.

METABOLIC CONTROL OF CHLORTETRACYCLINE BIOSYNTHESIS

COMPETITION BETWEEN METABOLIC PATHWAYS OF ACETATE

One of the fundamental prerequisites of high yields of CTC is a sufficient supply of building units for the synthesis of the naphthacene ring, *i.e.* acetyl- or malonylCoA. We investigated the activity of metabolic pathways that might compete with the production of the antibiotic: lipid synthesis and oxidation of acetate in the tricarboxylic -acid cycle.

LIPOGENESIS

The relationship between lipid synthesis and CTC production was followed in five strains of *S. aureofaciens*[28]. In spite of the fact that the strains produced different amounts of CTC in a synthetic medium, the lipid content in their mycelium was practically constant. Fatty acids in the lipid fraction were identified by gas chromatography. The main component of all the fractions were branched fatty acids (Table II). The character of the acids present did not point to the structure of the tetracycline nucleus.

S. aureofaciens does not synthesize lipids as reserve material, their formation being more characteristic of the anabolic phase. Using incorporation of labelled acetate it was shown that it reaches a maximum at about 12 h of cultivation, at the time of intense protein synthesis[18]. The rate of incorporation of acetate was higher in the production strain of *S. aureofaciens*, apparently because of the lower activity of the tricarboxylic-acid cycle (Fig. 5). The relative amount of fatty acids in the dry weight varies only negligibly during cultivation, rapid synthesis thus being accompanied by rapid utilization[29]. The results obtained indicate that lipogenesis and CTC biosynthesis proceed at a different period of development of the submerged culture and do not compete with each other. Another substantial difference between the two processes consists in different demands for reducing equivalents. While CTC (containing in its molecule a number of oxygen functions) requires only 3 moles NADPH[30], a saturated chain of, say, palmitic acid, requires 14 moles of NADPH to be synthesized.

TABLE II

Proportion of Fatty Acids in Lipid Fractions of *S. aureofaciens* (relat. %)

Fatty acid		Tri-glycerides	Di-glycerides	Free fatty acids	Mycelium hydro-lyzate[a]
Lauric	$C_{12}:0$	0.4	0.4	0.4	0.4
Myristic	$C_{14}:0$	2.0	0.8	1.6	0.9
12-Methyltridecanoic	$iC_{14}:0$	2.3	1.2	1.9	2.9
13-Methyltetradecanoic	$iC_{15}:0$	17.6	22.0	26.0	23.2
14-Methylpentadecanoic	$iC_{16}:0$	12.1	14.3	15.6	19.2
Palmitic	$C_{16}:0$	15.4	15.9	14.9	17.4
Palmitoleic	$C_{16}:1$	12.9	11.3	14.8	0
14-Methylhexadecanoic	$aC_{17}:0$	14.2	15.5	15.9	27.8
Stearic	$C_{18}:0$	2.7	2.0	1.8	0
Oleic	$C_{18}:1$	9.8	6.1	4.9	0
Linoleic	$C_{18}:2$	9.6	7.1	4.3	0

[a] Fatty acids remaining in mycelium after extraction, released by alkaline hydrolysis.

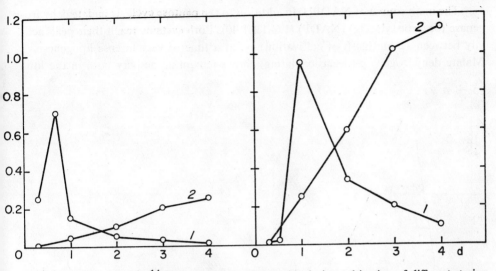

Fig. 5. Incorporation of 1-^{14}C-acetic acid into fatty acids during cultivation of different strains of *S. aureofaciens. Left:* strain RIA 57; *right:* strain 8425; *1* incorporation (µCi/g mycelial dry weight), *2* chlortetracycline (µg/ml × 10^{-3}).

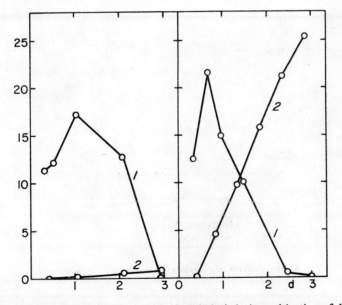

Fig. 6. Activity of malate dehydrogenase (decarboxylating) during cultivation of different strains of *S. aureofaciens. Left:* strain RIA 57; *right:* strain 8425; *1* specific activity (nmol/min per mg protein), *2* chlortetracycline (µg/ml × 2 . 10^{-2}).

The main source of NADPH, in addition to the pentose cycle, is malate dehydrogenase (decarboxylating) (NADP) (EC 1.1.1.40). Both systems reach their peak activity between 12 and 24 h of cultivation, *i.e.* at a time of very intense lipogenesis[19]. Malate dehydrogenase (decarboxylating) shows maximum activity both in the low-

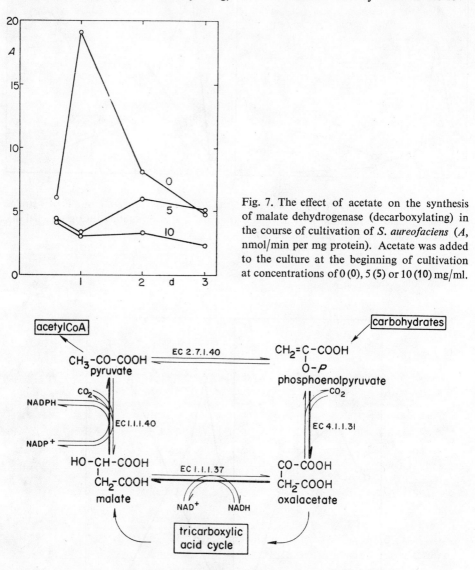

Fig. 7. The effect of acetate on the synthesis of malate dehydrogenase (decarboxylating) in the course of cultivation of *S. aureofaciens* (*A*, nmol/min per mg protein). Acetate was added to the culture at the beginning of cultivation at concentrations of 0 (0), 5 (5) or 10 (10) mg/ml.

Fig. 8. Metabolism of dicarboxylic acids in *S. aureofaciens*.

production and in the production variant (Fig. 6). In spite of substantial differences in the levels of antibiotic produced the course of activity of the enzyme has an equal tendency in both strains.

The enzyme synthesis is repressed by acetate. Fig. 7 shows that 1% acetate added to the medium at the beginning of cultivation will practically suppress its formation. This regulatory mechanism might control the transition from the anabolic to the katabolic phase. It is likely that malate dehydrogenase (decarboxylating) is synthesized by a repressible system, the function of which is controlled by acetate (or a near metabolite). Suppression of the formation of the enzyme after termination of the anabolic

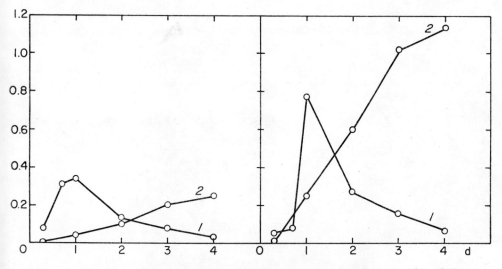

Fig. 9. Activity of acetylCoA carboxylase during cultivation of different strains of *S. aureofaciens*. *Left:* strain RIA 57; *right:* strain 8425; 1 acetylCoA carboxylase (c.p.m./mg protein × 10^{-4}), 2 chlortetracycline (µg/ml × 10^{-3}).

phase may be due to accumulation of acetylCoA which is oxidized more slowly in the tricarboxylic-acid cycle.

Malate dehydrogenase (decarboxylating) participates first of all in anabolic processes — its activity drops sharply with beginning production of the antibiotic. The role of the enzyme in the formation of NADPH is shown in Fig. 8. Malate is formed from oxalacetate through the action of cytoplasmic malate dehydrogenase[31] with simultaneous oxidation of NADH. NADPH is generated by decarboxylating malate dehydrogenase during oxidative decarboxylation of malate. An essential requirement of the transhydrogenation is sufficient oxalacetate, which is the substrate of malate dehydrogenase.

The activity of the key enzyme of lipogenesis, acetylCoA carboxylase (EC 6.4.1.2) in the high-production variant of *S. aureofaciens* is only twice as high as in the low-production strain[20]. The peak of enzyme activity (determined by incorporation of $^{14}CO_2$ into acetylCoA) is reached during the anabolic phase of cultivation while during intense formation of CTC the activity is fundamentally lower (Fig. 9).

In this connection it is of interest to note that the activity of the pyruvate dehydrogenase complex is also highest at the beginning of cultivation. At the time of intense biosynthesis only minimum activity of this system was demonstrated[21]. The possibility of formation of acetylCoA in this way during the katabolic phase is very small.

The formation of malonylCoA as an intermediate of CTC biosynthesis by carboxylation of acetylCoA[32] thus has not been demonstrated. The principal source of malonate, or malonylCoA, in the biosynthesis of CTC may lie in another reaction,

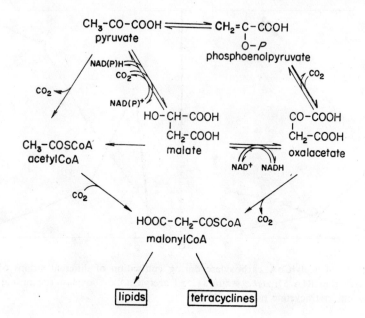

Fig. 10. Possibilities of formation of malonylCoA in *S. aureofaciens*.

e.g. decarboxylation of oxalacetate (Fig. 10), similar to *Phaseolus vulgaris*[33] or *Penicillium islandicum*[34]. The possibility of formation of malonate in this way is further suggested by the presence of phosphoenolpyruvate carboxylase[35].

The results thus make it possible to conclude that lipogenesis is most active during the anabolic phase where it is closely associated with growth and hence cannot substantially compete with the utilization of acetylCoA for CTC biosynthesis. The investigation of the changes in fatty acid content during fermentation supported this view[29].

TRICARBOXYLIC-ACID CYCLE

A number of metabolic differences exist between the low-production strain *S. aureofaciens* RIA 57 and the production variant *S. aureofaciens* 8425. Under submer-

TABLE III

Changes in Activities of Enzymes during Cultivation of S. aureofaciens (nmol/min per mg of protein)

Strain		RIA 57					8425				
Time of cultivation, h		12	24	48	72	96	12	24	48	72	96
Chlortetracycline produced, μg/ml		—	70	107	155	135	—	137	482	755	955
Enzyme	Citrate synthase	33.6	39.0	33.7	106	77	10.4	19.0	33.2	11.4	20.0
	Aconitate hydratase	105	89	82	54	47.0	31.8	31.5	29.0	28.5	22.3
	Isocitrate dehydrogenase	77	59	66	60	67	52	41.3	26.7	33.0	27.0
	Fumarate hydratase	70	10.0	70	94	116	40.3	39.0	45.5	51.5	53.5
	Malate dehydrogenase	655	840	1030	2380	2020	310	675	770	370	560

TABLE IV

Changes in Activities of Enzymes during Growth of S. aureofaciens in Medium with 50 μM Benzyl Thiocyanate (nmol/min per mg of protein)

Strain		RIA 57					8425				
Time of cultivation, h		12	24	48	72	96	12	24	48	72	96
Chlortetracycline produced, μg/ml		—	65	180	290	280	—	212	652	915	1280
Enzyme	Citrate synthase	14.8	28.8	16.6	22.7	59	22.5	14.9	10.2	3.6	10.0
	Aconitate hydratase	74	50	21.0	17.4	46.4	42.0	29.1	24.0	17.1	26.7
	Isocitrate dehydrogenase	49.8	30.0	13.0	10.0	40.5	65	58	46.3	18.9	25.4
	Fumarate hydratase	33.7	47.0	26.5	30.7	70	10.5	48.5	33.9	34.7	46.3
	Malate dehydrogenase	268	343	317	506	1190	513	613	452	458	682

ged conditions, the low-production strain will grow faster and the mycelial dry weight will be greater.

In cell-free extracts of both strains, the activity of the following enzymes was followed spectrophotometrically during cultivation: citrate synthase (EC 4.1.3.7), aconitate hydratase (EC 4.2.1.3), isocitrate dehydrogenase (NADP) (EC 1.1.1.42), fumarate hydratase (EC 4.2.1.2) and malate dehydrogenase (EC 1.1.1.37)[13]. It was found that the enzyme activity of the low-production strain is greater during cultivation than the activity of the production strain (Table III).

The anabolic phase of both strains is charaterized by high enzyme activity in the initial reactions of the cycle and by lower activity of the terminal enzymes. The tricarboxylic-acid-cycle intermediates are apparently utilized for the synthesis of cell components. The formation of citrate requires a supply of oxalacetate outside the cycle, apparently *via* phosphoenolpyruvate carboxylase (EC 4.1.1.31), an anaplerotic enzyme, catalyzing the irreversible carboxylation of phosphoenolpyruvate to oxalacetate. Pyruvate carboxylase (EC 6.4.1.1) has not been detected in *S. aureofaciens*[35].

Under all conditions, a high activity of malate dehydrogenase has been observed. Citrate synthase has a low activity at the beginning of cultivation. It is likely that during the anabolic phase a part of the oxalacetate is metabolized *via* malate dehydrogenase (decarboxylating) to pyruvate. The observed activity of malate dehydrogenase (decarboxylating), detected only during the anabolic phase, is in agreement with this[19].

The addition of benzyl thiocyanate substantially affected the activity of the individual enzymes of the tricarboxylic-acid cycle. In addition to retarded consumption of sucrose and increased production of CTC the activity of enzymes was lower than in the control (Table IV).

It is thus likely that in the production strain (or in the culture with benzyl thiocyanate) the low activity of the tricarboxylic-acid cycle may channel the utilization of acetate toward biosynthesis of CTC.

The quantity of the cell enzyme alone does not give a satisfactory picture of the metabolic activity of the culture. It shows the maximal capacity of the pathway. However, the intracellular concentrations of intermediates are mostly at levels that will not allow enzymes to function at maximal rate. In addition, the action of enzyme effectors must be considered. Determination of ATP in the mycelium[23] hence supplemented the data on the energy metabolism of the two strains obtained by following the activity of the tricarboxylic-acid cycle.

The ATP concentration follows the same trend in both strains during cultivation, the concentrations found in the low-production strain being 2—4 times higher than in the production strain (Fig. 11). In both cases the anabolic phase is characterized by a high concentration of ATP in the mycelium, attaining a peak after 12 h of fermentation. A second peak was observed after cessation of growth but only in the low-production strain.

The results are in agreement with the data on the activity of the individual enzymes of the tricarboxylic-acid cycle. Similarly, the activity of ATPase (EC 3.6.1.3) indicates that the low content of ATP in the production strain is not due to its greater degradation, the low-production strain having displayed a higher ATPase activity than the production variant (Fig. 11).

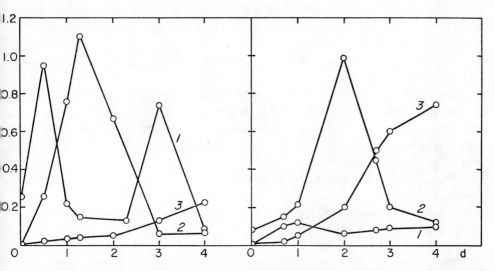

Fig. 11. Changes in intracellular concentration of ATP and ATPase activity during cultivation of different strains of *S. aureofaciens. Left:* strain RIA 57; *right:* strain 8425; *1* ATP (µg/ml), *2* ATPase (orthophosphate µg/mg protein $\times$ 5 . 10^{-3}), *3* chlortetracycline (µg/ml $\times$ 5 . 10^{-4}).

The concentration of ATP in the low-production strain corresponds to concentrations found in other microorganisms or animal tissues[36,37]. The production strain apparently possesses a certain defect in its energy metabolism (judging from the low activity of the tricarboxylic-acid cycle and the low ATP level), which results in an increased utilization of acetate for secondary biosynthesis during the katabolic phase.

The rate of oxidation of acetate in the tricarboxylic-acid cycle is affected by ATP which acts as a negative effector of the initial enzyme of the cycle, *i.e.* citrate synthase[38]. The rate of citrate formation in *S. aureofaciens* is also affected indirectly by ATP, the negative action being on the activity of phosphoenolpyruvate carboxylase[39].

Concentrations of ATP inhibiting the two enzymes (0.2—0.5 mM) were found only in the low-production strain. Modulation of their activity prevents the metabolic disturbance caused by overproduction of energy under conditions when acetate is metabolized predominantly by the tricarboxylic-acid cycle. A change of the genotype, to give the high production strain, results in a reduction of energy metabolism which completely excludes the possibility of action of these control mechanisms.

GENETIC CHANGES IN THE ENERGY METABOLISM AND RATE
OF CHLORTETRACYCLINE BIOSYNTHESIS

The principal difference between a production and a low-production strain may be seen in the intensity of energy metabolism during the katabolic phase. In the low-production strain, oxidation of acetate in the tricarboxylic-acid cycle and formation of ATP with subsequent conservation of energy in polyphosphate predominate:

$$x\ ATP\ +\ (PO_3^-)_n \longrightarrow x\ ADP\ +\ (PO_3^-)_{n+x}$$

The production strain with suppressed energy metabolism utilizes acetylCoA (or an equivalent building unit) for CTC biosynthesis. In contrast with the standard strain which utilizes about 0.3% of total sugar for the synthesis of the antibiotic, the high-producing strain under optimal conditions (at 3% sucrose in the medium) may transform as much as 30% of the sugar to CTC.

The quantitative role of energy metabolism varies from strain to strain. The intensity of CTC biosynthesis is thus inversely proportional to the rate of energy metabolism. This is determined by the expression of individual genes responsible for the synthesis of enzymes participating in the individual steps of the tricarboxylic-acid cycle and the respiratory chain.

These pathways represent a metabolic complex controlled by a large number of genes. Genes responsible for a given segment of metabolism (closely linked genes) are usually grouped in an operon which facilitates simultaneous regulation. An example here is the course of specific activities of citrate synthase and malate dehydrogenase during cultivation (Fig. 3).

This truly polygenic system allows a wide variability of the intensity of energy metabolism in different variants through mutational changes in some step of the respiratory chain. We may consider here changes of the structural or regulatory genes which may result in changes in the rate of enzyme formation. However, one must consider that such mutations may affect the activity of some enzyme reactions, participating only indirectly in the biosynthetic sequences (transport of the metabolite or availability of cofactors or effectors). A consequence of these genetic changes is a more or less pronounced retardation of energy metabolism and hence a corresponding change in the availability of building units for the synthesis of the antibiotic.

As an example we may cite the so-called phosphorus-resistant mutants of *S. aureofaciens*[40]. In most strains grown under submerged conditions the optimal content of inorganic phosphorus in the medium is about 30 µg/ml which is the rate-limiting concentration for oxidative phosphorylation. The production of the antibiotic is actually a phenotypic expression, since an increase in phosphate concentration results in more intense growth and energy metabolism, which is accompanied by a drop of secondary biosynthesis. In mutants, the rate of oxidative phosphorylation is controlled by a genetic block rather than by the concentration of orthophosphate itself, since an increase of phosphate concentration within a certain range has no effect on CTC bio-synthesis (Fig. 12).

Differences in metabolism were also found between the strain *S. aureofaciens* BMK (parent strain *S. aureofaciens* 8425) and the strain *S. aureofaciens* LS-B-16[41]. The LS-B-16 strain originates from another line and belongs to the so-called starch-type strains (it degrades sugars slowly, starch being the most suitable substrate for

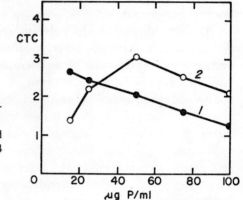

Fig. 12. Effect of various concentrations of ortho-phosphate on the production of CTC ($\mu g/ml \times \times 10^{-3}$) by *S. aureofaciens* 8425 (*curve 1*) and by the "phosphorus-resistant mutant" UVB I/84 (*curve 2*).

CTC production). In a medium with sucrose it will grow slowly, attaining low yields, and the biosynthesis of the antibiotic is not stimulated by benzyl thiocyanate. Using specifically labelled glucoses we could demonstrate a greater role of the pentose cycle in sugar katabolism than in the case of *S. aureofaciens* BMK.

Deeper knowledge of the metabolism of the producer will make it possible to determine the strategy during cultivation screening of variants obtained after muta-genic treatment; by suitable choice of screening medium mutants of desired properties can then be selected.

CONCLUSION

The yield of chlortetracycline obtained during submerged cultivation of *S. aureo-faciens* is the result of interaction of a number of genes, in the expression of which environmental conditions play their role. VANĚK, who analyzed the metabolic path-way of CTC biosynthesis, observed[42] that some 200 genes may be involved in the synthesis of the antibiotic. During the first segment of the pathway leading from glu-cose to acetylCoA, some 100 genes take part. In the second segment from acetylCoA to the tricyclic nonaketide, 28 reactions are involved. In the final phase, from the condensation of the tricyclic skeleton to the final product, 11 genes may participate, and be responsible for the formation of 72 compounds (27 isolated already and 45 presumed to exist). The last group of genes represents the so-called major genes, determining the biosynthesis of tetracyclines as a qualitative character.

The results discussed indicate that the genes responsible for the biosynthesis of chlortetracycline as a quantitative character, lie in the region of metabolism of acetylCoA. Genes of the tricarboxylic-acid cycle and respiratory-chain enzymes play a major role here. By slowing down this metabolic pathway (by affecting the rate of one of its reactions) acetylCoA begins to be metabolized preferentially by other pathways. One of them is the formation of the tricyclic nonaketide, which has relatively low energy demands.

One may thus speak of a polygenic character determining the yield of chlortetracycline but in a way which differs qualitatively from the previous concept. The production activity is the result of the interaction of a number of genes which control the individual steps of metabolic pathways, competing with biosynthesis itself. One need not assume the action of the so-called genes of specific effect. The polygenic system determining the production as a quantitative feature, represents both the structural and control genes of the biosynthetic pathway itself as well as of the alternative pathways competing for a common precursor. A decrease in the metabolic rate of a competing pathway caused either by a hereditary change at the level of some structural or a control gene, or by environmental changes, results in an intensification of secondary metabolism and increased antibiotic synthesis. Mutations affecting the rate of metabolic conversions of acetate, which compete with secondary biosynthesis may yield the same result, as a limitation in their rate by cultivation conditions. The high-production mutant during submerged cultivation then acts as a phenocopy of the standard strain cultivated under limiting conditions.

REFERENCES

1. MATHER K.: Biol. Rev. **20**, 32 (1943).
2. WAGNER R. P., MITCHEL H. K.: *Genetics and Metabolism*, 2nd Ed., p. 409. John Wiley, New York 1965.
3. DAVIS B. D.: Science **170**, 1279 (1970).
4. SERMONTI G.: *Genetics of Antibiotic-Producing Microorganisms*, p. 118. Wiley-Interscience, London 1969.
5. DiMARCO A., PENNELLA P., p. 45 in D. J. D. Hockenhull (Ed.): *Progress in Industrial Microbiology*, Vol. I. Heywood & Co., London 1959.
6. BIFFI G., BORETTI G., DiMARCO A., PENNELLA P.: Appl. Microbiol. **2**, 288 (1954).
7. BORETTI G., DiMARCO A., JULITA P., RAGGI F., BARDI U.: Giorn. Microbiol. **1**, 406 (1956).
8. HOŠŤÁLEK Z.: Fol. Microbiol. **9**, 78 (1964).
9. MATELOVÁ V., MUSÍLKOVÁ M., NEČÁSEK J., ŠMEJKAL F.: Preslia (Prague) **27**, 27 (1955).
10. HOŠŤÁLEK Z., JANEČEK J., DOSKOČIL J., KAŠPAROVÁ J.: Naturwiss. **45**, 396 (1958).
11. HOŠŤÁLEK Z., HEROLD M., NEČÁSEK J.: Naturwiss. **45**, 543 (1958).
12. HOŠŤÁLEK Z.: Fol. Microbiol. **9**, 96 (1964).
13. HOŠŤÁLEK Z., et al.: Biotechnol. Bioeng. **11**, 539 (1969).
14. VANĚK Z.: Čs. mikrobiol. (Prague) **2**, 275 (1957).

15. VANĚK Z.: Čs. mikrobiol. (Prague) **3**, 364 (1958).
16. DOSKOČIL J., HOŠŤÁLEK Z., KAŠPAROVÁ J., ZAJÍČEK J., HEROLD M.: J. Biochem. Microbiol. Technol. Eng. **1**, 261 (1959).
17. MIKULÍK K., KARNETOVÁ J., QUYEN N., BLUMAUEROVÁ M., KOMERSOVÁ I., VANĚK Z.: J. Antibiotics, **24**, 801 (1971).
18. BĚHAL V., CUDLÍN J., VANĚK Z.: Fol. Microbiol. **14**, 117 (1969).
19. JECHOVÁ V., HOŠŤÁLEK Z., VANĚK Z.: Fol. Microbiol. **14**, 128 (1969).
20. BĚHAL V., VANĚK Z.: Fol. Microbiol. **15**, 354 (1970).
21. JECHOVÁ V., BĚHAL V., HOŠŤÁLEK Z., VANĚK Z.: Fol. Microbiol. *to be published.*
22. ONDRÁČEK P.: *Thesis.* Charles University, Prague 1970.
23. ČURDOVÁ E., HOŠŤÁLEK Z., VANĚK Z.: Fol. Microbiol., *to be published.*
24. MIKULÍK K., BLUMAUEROVÁ M., VANĚK Z., LUDVÍK J.: Fol. Microbiol. **16**, 24 (1971).
25. BU'LOCK J. D. p. 61 in Z. Vaněk, Z. Hošťálek (Eds.): *Biogenesis of Antibiotic Substances.* Academic Press, London 1965.
26. MIKULÍK K., KARNETOVÁ J., KŘEMEN A., TAX J., VANĚK Z., p. 201 in *Radiation and Radioisotopes for Industrial Microorganisms.* Internat. Atomic Energy Agency, Vienna 1971.
27. HOŠŤÁLEK Z., BARNES M. F., BU'LOCK J. D.: Fol. Microbiol. **15**, 209 (1970).
28. BĚHAL V., PROCHÁZKOVÁ V., VANĚK Z.: Fol. Microbiol. **14**, 112 (1969).
29. BĚHAL V., JÍLEK M.: Fol. Microbiol. **14**, 211 (1969).
30. MILLER P. A., HASH J. H., LINCKS M., BOHONOS N.: Biochem. Biophys. Res. Comm. **18**, 325 (1965).
31. VANĚK Z., HOŠŤÁLEK Z.: Adv. Hyg. & Exp. Medicine (Warsaw), *in press.*
32. McCORMICK J. R. D., p. 566 in M. Herold, Z. Gabriel (Eds.): *Antibiotics. Advances in Research, Production and Clinical Use.* Butterworths, London 1966.
33. DE VELLIS J., SHANNON L. M., LEW J. Y.: Plant Physiol. **38**, 687 (1963).
34. GATENBECK S., MAHLÉN A. K.: Acta Chem. Scand. **22**, 1696 (1968).
35. VOŘÍŠEK J., POWELL A. J., VANĚK Z.: Fol. Microbiol. **14**, 398 (1969).
36. POLAKIS E. S., BARTLEY W.: Biochem. J. **99**, 521 (1966).
37. BERRY N. N.: Biochem. J. **95**, 587 (1965).
38. HOŠŤÁLEK Z., RYABUSHKO T. A., CUDLÍN J., VANĚK Z.: Fol. Microbiol. **14**, 121 (1969).
39. VOŘÍŠEK J., POWELL A. J., VANĚK Z.: Fol. Microbiol. **15**, 153 (1970).
40. HEROLD M., SIKYTA B., HOŠŤÁLEK Z.: 2nd Ann. Meeting Czech. Biochem. Soc., Prague 1960. Abstracts p. 41.
41. HOŠŤÁLEK Z.: 9th Internat. Congr. Microbiol., Moscow 1966. Abstracts p. 178.
42. VANĚK Z., CUDLÍN J., BLUMAUEROVÁ M., HOŠŤÁLEK Z.: Fol. Microbiol. **16**, 227 (1971).

6/ Biosynthesis of Alkaloids

The Inheritance of the Alkaloid Content of Ergot in Parasitic and Saprophytic Cultures

NICOLAS VON BÉKÉSY

Research Institute for Medicinal Plants, Budapest, Hungary

Dedicated to Prof. Dr. Dr. h. c. Kurt Mothes on his 70th birthday in token of my respect and friendship.

A short survey of ergot research shows that ergot has been for a long time past a hard nut to crack. The mystification about ergot diseases, the different names for the different forms of appearance of the same fungus and the difficult and complicated way which led to ergot alkaloids becoming medicaments, shows the difficulties involved and as one proceeds from the cultivation of ergot on rye to the fermentative production of alkaloids, one is faced with further obstacles.

Essentially, each ergot which forms alkaloids on rye must be able to produce alkaloids in a saprophytic culture. Consequently, the problem of a saprophytic alkaloid production is not a problem of genetics but one of the nutrition and the phase of development. Up to the present this method has had little success but occasionally ergot strains were found which were able to produce alkaloids. It was possible to improve them by selection and to cultivate them for a long time. Consequently, the problem is more complicated and cannot be split into a purely genetic, nutritional or developmental problem. Now we are going to discuss the genetic part of this question relating to the parasitic and saprophytic ergot cultures.

INHERITANCE BY ASCOSPORE PROPAGATION

It is difficult to follow the generative events in ergot fungi. The sexual behaviour has already been examined 50 years ago[1] but it was not possible to look into all the details. Even experiments done recently lack an unambiguous interpretation. The details reported by KILIAN[1] were queried by KNIEP[2] but new research of KULKARNI[3] on *Claviceps microcephala* seems to verify his findings.

The fungus is homothallic. It is possible with a single ascospore germinating and growing well and saprophytically propagated to infect a carefully isolated rye, to follow the whole life cycle. But not all ascospores behave in the same way. There are spores which germinate normally but form only a small mycelium in the hanging drop. Instead of long threads of hyphae, crumbled unequal hyphae will grow and their development soon stops[4]. At that time, we believed that they were heterothallic and able to mate with a corresponding partner through an anastomosis which turns into a normal mycelium. Anastomoses had been described by JUNG and ROCHELMEYER[5] and proved by STRNADOVÁ[6]. The latter mated auxotroph mutants and obtained hetero-karyotic mycelia. The frequency of the anastomoses on *Claviceps purpurea* is cert-ainly not comparable with that on other fungi, *e.g. Aspergillus*. JUNG and ROCHEL-MEYER found also anastomoses in cultures of monoconidial isolates. This may lead, however, only to homokaryosis as most normal conidia are uninucleate. We believe, therefore, that in most cases ergot is homothallic.

Although the sexual development of the fungus is unipolar and therefore self-fertilized, it is of great importance for the vitality of ergot. This is represented by our ergot cultivation which has been carried out for many years. We have used two strains in our ergot cultures on a large scale. One of them was cultivated in the field for 16 years and was labelled "OY178"; it produced ergocornine and ergokryptine. The other has been cultivated for 12 years and was labelled "A20" and yielded ergotamine. Both strains have been yearly asexually propagated by selecting an appropriate scle-rotium with high alkaloid content. The inoculum made from the sclerotium was put on a malt-agar tube where it turned into a saprophytic culture. The ergotoxine-con-taining strain showed a more woolly form in the culture, whilst the ergotamine strain became rather mucous. To be able to distinguish between the two cultures readily we made a selection in the direction of the two markers. Then we continued this selection for several years but at the end the woolly, ergotoxine-producing strain degenerated. Conidial productivity and vitality in the field have dropped. The alkaloid content of the fungus decreased somewhat. The productivity of conidia was only one-fourth that of the original strain. The alkaloid content has fallen from $0.6-0.5\%$ to $0.5-0.4\%$. This strain selected from a wild one showed fine vitality during the first years. We wan-ted to regenerate this strain and began to propagate it sexually by ascospores. Unexpec-tedly, a great part of the sclerotia which were cultivated asexually for many years did not produce stromata. Even if some sclerotia did, the number of the stromata was not more than four, seldom five and they formed only a small head on a thin stem. After

TABLE I

Harvest Results of Sexually Propagated Ergot under Normal Climatic Circumstances

Strain	Yield kg/ha	Alkaloid content %
OY 178/Bk 83584	270	0.38
OY 178/Bk 83592	270	0.36
OY 178/Bk 83593	415	0.30
OY 178/Bk 83596	440	0.43
OY 178/Bk 83597	354	0.41

repeated cultivation through ascospores we got an ergot which now easily produces stromata with great heads and from which we are able to obtain ascospores without difficulties. The vitality of the strain has been also improved as shown in Table I.

The experiment was set in the wooded, hilly Bakony. It was not possible to have control plots but the yields were high and showed good vitality. The alkaloid contents in this district are usually lower than those in the vicinity of Budapest but this may be due to a different climate, the district lying 520 m above sea level while Budapest has an elevation of only 110 m.

Another experiment shows that the aggressivity of this generatively propagated ergot has been increased, in comparison with vegetatively cultivated ergot.

In May 1969, the surroundings of Budapest suffered a drought spell which caused the rye to be stunted. It was blooming just after earing so that we were not able to infect it at the right time but this late infection showed the increased aggressivity of ergot (see Table II). As the table shows, all yields were very low. That happened not only because of a late inoculation but the temperature during the inoculation was abnormally high (about 27 to 30 °C) and the air was very dry. But even this shows the differences in the yields as the yields of the generative cultures were about five times greater than the vegetative ones.

TABLE II

Increased Aggressivity of Ergot Strains Three Times Generatively Subcultivated, in Comparison with the Vegetatively Cultivated Ergot of a Consecutive 12-year Period in a Dry Year

Generative			Vegetative		
strain	yield[a] kg/ha	alkaloid %	strain	yield[b] kg/ha	alkaloid %
OY 178 Bk 83[c]	117	0.43	OY 178 Bk 81[e]	14	0.44
	45	0.46		9	0.45
	87	0.58		19	0.45
	98	0.29		19	0.46
				15	0.44
OY 178 Bk 83[d]	106	0.46		23	0.45
	108	0.37			
	75	0.52			
	115	0.42			
OY 178 L 641	151	0.43			
	108	0.57			
	50	0.54			

[a] Average 96 kg/ha.
[b] Average 17 kg/ha.
[c] From mass ascospores.
[d] From single ascospores.
[e] From conidia.

The behaviour of this strain by sexual propagation is very characteristic for the inheritance and it seems to be generally valid. If the distribution curve of the alkaloid content in the sclerotia is charted the results of Fig. 1 are obtained.

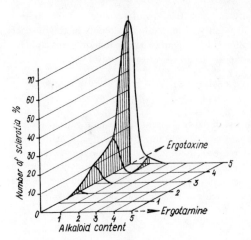

Fig. 1. Distribution curves of the alkaloid content of an asexually propagated ergo-toxine-containing ergot strain.

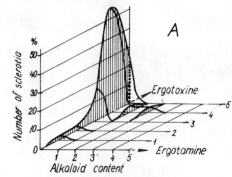

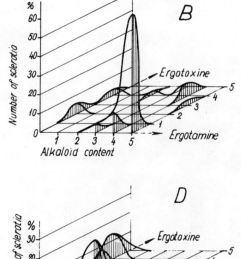

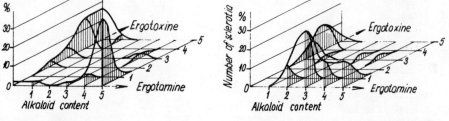

Fig. 2. Distribution curves of the alkaloid content after 1st generative propagation of the ergoto-xine-containing strain.

The ordinate shows the numbers of sclerotia expressed in per cent which belong to a given alkaloid content. On the horizontal plane, the alkaloid contents are plotted. The values from *1* to *5* are visual estimates of the alkaloid spots on a paper chromatogram which were compared with a known standard alkaloid spot. "*1*" means a very low alkaloid content, "*5*" a high content. When the alkaloid content is plotted on the outer line from the starting point "*0*" in the direction of the arrow to ergotoxine, it means that the sclerotia contain only ergotoxine and have no ergotamine content. On the other hand, if it lies on the outer line in the direction of ergotamine the ergot contains only ergotamine. All the other values which lie between the two outer lines belong to ergot sclerotia which have a mixed content of ergotamine—ergotoxine.

Fig. 1 relates to asexually propagated ergot. The greatest part of the sclerotia contain ergotoxine and are marked with "*5*". About 4% of the sclerotia contain in addition to ergotoxine a low amount of ergotamine which is marked with "*1*". One must not forget that all distribution curves of the alkaloid content show a bell-shaped form. If such a curve belongs to a culture which has a prevailing ergotoxine content and only a very small content of ergotamine, the ergotamine contents of the sclerotia can only be expressed if they lie on the side of the maximum alkaloid content of the bell-shaped curve.

The distribution curve of Fig. 2 is obtained by active discharge of the ascospores from the stromata of the sclerotia which belonged formally to the asexually cultivated ergot as shown in Fig. 1. We obtained many results in this experiment but we show only some of the most typical distribution curves. The form of the distribution curves depends to a great extent on the stromata from which the ascospores were obtained,

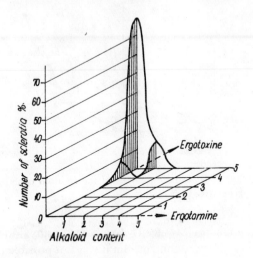

Fig. 3. Distribution curves of the alkaloid content of the ergotoxine-containing strain asexually propagated after it has passed a generative propagation.

but sometimes we get from the same stroma very different kinds of ascospores. In the figure, *A* is very similar to the starting culture but the alkaloid content has been somewhat diminished and the proportion of ergotamine-containing sclerotia has been increased. We used this culture for further cultivation and selected the pure ergotoxine-containing sclerotia which are marked with points. The other distribution curves are double-peaked and it is possible to get an ergot which contains only ergotamine but it is also possible to obtain cultures with mixed ergotoxine—ergotamine content. Repeated experiments proved that it is not possible to get rid of a small amount of ergotamine in these cultures. As we started from the pointed sclerotia with pure ergotoxine, we obtained by asexual propagation the distribution curve shown in Fig. 3. This culture consists of sclerotia with high ergotoxine content but a small percentage shows a contamination by ergotamine. If we propagate this strain in the sexual way, we get

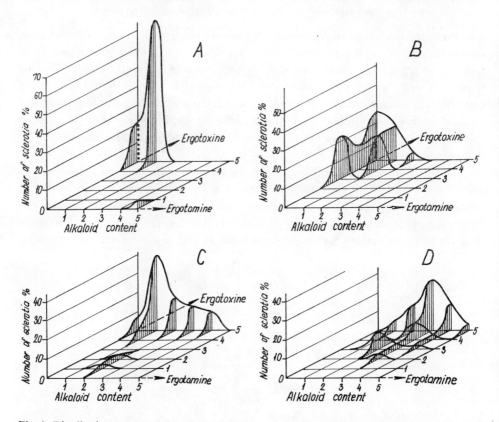

Fig. 4. Distribution curves of the alkaloid content of the ergotoxine-containing strain after 2nd generative propagation.

a minor cleavage. There are hardly any double-peaked curves and sclerotia with pure ergotamine content. But it was not possible to get rid of the small ergotamine content as all sclerotia which comprised a higher ergotoxine content included some ergotamine. In the following figures we present some typical cases. In Fig. 4, *A* shows some cultures with an alkaloid content of 0.68—0.71%, ergotamine participating with 15%. *B* shows some more flattened curves, *C* and *D* show curves with a considerable amount of ergotamine. After the third generative propagation we found no more double-peaked distribution curves and we found no more sclerotia with pure

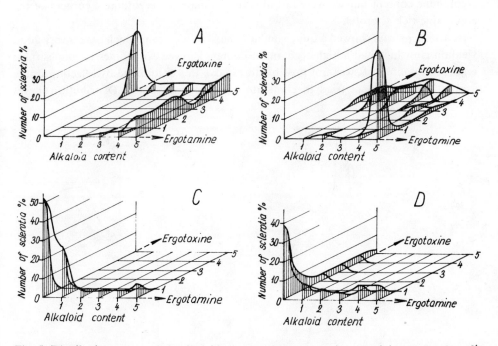

Fig. 5. Distribution curves of the alkaloid content of an ergotamine-containing strain after 1st generative propagation.

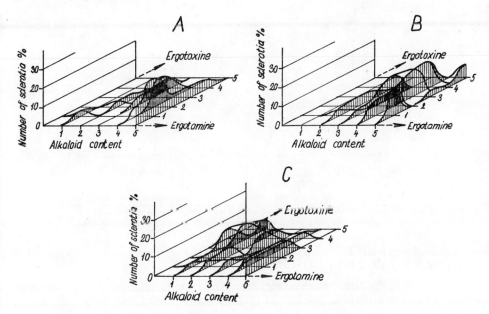

Fig. 6. Distribution curves of the alkaloid content from an ergotamine-containing strain after 2nd generative propagation.

ergotamine content but we were not able to eliminate the ergotamine content of the ergotoxine-rich sclerotia.

Another strain showed a quite similar behaviour. A strain which gave a very good yield on rye which had a high ergotamine content showed sometimes an undesirable ergotoxine content. It was not possible to remove this ergotoxine by selection in the asexual cultures, so we tried to eliminate it in a generative way. Distribution curves

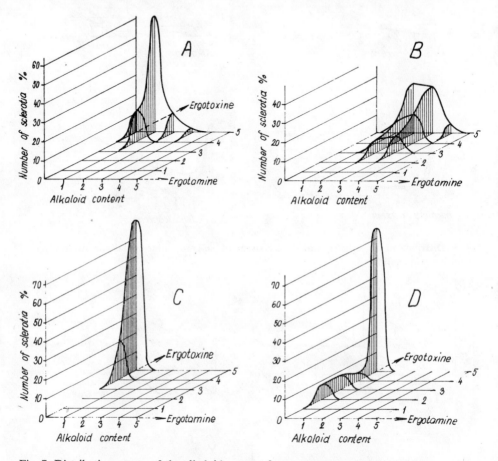

Fig. 7. Distribution curves of the alkaloid content from one-ascospore culture from an ergotoxine-containing strain. *A, B* after 2nd propagation; *C, D* after 3rd propagation.

from these sexual propagations are shown in Fig. 5. In *A*, there are about 30% sclerotia with pure ergotoxine, in *B* there are many sclerotia containing ergotamine, while others are sclerotia with a mixed alkaloid content. *C* shows only ergotamine but a great part of the sclerotia are free of alkaloids. *D* is somewhat similar but shows pure ergotamine and ergotoxine containing sclerotia. After the second generative propagation, when selection for pure ergotamine took place, sclerotia with pure

ergotoxine content were not found but sclerotia with mixed ergotamine and ergotoxine content became prevailing, as shown in Fig. 6. In some of these sclerotia the
sclerotia with mixed alkaloid content may belong in a small extent to the pseudo-
chimaerae but our previous examination showed that the contribution of chimaerae
cannot be more than 6%.

The intricacy of the heredity may be seen in the single-ascospore cultures. In
many cases, single-spore cultures behave as one expects them to. Fig. 7 shows cultures
after the second generative propagation of the earlier mentioned ergotoxine-contain-

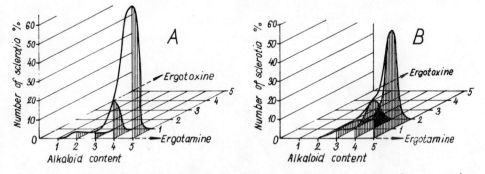

Fig. 8. Distribution curves of the alkaloid content from one-ascospore cultures of an ergotamine-
containing strain. A, B 2nd propagation.

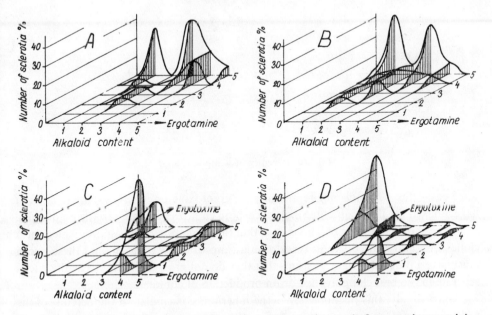

Fig. 9. Two-peak distribution curves from one-ascospore cultures. A, B ergotoxine-containing
strain; C, D ergotamine-containing strain.

ing strain. *A* is almost purely ergotoxine-containing, *B* contains mixed alkaloids but both are single-peaked. After the third sexual propagation the distribution curves showed a better distribution as shown in *C* and *D*. A similar pattern is shown by the ergotamine strain in Fig. 8. *A* is with pure ergotamine content, *B* has a small contamination of ergotoxine but the distribution curves are not always single-peaked. The ergotoxine-strain curves shown in Fig. 9 yielded curves with two peaks *A* and *B*. In

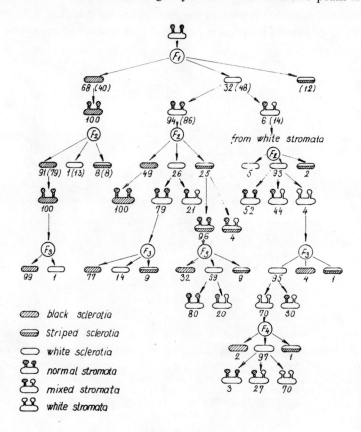

Fig. 10. Heredity of white sclerotia.

these cases, the peaks do not represent pure ergotoxine or pure ergotamine as both have a mixed alkaloid content. Single-ascospore cultures belonging to the ergotamine strain are shown in *C* and *D*. This means that the formation of ascospores which form clones with two quite different types of ergot must bear two nuclei, and these nuclei must have a different origin. It seems that they do not grow as dikaryons as the nuclei separate in the branching hyphae or on production of conidia.

Surprisingly, this phenomenon in the inheritance is similar to the heredity of the white sclerotia. One may recognize in Fig. 10 that even after the fourth sexual propagation by continuous selection in order to obtain pure, white sclerotia, with white stromata, it may happen that in the yield some sclerotia are black, others are white but have mixed stromata and only about 70% of the sclerotia are pure white with white stromata. At this time, it was very important for us to get albino sclerotia with pure white stromata in order to mate such white strains without alkaloid content with black, alkaloid-rich strains. But we were not able to do this experiment because we did not have a strain which gave in each case only white sclerotia with white stromata.

But even single ascospore clone cultures from a white sclerotium behave quite similarly as the single ascospore cultures with double-peaked distribution curves. We found in our earlier work sclerotia originating from mono-ascospore cultures which showed a white and black streaked colour. This means that a white and a black strain were growing together. As they originated from a single ascospore, we believed that such a phenomenon must be due to a back mutation.

In estimating the alkaloid content one must take in consideration that climatic conditions may cause a change not only in the total alkaloid content but also in the proportions of the single alkaloids. We found at the same type of clone culture that in the hilly region a strain with 0.35% peptide alkaloid content, which consisted of 75% ergotoxine and 25% ergotamine, showed in the plain an alkaloid content of 0.55%, the distribution being 90% ergotoxine and 10% ergotamine. Thus it may happen that a seemingly satisfactory result will become dubious next year.

From the genetic point of view it must be emphasized that the strain being discussed has been cultivated for 16 years asexually. During this long time, the alkaloid content showed a small decrease by about 20% but the relative amount of the different alkaloids remained constant. The strain showed among the peptide alkaloids an ergocornine content of 50% and an ergokryptine content of 30% with about 10% ergotamine. When this strain was selected by sexual propagation in the direction toward ergotamine an ergotamine strain with a small amount of ergotoxine was obtained, but in this ergotoxine the ratio of ergocornine to ergokryptine was the same as mentioned before. We never obtained in all our experiments an ergot with pure ergocornine or pure ergokryptine content. As the capacity of alkaloid biosynthesis is genetically coded in the chromosomes, ergocornine and ergokryptine must be closely linked. Perhaps the ergotamine production is also in connection with this locus, as we could not eliminate the small ergotamine content. Other ergot strains showed quite a similar behaviour although the relative amount of the alkaloids was quite different.

INHERITANCE OF THE ALKALOID CONTENTS BY ASEXUAL PROPAGATION

Previously we have seen some features of ergot in saprophytic culture which may be influenced by selection. We were able to select a woolly form and it is possible, by using a special nutrient substrate, to change through selection the colour of the cultures, and to control the number of conidia in the cultures. As we have seen, the number of conidia diminished when the selection proceeded in the direction of a woolly form but it is possible to increase the number of the conidia. This was needed for cultures to be utilized for mass production of ergot in the field, to give the farmer an economical material for inoculation. All these phenomena are well known. JINKS[7,8] has shown that variations in such characters are under cytoplasmic control during differentiation and ageing. He was able to increase by nearly two-fold the growth rate between the two extremes of high and low growth, in a homokaryon of *Aspergillus*

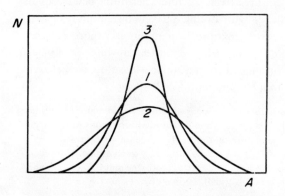

Fig. 11. Distribution curves of the alkaloid content of a pure strain of ergot. *A* alkaloid content, *N* number of sclerotia (%); *1* normal behaviour, *2* sensitive to environmental factors, *3* indifferent to environmental factors.

glaucus through repeated selection. But there are limits imposed be the nonviability of the most extreme forms. But the problems of breeding a strain of ergot with high alkaloid contents are much more difficult to survey than the problems associated with the physiology of nutrition and of development.

In the first place, let us observe the behaviour of ergot alkaloids on rye and on grass, respectively. If we infect rye with a pure culture we get an ergot which will show a distribution curve of the alkaloid content of the sclerotia with an entirely normal bell-shaped form, like Fig. 11 (Refs[9,10]). This distribution curve may move according to the climatic conditions to a lower or to a higher alkaloid content. If the curve flattens, as is shown by line *2*, it means that the ergot is sensitive to environmental factors. However, if the curve shows a steep slope, such as line *3*, then the culture is more indifferent. If we make a propagation from on esclerotium with the lowest alkaloid content, and another from a sclerotium with the highest alkaloid content, both ergots will show the same distribution curves. But if we inoculate with this ergot different grasses we shall find very great differences in the alkaloid content. This happened to HARTWICH[11] who found a very high alkaloid content of 1% in one ergot strain which was grown on *Molinea coerulea* Moench. This high alkaloid content was confirmed with the new analytical methods of ROCHELMEYER[12] and by BLAŽEK and BÖSWART[13]. According to KYBAL[14] this is caused by an optimum nitrogen supply together with a limited proportion of phosphorus. A considerably lower alkaloid content was found by GRÖGER[15] on different kinds of *Bromus* and *Calamagrostis* and by us[16] on *Lolium aristatum*, *L. perenne*, on *Festuca sulcata*, etc. If we inoculate rye with this grass-grown ergot we get an alkaloid content which corresponds to that of the ergot grown on rye. It is not possible to explain this phenomenon by saying that the nutrition on grass was insufficient, as the sclerotia are well developed on grass. Besides, if the nutrient transport is slower on rye, such as by breaking the stem under the ears, this will hardly influence the alkaloid content[17]. Even other factors, such as shading or heavy rust infection, has no effect on the alkaloid content. There must be some specific substances which affect the regulating system by converting the spha-caelia into sclerotium and by producing alkaloids.

Last year we re-examined the ergot grown on grass to evaluate the relative amount of different alkaloids. We used for this purpose the above-mentioned ergo-toxine-containing culture. Table III shows that the special alkaloids are the same as they were on rye but the ratio of ergotoxine to ergotamine was changed and one of the grass ergots contained no alkaloid at all. Unfortunately, the quantity of this grass ergot was very small so that we could only estimate the spots on a paper chromatogram.

Sometimes it may happen that the ergot of grass which we used for inoculating rye, shows on rye a diminished alkaloid content. If this diminished alkaloid content is permanent, the grass must have had a genetic influence on ergot. This may happen to an inhomogeneous strain by selection or by mutation. But if we have inoculated with the same strain the same grass in different plots several times and all ergot of these plots shows the same diminished alkaloid content, this diminished alkaloid content was certainly not caused by mutation.

It is well known that ergot contains none or only traces of alkaloids in the spha-caelia stage[10,18]. Only when the hyphae begin to form the compact mycelium of the sclerotium do the alkaloids come into appearance. This conversion from the conidia-producing sphacaelia stage to the compact-sclerotium stage is connected with a change in the number of nuclei. MILOVIDOV[19] and JUNG and ROCHELMEYER[5] found that the conidia of honeydew are uninucleate. The sphacaelia itself should be, after MILO-VIDOV, binucleate but JUNG and ROCHELMEYER suggest that the conidiophores are binucleate only if the conidia are not yet abstricted from the conidiophores and the nucleus has not moved into the conidia. But the closely packed parallel hyphae of the

TABLE III

Alkaloid Content of Grass Ergot Inoculated with the Ergotoxine Strain

Species	Ergometrine	Ergotamine	Ergotoxine
Aneurolepidum angustum	trace	1	5
Secale kuprijanovii	—	trace	5
Secale montanum	—	trace	5
Dasypyrum villosum	trace	trace	5
Roegneria fibrosa	—	4	2
Brachypodium pinnatum	—	—	—

sclerotium are, after MILOVIDOV and JUNG and ROCHELMEYER, always multinucleate. Therefore, it seems that if the hyphae begin to be multinucleate they do not produce any more conidia but form the pseudoparenchyme of the sclerotium. The sclerotium itself in the beginning is not composed of growing hyphae. The hyphae become only divided by transverse walls and increase in thickness and are later filled with protein and oil.

Let us now discuss the behaviour of ergot in saprophytic culture. Young cultures with growing hyphae have no alkaloid content, especially if they are able to abstract conidia in large quantities. This stage of the culture corresponds to the sphacaelia stage. If the cultures grow old and form a mycelial web, like a sclerotium, they may form alkaloids[20] but only if we use a suitable nutritive material[21,22]. We found that the nutrient of ABE modified after HOFMANN et al.[23] is adequate for this purpose.

The cells of these sclerotium-like webs are in many respects very similar to the true sclerotia and are according to JUNG and ROCHELMEYER[5] also dikaryotic. But it is not possible to obtain alkaloids if one uses for cultivation a material rich in nutrients, e.g. malt extract. The fungus grows well on this medium and forms microscopic and macroscopic sclerotium-like structures and it is very difficult to distinguish them from structures which form alkaloids. However, these do not have any traces of alkaloid content. This influence of the nutrient on the alkaloid content is reminiscent of the parasitic ergot, the alkaloid content of which is changed on different grasses. Accordingly, the fungus may produce alkaloids only if it has developed much further than the conidia-producing and growing stage and it is in the stage of sclerotium formation. In addition, it must have a nutrient which favours the formation of alkaloids.

If we use conidia from dilute honeydew for plating it is noteworthy that all single-conidium colonies are exactly of the same shape. But if we use for plating conidia from a saprophytic culture, we will find some variations in the phenotype. We used in both cases asexual propagation, but the conidia of the honeydew represent a natural form of propagation, all conidia are abstricted from the conidiophores and are quite similar. When honeydew is diluted, germination takes place, and in about 12 h almost simultaneously all conidia will germinate. The conidia of the saprophytic culture are not so much alike as they may be abstricted from hyphae growing under different environmental factors. Some of them grow deep in the medium, others on the surface, some of the hyphae form conidia, others form atrospores, and so on. The germination of such spores is prolonged, the germs are very different and one finds always spores which do not germinate at all. This different behaviour belongs to the category of cytoplasmic variations as JINKS[24] has shown with other fungi.

Having a colony which is able to form alkaloids, the productivity is very unstable in the greatest part of subcultures. From the ergotoxin culture which was several times mentioned before, MOLNÁR and TÉTÉNYI[25] got a subculture which formed alkaloids under saprophytic conditions. This subculture produced, under favourable conditions, in surface cultures 1 g, in shake cultures 0.5 g alkaloid per liter, while the dry hyphae contained 2% of alkaloids. This culture was at the beginning very unstable.

It showed that the antecedent of the inoculum in propagating plays an important role in the formation of alkaloids. A similar observation was previously reported by TABER and VINING[26,27] in a series of experiments carried out to clear up the matter. It was demonstrated by MOLNÁR[28] that if one inoculum is taken from the surface of a 7-day-old culture and another is taken from a 56-day-old surface culture, and both are used for inoculation, under the same circumstances, after a period of 21 days the culture which was inoculated with the young mycelium, will show an alkaloid content of 0.21%, while the other culture inoculated with the old mycelium showed an alkaloid content of 0.99%.

If we prepare shake cultures from the same strain and use for inoculation a mycelium which has developed in 5−7 days and transfer a part of this young mycelium to inoculate a next shake culture, and if we repeat this procedure four times and use the fourth shake culture as surface culture, we get after 21 days in the surface culture the very small alkaloid content of 0.05%. It is possible, by using young cultures as inocula, to decrease the alkaloid content but it is also possible to reverse this repression by using repeated inoculations with old mycelia. Our own experiments showed that if the fungus is growing on a nourishing medium, such as malt extract, the alkaloid-producing ability may be lost very easily.

Apparently, a young mycelium accommodates itself to the new environmental conditions much more easily than an old mycelium. Enzymes needed for growth may be constitutive and are always present in the mycelium. They are functioning in the fresh nutrient easily and this is promoted by repeated inoculations. But as we have seen, growing mycelia and conidia abstricting hyphae do not produce alkaloids. The enzymes which are needed for producing alkaloids are adaptive and are present in mycelia only in a very small amount. By inoculation with old hyphae the suppression of the alkaloid-synthesizing enzymes goes on only slowly, even if in the fresh nutrient the hyphae are able to grow, the nutrient is sooner exhausted than the enzymes are repressed. All these changes in the hyphae are of a quantitative rather than qualitative nature and they show a continuous variation which means that they are under cytoplasmic control.

DELBRÜCK discussed[29] (cited after BRESCH[30]) a "pseudoheredity" to explain the extranuclear inheritance. His fundamental idea was a reciprocal blocking of two-enzyme reactions. This concept, expressed by the terminology of the operon theory[31] and adapted to our complex problem, is represented by a regulatory system as shown in Fig. 12. We will suppose that operon 1 contains all genes which control the growth of hyphae. Operon 1 is in a nonrepressed state and forms the 1 enzymes. These enzymes act if a suitable substrate, the culture medium, is present. These enzymes, marked as 1, are located in the cytoplasm. A part of the molecules are used up for the growth of the cells, another part will accumulate in the cell as substrates for enzyme group 2, and a small part of the molecules will act as corepressor 1. Corepressor 1 reacts with repressor 2. If enough substrate 1 is present the enzymes marked as 2, controlled by the operon, are inactive since operon 2 is blocked. In this case, no corepressor 2 is synthesized, repressor **1** remains inactive and cannot stop the activity of operon **1**. Therefore, operon **1** is working continuously. But if substrate **1** is exhausted, or if there appears a metabolite which interacts so that the enzymes are not able to form corepressor **1**, repressor **2** remains inactive and operon **2** will begin to work. Operon **2** contains genes coding for enzymes **2** which convert the thin hyphae into sclerotium and form the colour of the mycelium. Enzymes of group **2** may be probably adaptive

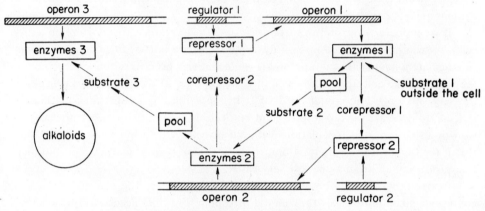

Fig. 12. Sketch of a regulating system.

enzymes which are only synthesized by the active operon **2**. When the enzymes of group **2** are working, they need substrate **2**, which was produced by enzyme **1** and which has already accumulated in the cell. But enzymes of group **2** also synthesize substrate **3** which is needed to form the alkaloids. Moreover, the enzymes of group **2** produce corepressor **2** which activates repressor **1** and stops the action of operon **1**. In this way, the product which is accumulated in the cell cannot be used again for growth but the hyphae are transformed into the sclerotial stage. The action of operon **3** containing genes for the alkaloid-forming enzymes of group **3** proceeds along similar lines. These enzymes must also be adaptive. Their substrates are produced by enzymes of group **2** and if there is no corepressor present that would interact with the repressor of operon **3**, it can work and alkaloids are produced.

The growth of ergot in a saprophytic medium is a non-organized loose growing form of filamentous hyphae. These unorganized forms of hyphae are regulated in the parasitic form on the host by the host plants. In the first step the sphacaelia is built. There must be a substance arising in the host plant, or from the mutual interaction between the fungus and the host plant, which regulates this growing mycelium. Figs 13 and 14 from my earlier work show this for the rye plant. The *right-hand* side shows a healthy ovary on the pedicel, the embryonal sac is shrivelled. On the *left* there is an attacked plant, on the *upper* part one may see the sphacaelia, *below* the beginning of the sclerotium. The sclerotium is sitting on the pedicel, but there is a cavity where the fungus has hydrolyzed the center of the pedicel. This part must be responsible for the substances which are necessary for organizing the sphacaelia. If the supply of these substances is waning the unorganized loose mycelium comes into appearance. This may be shown in Fig. 15 which shows a stem infection. The mycelium web has a woolly form which no organized mycelium possesses.

Let us now return to the regulating system. Substrate 1 on the host plant supplies the fungus not only with the substances which are needed for growth but also with substances needed for the conversion of the loose mycelium to the sphacaelia stage. But to be able to pass from the sphacaelia stage to the sclerotium stage, another substance is required, perhaps a metabolite which stops the production of sphacaelia and leads to the formation of the sclerotium. That must be able to stop the synthesis on corepressor 1 as we have seen previously. We know the phenomenon of degeneration which may cause a loss of the ability to form sclerotia[32]. Only honeydew is produced with or without conidia[33] so that ergot must comprise in its regulating system a sensitive group which may be blocked and this is reminiscent of the group of regulator 2, repressor 2, and corepressor 1.

One does not find this system in a saprophytic medium, when alkaloids are formed. In a young culture there are no alkaloids but if the hyphae are converted to the sclerotia-like mat, alkaloids are produced. Such strains may easily lose the conidia-producing capacity. In the second case, the strain can form sclerotia-like mat, but alkaloids are not constituted. This is often found with inoculations of wild ergot. In this case substrate 3 is not convenient for enzymes 3, or operon 3 is somehow blocked. In the third case, the strain is able to produce alkaloids just after inoculation*. The filamentous hyphae are scarcely formed as they soon pass to the sclerotia-like web. There is no difference between a young 10-day-old and a 70-day-old inoculum, both producing alkaloids in the same quantity. We suppose that in this case both the regulators 2 and 1 are out of function, or corepressor 1 and 2 are not formed. The growing mycelium controlled by operon 1 is handed over to operon 2 and 3, as none of the operons are blocked. Such mycelium is growing in most cases slowly. At least MANTLE and TONOLO[33] demonstrated a strain where the plectenchymatic sclerotial mat can produce conidia and alkaloids, too, but only in a surface culture. I believe that in this case the aerial part and the submerged part on the surface cultures act with the same, but differently blocked, regulating system.

This regulating system is based, in the morphological sense, on the different kinds of hyphae which belong to different groups in the regulating system and I tried to outline those groups in the sketch. It shows that though everything is coded in the chromosomes, the action of the genes is controlled by the cytoplasm which is affected, on other hand, by environmental factors.

This investigation was supported by Chemical Work Gedeon Richter, Budapest.

---

* About such a strain the author was kindly informed by Mrs Borowski (Arzneimittelwerk Dresden, Biologisches Institut, Radebeul 1).

REFERENCES

1. KILLIAN C.: Bull. Soc. Mycol. France **35**, 182 (1919).
2. KNIEP H.: *Die Sexualität der niederen Pflanzen.* Jena 1928.
3. KULKARNI U. K.: Mycopathol. Mycol. Appl. **20**, 19 (1963).
4. VON BÉKÉSY N.: Z. Pflanzenzücht. **35**, 461 (1956).
5. JUNG M., ROCHELMEYER H.: Beitr. Biol. Pflanzen **35**, 343 (1960).
6. STRNADOVÁ K.: Herba Hung. **7**, 45 (1968).
7. JINKS J. L.: Proc. Roy. Soc. B **146**, 527 (1957).
8. JINKS J. L.: J. Gen. Microbiol. **20**, 223 (1959).
9. MOTHES K., SILBER A.: Fortschr. Forsch. **26**, 101 (1954).
10. SILBER A., BISCHOFF W.: Pharmazie **9**, 46 (1954).
11. HARTWICH C.: Schweiz. Wochenschr. Chem. Pharm. **50**, 281 (1895).
12. ROCHELMEYER H.: Pharmazie **7**, 326 (1949).
13. BLAŽEK Z., BÖSWART J.: Pharmazie **8**, 851 (1953).
14. KYBAL J.: Phytopathology **53**, 363 (1963).
15. GRÖGER D.: Abh. Deut. Akad. Wiss., Kl. Chem. Geol. Biol. 243 (1956).
16. VON BÉKÉSY N.: Phytopathol. Z. **26**, 49 (1956).
17. VON BÉKÉSY N.: Biochem. Z. **30**, 368 (1940).
18. STERNON M. F.: Bull. Roy. Acad. Med. Belg. **1**, 463 (1936).
19. MILOVIDOV P.: Acta Histochem. **4**, 41 (1957).
20. ABE M.: Ann. Report Takeda Res. Lab. **10**, 152 (1951).
21. ABE M.: Ann. Report Takeda Res. Lab. **10**, 110 (1951).
22. ABE M.: Ann. Report Takeda Res. Lab. **10**, 129 (1951).
23. HOFMANN A., BRUNNER R., KOBEL H., BRACK A.: Helv. Chim. Acta **40**, 1358 (1957).
24. JINKS J. L.: Compt. rend. Trav. Lab. Carlsberg, Ser. Physiol. **26**, 183 (1956).
25. MOLNÁR G., TÉTÉNYI P.: Hung. Pat. 150,631 (1962).
26. TABER W. A., VINING L. C.: Can. J. Microbiol. **5**, 418 (1959).
27. TABER W. A., VINING L. C.: Can. J. Microbiol. **4**, 611 (1958).
28. MOLNÁR G.: Herba Hung. **4**, 90 (1965).
29. DELBRÜCK M.: Edn. Centre Nat. Rech. Sci. 25 (1949).
30. BRESCH C.: *Klassische und molekulare Genetik.* Springer-Verlag, Berlin—Heidelberg—New York 1965.
31. JACOB F., MONOD J.: J. Mol. Biol. **3**, 318 (1961).
32. VON BÉKÉSY N.: Acta Agric. Hung. **2**, 125 (1952).
33. MANTLE P. G., TONOLO A.: Trans. Brit. Mycol. Soc. **51**, 499 (1968).

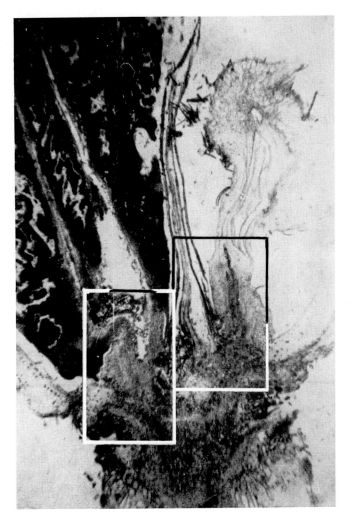

Fig. 13. Cross section of two flowers on the ear. *Right:* a healthy ovary on the pedicel; *above:* the sphacaelia stage; *below:* beginning of a sclerotium.

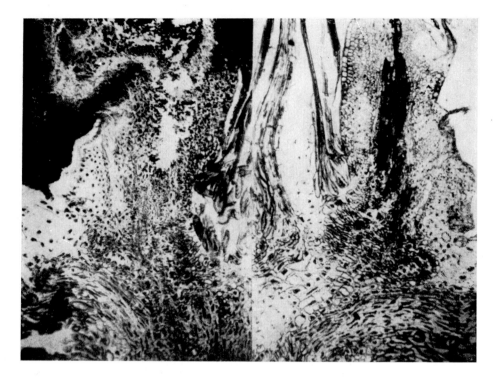

Fig. 14. The framed part of Fig. 13 enlarged. *Right:* healthy pedicel with vascular bundle; *left:* vascular bundle replaced by a hydrolyzed cavity. On the side of the cavity the nuclei of the cells are swollen and heavily stained.

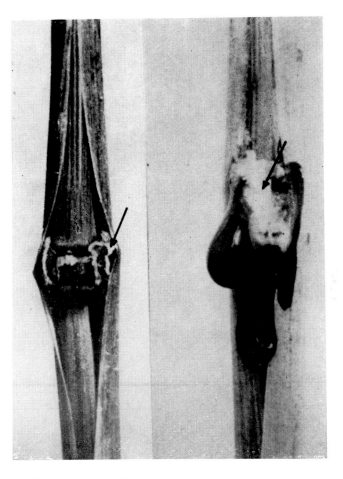

Fig. 15. Stem infection on a node. *Right:* sphacaelia stage with honeydew drops. Above the drops unorganized woolly mycelium. *Left:* The same node in sclerotial stage. Arrow shows the woolly form of mycelium.

Genetic Problems of Production of Ergot Alkaloids in Saprophytic and Parasitic Conditions

C. Spalla

Farmitalia — Istituto Ricerche, Milano, Italy

INTRODUCTION

The problem of the production of ergot alkaloids in submerged culture using the techniques traditionally employed in industrial fermentation, can be considered as solved. In fact, with these techniques lysergic acid and its amide, as well as all the alkaloids of both groups of ergotamine (ergotamine and ergosine) and of ergotoxine (ergocristine, ergocornine and ergokryptine) can be economically produced nowadays. The realization of these production processes required a long period and much effort: the progression of the various stages that allowed to obtain the results and the features that characterize these processes make the study of this particular group of fermentations very interesting.

I have to say right now that, from the metabolic point of view, these fermentations, even though not very different in their main features from other fermentations for obtaining secondary metabolites, show particular characters such as the frequent need of two carbon sources, the capacity of producing in synthetic media, the need for high amounts of sugars in the medium, the accumulation of fats and sterols together with alkaloids and, finally, the somewhat long span of the producing phase.

I think, however, that the recent and most interesting aspect of this kind of fermentations is not represented by these features but by the genetic characteristics of the producing strains. It is from this point of view that I want to examine the production of alkaloids with strains of the genus *Claviceps* in this paper.

The life cycle of *Claviceps* is well known (Fig. 1). Alkaloids are found in sclerotia; these are formed by mycelium that can be grown in saprophytic conditions. Normally in these conditions, however, the mycelium does not produce alkaloids and all the numerous attempts to obtain production gave only very scant or doubtful results until Abe in 1948[1] succeeded in obtaining production of alkaloids of the group of clavines in surface culture. Unfortunately, alkaloids of this kind are pharmacologically inactive. A few years later, the same author and his coworkers[2] could obtain production of peptide alkaloids in surface culture. Analogous results were obtained by Stoll *et al.*[3,4], Taber and Vining[5], Kybal and Starý[6], Tyler[7] and Rassbach[8]. The yields, however, were low and the processes not for industrial purposes. The problem had not been solved for several years until ARCAMONE et al. in 1961[9] isolated from *Paspalum distichum* a strain of *Claviceps paspali* able to produce in submerged culture high amounts of a new alkaloid, α-hydroxyethylamide of lysergic acid. ARCAMONE'S strain was characterized by a high virulence on rye and by lack of conidia. Shortly later, other workers isolated other strains of *C. paspali* producing the same alkaloid[10,11] or a derivative of it with a double bond in position 8[12]. All these strains were characterized, like that of ARCAMONE et al., by the fact that they failed to produce conidia. Only the strain by KOBEL et al. produced a few and under special conditions only.

From the pharmacological point of view, however, the most important alkaloids are the peptide derivatives of lysergic acid and the production of these cannot be achieved with *C. paspali*. The problem has been studied for several years at this Institute: many strains of *Claviceps* were isolated and studied under various conditions until strains of *Claviceps purpurea* producing ergotamine[13,14] and others of the same species producing, respectively, ergokryptine and ergotamine, ergocornine and ergosine, and ergocristine, were obtained[15].

These strains are characterized by the fact that they do not produce conidia or only a few and under particular conditions only. All of them produce high amounts of alkaloids in submerged culture. The problem of producing all ergot alkaloids by fermentation thus appears to be definitively resolved.

From the first paper on this problem, published by BONNS in 1922[16], to that of AMICI et al. of 1965 which announced the solution of the problem, there is a span of 43 years. In a survey on the topic[17], published in 1964, ABE and YAMATODANI quote 135 scientific papers. From 1964 up to now many other papers have been published. These data clearly show the endeavour required for the solution of the problem. It is evident that the problem, though apparently easy, concealed big difficulties. Examining the results obtained in the light of our present knowledge, we can say that the greatest difficulties to be overpassed were of genetic nature. A review of the main data known up to now on the genetic characteristics of the various *Claviceps* both in saprophytic culture and parasitic conditions will demonstrate the previous statement and allow to draw a general picture of some genetic factors that govern the production of alkaloids in *Claviceps*.

VARIABILITY OF *Claviceps* SPECIES

One of the most evident characteristics of the species of the genus *Claviceps* studied is their enormous variability. This variability in cultures of *C. purpurea* freshly isolated from sclerotia has been studied at this Institute[18]. The cultures obtained from sclerotia collected from ears of rye at Brunico (Italy), Usseglio (Italy) and Spain, were studied on slants of medium T_2 (Table I) and on Petri dishes with the same medium in which the strains were left to grow until giant colonies were obtained. The following features of the strains were considered: type of growth of the vegetative mycelium

TABLE I

Composition of the media

Component	T_2	T_S	T_{25}
Sucrose	100 g	100 g	300 g
Asparagine	10 g	10 g	—
Citric acid	—	—	15 g
Yeast extract	0.1 g	0.1 g	0.1 g
KH_2PO_4	0.25 g	0.5 g	0.5 g
$MgSO_4 \cdot 7\,H_2O$	0.25 g	0.3 g	0.5 g
KCl	0.12 g	—	0.12 g
$Ca(NO_3)_2 \cdot 4\,H_2O$	1 g	—	—
$FeSO_4 \cdot 7\,H_2O$	0.020 g	0.007 g	0.007 g
$ZnSO_4 \cdot 7\,H_2O$	0.015 g	0.006 g	0.006 g
NH_4OH	—	—	to pH 5.2
Tap water	to 1000 ml	to 1000 ml	to 1000 ml
pH	5.2	5.2	5.2
Agar	18 g	—	—

(flat, folded, wrinkled, bulky, clotted), features of the aerial mycelium (velvety, cottony, formation of synnemata) and color. After five transfers on slants of medium T_2, 55 strains out of 61 did not show any more characteristics of the original parent strain. For some of the cultures the changes were less striking and might consist in the appearance of some new color shades, as well as in changes of the aerial mycelium; for others dramatic changes occurred with all the characters examined: some of the cultures lost completely their ability to form aerial mycelium and assumed a much more folded or bulky growth; some others originally showed a velvety ivory aerial mycelium turned to cottony violet. Analogous results were obtained when the cultures were maintained on potato agar or Bennett's agar. The observation of giant colonies demonstrated that about 60% of the colonies produced sectors. These would vary in their color, aspect, presence or absence of aerial mycelium. In other cases, sectors of overgrowth were also observed (Figs 2—5).

High variability was also observed in *C. paspali* by PACIFICI et al.[11]. In fact, submerged cultures in flasks seeded with inocula from the very same slant showed large variation of the quantitative growth and of its type (feathery to compact mycelial spheres), of the pigment (from none to yellow or brown), and finally, of the production of alkaloids (from 0 to 300 µg/ml). This variability was partially removed by using 5% homogenized inoculum. The use of 1% inoculum, even though well homogenized, gave, however, high variability. Interesting results are reported by PACIFICI et al.[19] in a subsequent paper. They made selections by isolating fragments of hyphae, testing their producing capacity, choosing the most active colonies and making new isolations and selections repeated several times. After the first selection, 11 out of the 29 examined strains were found unproductive and 18 productive. A high degree of variation of pigment and of growth was also observed. The second and third selection gave always producing strains showing uniform pigmentation and growth. The average production obtained rose considerably within the three selections; 37 μg/ml in the first selection (the producing strains only have been considered); 209 μg/ml in the second; and 408 μg/ml in the third. Furthermore, with the increase of the production the variability among the strains decreased very much. In fact, the mean deviations of the results of the three selections, expressed as per cent of the mean itself, are 81, 49, and 24, respectively.

A high degree of variability characterized also the strains of *C. purpurea* producers of peptide alkaloids which we isolated. Such variability progressively decreased with the increase of the production obtained through selection. Another feature that characterizes our strains is the more or less rapid loss of activity as a consequence of the successive tranfers on solid media. On medium T_2, for instance, after three to five transfers, the production drops to 50% and after six to eight, it is practically abolished.

HETEROKARYOSIS AND PRODUCTION OF ALKALOIDS

Another fact that we observed in all our producing strains is the formation of sectors in giant colonies [20,21]. These sectors frequently start from a point distant from the center and appear also when the giant colony originates from a single fragment of the hypha. The giant colonies frequently show also marked irregularities at their edges. These characteristics are in agreement with the hypothesis that the strains are heterokaryons[22,23]. The problem has been studied with strain 275 FI[20], FI 32/17 and FI 82[21], and FI S 40/T[24].

Strain 275 FI produces in submerged culture in medium T_2[25] (Table I) about 1000 μg/ml of ergotamine[25] and in giant colonies on medium T_2 it gave rise to three types of sectors characterized by different colors. The cultures obtained from these sectors were unproductive and they did not produce sectors in giant colonies. It has been therefore thought that strain 275 FI was a heterokaryon formed by the cooperation of the three cultures obtained from these sectors. This conclusion received strong support from the observation that in mixed cultures of the three strains obtained from sectors, numerous anastomoses were formed. As is well known, anastomoses represent one of the most efficient means showed by fungi to allow the migration of nuclei from one hypha to another and to reach the heterokaryotic condition. It is also known that the mycelium of ascomycetes forms septa which have only a mechanical support function; they permit, however, not only the passage of cytoplasm, but also of nuclei: the condition of plurinuclearity evidently necessary for heterokaryosis is therefore a rule in these organisms. The three strains isolated from the sectors were studied from the metabolic point of view in submerged culture[25]. We saw that whereas the original strain utilized high amounts of both sucrose and citric acid, the first of the strains obtained from sectors utilized high amounts of sucrose only, the second high amounts of citric acid only and the third utilized both substrates very little. The first two, therefore, behaved in a complementary way. In heterokaryons, the phenomena of complementation are known: *i.e.* two strains with complementary deficiencies annul, at least in part, their deficiencies if their nuclei are associated in a heterokaryon[22]. The usefulness of the association is then evident in the first two strains obtained from 275 FI which, as said before, are complementary. The study was further extended by the synthesis of a heterokaryon from the above two complementary strains. A new strain was obtained in this way with cultural characteristics more or less similar to those of strains 275 FI, able to produce alkaloids (even though in reduced amounts as compared to the production of strain 275 FI) and which gave origin to sectors when grown in giant colonies. It was therefore considered as demonstrated that strain 275 FI was a heterokaryon and that the heterokaryotic condition was indispensable in this strain for the production of alkaloids.

Strain 275 FI, as has been said before, is characterized by the absolute inability to form conidia. Conidia of *C. purpurea* are uninucleate and, therefore, in conidiating strains the heterokaryon breaks up and separates into the various conidial forms. The possibility of obtaining conidia would represent therefore a useful means for the analysis of heterokaryons. Strain FI 32/17 that produces in medium T_{25} in submerged culture 1800—2000 µg/ml of a mixture of ergotamine and ergokryptine, and which, as the former strain, shows sectors in giant colonies, generally does not form conidia. Sometimes, however, some of the cultures for reasons not yet explained, produce conidia. These cultures, in the normal condition of fermentation, give a very low production (200—500 µg/ml). If an isolation is carried out starting from nonsporulated cultures, colonies are obtained which give productions varying from 700 to 2000 µg/ml; if the isolation is carried out starting from conidia, only unproductive colonies appear. In strain FI 32/17 too, therefore, the production of alkaloids is dependent on the heterokaryotic condition.

The disjunction of a heterokaryon can be obtained in two ways: through an unbalanced growth due to a different ratio of multiplication of the various nuclei that constitute it, and that is made evident by sectors and various irregularities of the cultures[22,23] or, as pointed out before, through formation of uninucleate reproductive structures, in particular conidia. The final result is always the same; but, whereas in the case of uninucleate conidia the heterokaryon separates immediately into the forms which constitute it, giving rise to homokaryons, in the case of breaking due to unbalance, the appearance of other kinds of heterokaryons characterized by different quantitative ratios among the various nuclear types is possible. The evident result of these possibilities is that whereas from isolation of conidia we can obtain as many variants as there are nuclear types of the heterokaryon only (usually not many), in the case of imbalance of one strain that does not produce conidia a great deal of forms can be obtained because of the high number of the possible ratios among the nuclei. In this latter case too, however, homokaryons can be obtained at once and this happens if the conditions are such that the homokaryons find high advantage as compared with heterokaryons. In the case of heterokaryons of alkaloids-producing *Claviceps* we found examples of both types of behavior.

Strain FI 82[21] produces 150—200 µg/ml of chanoclavine in stationary culture in medium TS (Table I). It seldom forms conidia in medium T_2. The giant colonies form two types of sectors well characterized by their color and aspect. Cultures obtained from these sectors are practically unproductive. If an isolation from conidia is carried out, two types of colonies corresponding to those of the original sectors appear. In this case then, in the giant colonies, the disjunction was complete through the formation of sectors, at least for the considered characters (aspect and alkaloid production).

In another case, on the contrary, the disjunction in the giant colonies gave results different from those with conidia. This happened with strain FI S 40/T, able to produce, in submerged culture in medium TS, about 1200 µg/ml of ergocristine[24]. This strain has been obtained from strain FI S 40[15] through selection of cultures obtained by isolation of fragments of hyphae. Normally on medium T_2 it does not form conidia; sometimes, however, for reasons unknown to us, under the same conditions, it does form some. Starting from a culture surely not sporulated, we prepared giant colonies on plates of medium T_2. From the most characteristic sectors that appeared, cultures were isolated and tested in fermentation. Furthermore, a fragment of each sector was utilized for preparing a new series of giant colonies. From the sectors of this second series, analogously a third series was prepared and from this a fourth one. The obtained results can be summarized as follows: strain FI S 40/T gave rise to colonies with many sectors. In the course of the following subisolations the tendency to produce sectors decreased without disappearing completely. The productivity of some of the sectors was nil whereas that of many others was rather high. The productivity showed a tendency to decrease in the successive series and in the last one most of the sectors were unproductive. During these subisolations about 30 colonies were obtained which differed in one or more features. The production of these colonies varied from zero to 1100 µg/ml. In the meantime, an isolation was carried out from a colony of FI S 40/T that produced conidia. To make sure to isolate conidia only, a suspension of them was prepared and filtered through sintered glass. The filtrate observed under the microscope contained conidia and was free of fragments of hyphae. The isolated colonies were tested in fermentation for their capacity to produce alkaloids and underwent the test of the giant colony. The production was absent in all the cases. No colony produced sectors. The aspect of the colonies allowed one to group them into four morphological groups. Strain FI S 40/T is therefore an example of different behavior in the disjunction by unbalance and by segregation into conidia. In other words, in this strain the complete disjunction of the heterokaryon by means of unbalanced growth is not reached at once but through a large number of other heterokaryons.

The reported data are without doubt sufficient to draw the conclusion that the production of alkaloids is generally correlated to the heterokaryosis of the producing strains and that heterokaryosis is frequent in the genus *Claviceps*. In fact, also the high morphological variability observed by GREIN[18] in freshly isolated strains can find its explanation in their heterokaryotic condition. The same can be probably said for the high variability observed in *C. paspali* by PACIFICI *et al.*[11,19]. Heterokaryosis is not a very stable condition and depends on environmental conditions for its preservation. The heterokaryons can react by getting completely out of balance and forming the homokaryons or by giving rise to new heterokaryons characterized by different nuclear ratios. Particularly important is the fact that all the producing strains of both *C. purpurea* and *C. paspali*, normally do not form conidia. It is evident, of course, that their existence as heterokaryons is not compatible with reproduction through uninucleate conidia.

HETEROKARYOSIS IN THE LIFE CYCLE OF *Claviceps*

As I said at the beginning, speaking of the cycle of *C. purpurea*, the only phase in which alkaloids are produced is the sclerotial one. In the light of what I said before about the widespread necessity of a heterokaryotic condition for the production of alkaloids, it is logical to suppose that sclerotia are constituted by heterokaryotic mycelium. A convincing proof of this supposition has been given by the fact that, by inoculating plates of medium T_2 with fragments of sclerotia, nearly always giant colonies are obtained that show evident sectors[21]. Once it has been demonstrated that sclerotia are mostly heterokaryotic, another problem arises. In the cycle of *Claviceps* there are two stages represented by uninucleate forms: conidia and ascospores. Besides the conidia, ascospores also give obviously rise, when isolated, to homokaryotic colonies[21]. It is from the ascospores that the infection starts and the mycelium of the sclerotium itself must, sooner or later, produce again ascospores. They represent therefore a compulsory step through which the heterokaryons must necessarily split into the forms that constitute them. The same can be said for conidia. The coexistence in a biological cycle of a heterokaryotic stage and uninucleate stages can only be possible, evidently, if in the meantime one or more phases exist during which the uninucleate forms fuse to form plurinucleate structures. This can happen in practice, if conidia are disseminated *en masse*. As a consequence, during germination, the numerous promycelia can form anastomoses and join the nuclei. It is known that a very young mycelium forms anastomoses very easily. In the genus *Claviceps* the conidia are sunk into a sticky substance, the honeydew, and consequently dissemination cannot occur in other way than *en masse*. The presence of a substance similar to honeydew and the formation of clusters of conidia is known to occur also in agar cultures[18] (Figs 6, 7). It has been observed that the dissemination of ascospores occurs *en masse*, too: ascospores are violently ejected in enormous numbers around the sphaeridium when this is ripe and they germinate quickly, forming a mat of hyphae in which the formation of anastomoses can take place easily[21].

These observations and considerations make acceptable the hypothesis that it is possible to obtain plurinucleate heterokaryotic structures starting from uninucleate ones. In order to be able to prevail, however, the first ones must possess some advantages as compared with the homokaryons. In fact, the mass of conidia or ascospores that reach the organ of the plant to be infected by their germination give rise, as already said, to heterokaryons through fusion of the hyphae, but give also rise to numerous homokaryons when the hyphae do not fuse. A mixture of homokaryotic and heterokaryotic forms is thus obtained and the prevailing of one or the other form will depend on the relative degree of competitivity only. The coexistence in the honeydew of two reproductive forms has been demonstrated by KYBAL[26]: the first form seems to give rise to a thin mycelium of the sphacaelial type; the second one to a thicker mycelium of the sclerotial type. Now, since the heterokaryotic sclerotial type is normally prevalent, and in this way the sclerotia are formed, we must admit that heterokaryotic mycelium possesses some advantage as compared with the homokaryotic one. This advantage could be connected with two facts: first, the heterokaryotic mycelium possesses a higher degree of virulence; second, the environment (honeydew) in which the competition between the two forms occurs, is favourable to the heterokaryon.

That the alkaloid-producing strains are more virulent than the nonproducing ones has been demonstrated by ARCAMONE et al.[9]. It is known that in some cases heterokaryosis increases the virulence[27,28]. As far as Claviceps is concerned, a hypothesis could be suggested from what is known to occur in the infection of rice known as "seedling blight" caused by Fusarium fujikuroi. This fungus produces high amounts of growth factors known as gibberellins. It has been observed that both the severity of infection and the virulence of the various strains of the above species are proportional to the quantity of gibberellins they synthesize[29]. Also in the case of Claviceps one could think that a correlation exists between virulence and production of some kind of growth factors and that the production of the latter is for some metabolic reasons bound to heterokaryosis. In this way, the ability of the heterokaryotic forms to prevail over the homokaryotic ones under parasitic conditions could be explained. The production of growth factors (indolylacetic acid, 2-indolylpropionic acid) by strains of the genus Claviceps is well known[30-32]. It would be possible to suppose that the heterokaryotic strains, besides their production of alkaloids, also produce high amounts of these growth factors. This kind of hypothesis is attractive also because both indolylacetic or 2-indolylpropionic acid and alkaloids contain an indole group and, therefore, at least one part of their biosynthetic pathways is common. Another hypothesis would be that since the derivatives of lysergic acid possess an indole structure, they can display an auxinic activity of their own. With the purpose of proving this last hypothesis some derivatives of lysergic acid have been tested in comparison with indolylacetic acid for their ability to stimulate the elongation of rye coleoptiles[33]. We preferred to employ rye instead of oats, because it is the most common

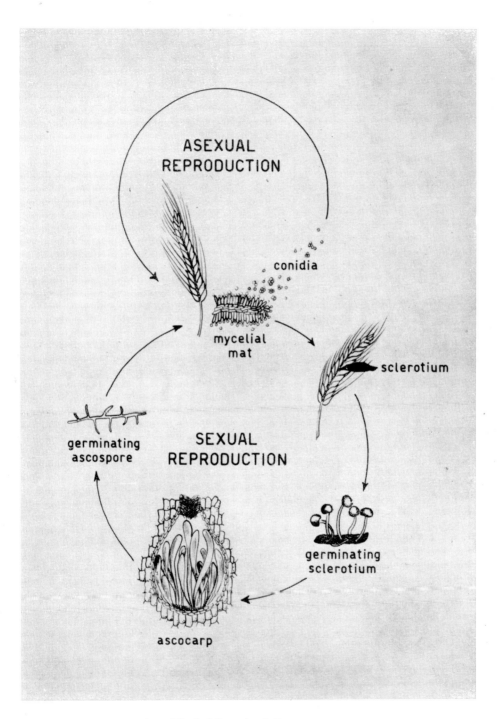

Fig. 1. Life cycle of *C. purpurea*.

Fig. 2. Giant colony of strain FI 640.

Fig. 3. Giant colony of strain FI 642.

Fig. 4. Giant colony of strain FI 537.

Fig. 5. Reverse of a giant colony of strain FI 539.

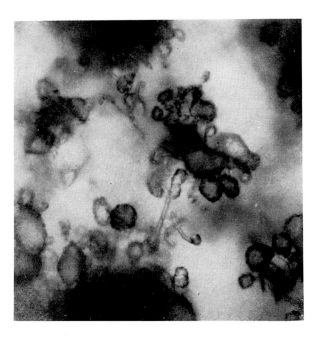

Fig. 6. Globose bodies of honeydew holding together masses of conidia. Strain B12,   × 670.

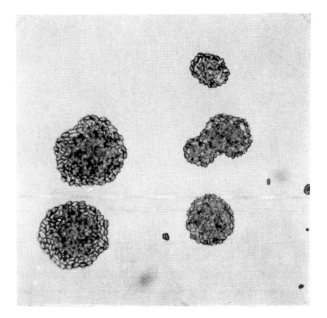

Fig. 7. Globose bodies of honeydew holding together masses of conidia. Strain B12,   × 1200.

TABLE II

Effect of Ergot Alkaloids on Rye Coleoptile

Substance	Concentration μg/ml	Length of coleoptile mm[a]
None	—	10.9
Indolylacetic acid	2	14.1
Lysergic acid	2	12.4
Ergotamine	4	12.2
Ergokryptine	4	12.4
Ergocornine	4	12.3

[a] Mean of 40 coleoptiles.

host for *Claviceps*. The growth of 10 mm elongating segments of rye coleoptile was studied when floated for 20 h in the dark on solutions that contained 20 g/l of sucrose and 2 mg/l of indolylacetic acid buffered to pH 4.8 with McIlvain's standard buffer diluted 30-fold. The test was carried out on 40 coleoptiles for each substance. The obtained results are reported in Table II and demonstrate that lysergic acid, ergotamine, ergokryptine and ergocornine, though less active than indolylacetic acid, show some auxinic activity. It would be therefore reasonable to suppose that the heterokaryons show a higher virulence also because they produce alkaloids and for this reason they prevail over the homokaryons in the plant.

As for the second possibility, *i.e.* that the heterokaryon prevails because it is favored by the environment, it is known that the environmental conditions influence the stability of the heterokaryons not only by conditioning their existence but also by provoking variations of the ratios between the different nuclear components[34]. In particular, the concentration of sugars in the medium is very important from this point of view. If we recall that honeydew represents the medium in which the sclerotium is formed and that it is particularly rich in sugars, it does not seem illogical to suppose that in such particular conditions the heterokaryon finds ideal conditions for its development. This hypothesis seems to be supported by the fact that media for producing alkaloids in submerged culture are extremely rich in sugars. In the majority of cases in fact sugars may reach concentrations of 25—30%[15,35].

CONCLUSIONS

The above-said can be summarized in the following. In nature the production of ergot alkaloids in *Claviceps* occurs in the sclerotium only. This sclerotium is mainly composed of heterokaryotic mycelium. The formation of ascospores provokes the disjunction of the heterokaryon which probably is formed again as the ascospores are disseminated *en masse*, and, when they germinate, can form anastomoses. The formation of conidia can also provoke the disjunction of the heterokaryon. In this case, too, the heterokaryon can be formed again since also the conidia are disseminated *en masse*. The prevalence of the heterokaryotic mycelium over the homokaryotic one in the infected organ may be bound to the higher virulence of the former which could be correlated to the production of plant growth factors such as indolylacetic acid or the alkaloids themselves or to the advantage that is given to the heterokaryons by the particular hypertonic conditions of the environment.

The production of alkaloids in saprophytic culture is generally bound to the heterokaryotic condition of the producing strains. The heterokaryons able to produce alkaloids in culture, even though theoretically obtainable in various ways (synthesis starting from homokaryons, mutations), are obtained practically by isolation from sclerotia. However, heterokaryons are not always obtained from sclerotia; in fact, sclerotia are usually covered by conidia that are residues of the first phase of infection (honeydew). Furthermore, the heterokaryons are unstable in saprophytic culture because they form uninucleate conidia which provoke the segregation of the various nuclear forms, and because in culture they show the tendency to get out of balance, giving rise both to the homokaryotic forms that constitute them and to other heterokaryons with different nuclear ratios. Some strains get out of balance very easily, others are more stable.

In conclusion from what is known up to now, for producing alkaloids in saprophytic culture, a strain must be heterokaryotic with a certain ratio among the component nuclei, must not form conidia and must be stable under the culture conditions.

It is evident that a strain that sums up all these characteristics is not so easily found. If we add that the media in which alkaloid production occurs are not of common use (for instance, they contain sugars in very high concentration and an organic acid) we can understand why the discovery of methods that would allow alkaloid production under submerged culture required so much effort.

REFERENCES

1. ABE M.: J. Agric. Chem. Soc. Japan 22, 2 (1948).
2. ABE M., YAMANO T., KOZU Y., KUSUMOTO M.: J. Agric. Chem. Soc. Japan 25, 458 (1952).
3. STOLL A., BRACK A., KOBEL H., HOFMANN A., BRUNNER R.: Helv. Chim. Acta 37, 1815 (1954).
4. STOLL A., BRACK A., HOFMANN A., KOBEL H.: U. S. Pat. 2,809,920 (1957).
5. TABER W. A., VINING L. C.: Canad. J. Microbiol. 3, 55 (1957).
6. KYBAL J., STARÝ F.: Planta Medica 6, 404 (1958).
7. TYLER V. E., Jr.: J. Am. Pharm. Assoc. Sci. Ed. 47, 787 (1958).
8. RASSBACH H., BÜCHEL K. G., ROCHELMEYER H.: Arzneim. Forsch. 6, 690 (1956).

9. ARCAMONE F., CHAIN E. B., FERRETTI A., MINGHETTI A., PENNELLA P., TONOLO A., VERO L.: Proc. Roy. Soc. B **155**, 26 (1961).
10. GRÖGER D., TYLER V. E., Jr.: Lloydia **26**, 174 (1963).
11. PACIFICI L. R., KELLEHER W. J., SCHWARTING A. E.: Lloydia **25**, 37 (1962).
12. KOBEL H., SCHREIER E., RUTSCHMANN J.: Helv. Chim. Acta **47**, 1052 (1964).
13. AMICI A. M., MINGHETTI A., SCOTTI T., SPALLA C.: French Pat. 1,404,533 (1965); Brit. Pat. 998,254 (1965); Chem. Abstr. **63**, 9026f (1965).
14. TONOLO A.: Nature **209**, 1134 (1966).
15. AMICI A. M., MINGHETTI A., SCOTTI T., SPALLA C., TOGNOLI L.: Appl. Microbiol. **18**, 464 (1969).
16. BONNS W. W.: Amer. J. Bot. **9**, 339 (1922).
17. ABE M., YAMATODANI S.: Progr. Ind. Microbiol. **5**, 203 (1964).
18. GREIN A.: Giorn. Microbiol. **15**, 217 (1967).
19. PACIFICI L. R., KELLEHER W. J., SCHWARTING A. E.: Lloydia **26**, 161 (1963).
20. AMICI A. M., SCOTTI T., SPALLA C., TOGNOLI L.: Appl. Microbiol. **15**, 611 (1967).
21. SPALLA C., AMICI A. M., SCOTTI T., TOGNOLI L., p. 611 in *Fermentation Advances*. Academic Press, New York 1969.
22. DAVIS R. H., p. 567 in G. C. Ainsworth, A. S. Sussman (Eds.): *The Fungi*, Vol. II. Academic Press, New York—London 1966.
23. HEALE J. B.: J. Gen. Microbiol. **45**, 419 (1966).
24. SPALLA C., TOGNOLI L., *unpublished data*.
25. AMICI A. M., MINGHETTI A., SCOTTI T., SPALLA C., TOGNOLI L.: Appl. Microbiol. **15**, 597 (1967).
26. KYBAL J.: Planta Medica **12**, 166 (1964).
27. BUXTON E. W.: J. Gen. Microbiol. **10**, 71 (1954).
28. BUXTON E. W.: J. Gen. Microbiol. **15**, 133 (1956).
29. MING Y. N., LIN P. C., YU T. F.: Sci. Sinica **15**, 371 (1966).
30. YAMANO T.: J. Agric. Chem. Soc. Japan **35**, 1284 (1961).
31. YAMANO T., KISINO K., YAMATODAMI S., ABE M.: Ann. Report Takeda Res. Lab. **21**, 83 (1962).
32. YAMANO T., KUSUMOTO M., YAMATODANI S., ABE M.: Ann. Report Takeda Res. Lab. **20**, 33 (1961).
33. MARNATI M. P., SPALLA C., *unpublished results*.
34. BUXTON E. W., p. 372 in J. G. Horsfall, A. E. Dimond (Eds.): *Plant Pathology*, Vol. II. Academic Press, New York—London 1960.
35. CASTAGNOLI N., TONOLO A.: Symp. 9th Internat. Congr. Microbiol., p. 31. Moscow 1966.

Physiological Aspects of Alkaloid Production by *Claviceps* Species*

L. C. Vining

> *Atlantic Regional Laboratory, National Research Council of Canada, Halifax, Nova Scotia, Canada*

As a group of natural products the ergot alkaloids are biosynthetically related in having ergoline (I), or a potential ergoline ring system, as a common structural element. The primary precursors of this moiety are tryptophan and mevalonic acid,

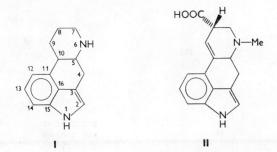

I   II

and the initial reaction characterizing the secondary metabolic pathway is 4-isoprenylation of the indole ring[1]:

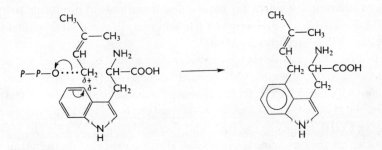

The classic ergot alkaloids are amide derivatives of lysergic acid (II) or its C8 isomer and, until quite recently, they had been found only in *Claviceps* species. It was possible to believe that they were produced by reactions peculiar to these fungi, that they played some unique role in their economy, and that evolution of the bio-

---

* Issued as N.R.C.C. 12599.

synthetic pathway had conferred on the genus some advantage connected with its ecological adaptation. With the discovery of ergoline derivatives in other fungi and in a higher plant Family, the basis for such a hypothesis has been weakened; the arguments must now be reexamined from alternative, if equally insecure, viewpoints, and we should consider the wealth of new recent information on ergot physiology to see if it can shed any new light on the role of these alkaloids in the organisms which produce them.

DISTRIBUTION OF THE 4-ISOPRENYLTRYPTOPHAN PATHWAY

The impression that ergoline biosynthesis was an exclusive property of the genus *Claviceps* now seems a natural consequence of the way in which interest in this subject developed. Ergot alkaloids owe their early detection to the powerful physiologic activity of lysergic acid depsipeptides (III) produced by *Claviceps purpurea*. Since *Clavi-*

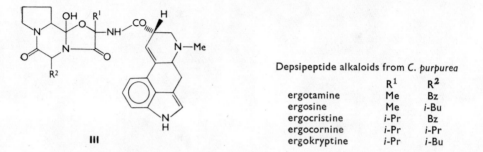

Depsipeptide alkaloids from *C. purpurea*

	R^1	R^2
ergotamine	Me	Bz
ergosine	Me	i-Bu
ergocristine	i-Pr	Bz
ergocornine	i-Pr	i-Pr
ergokryptine	i-Pr	i-Bu

III

ceps are parasitic fungi with a host range including common fodder grasses and cereal crops[2], their propensity to accumulate these mycotoxins has received widespread attention. In addition, the search for more efficient methods of producing therapeutically useful derivatives has maintained a lively interest in the alkaloid content of natural ergot as well as saprophytic cultures.

The first conclusive evidence for the wider distribution of ergolines was provided by SPILSBURY and WILKINSON[3] who isolated festuclavine (IV) and two new bases, fumiclavines A (V) and B (VI), from cultures of *Aspergillus fumigatus*. YAMANO and co-workers[4] later detected these and the additional clavine alkaloids agroclavine (VII), elymoclavine (VIII), chanoclavine (IX), and fumigaclavine C in all eight isolates of the fungus which they examined. Clavine alkaloids have been discovered in several *Penicillium* species[5-7], and *Penicillium cyclopium* isolated from stored grain produces a toxic substance, cyclopiazonic acid (X)[8], which radiotracer experiments have proven to be biosynthesized from L-tryptophan and mevalonate. Accumulation of β-cyclopiazonic acid (XI) in cultures lacking trace metals suggests that 4-dimethylallyltryptophan is an intermediate. A series of compounds recently isolated from a *Penicillium* species appear to be dimeric clavine alkaloids[9].

Bases giving indole colour tests have been detected in other fungi[3] and one survey of 1000 species uncovered 20 cultures producing metabolites of this type[7]. *Basidiomyi cetes* and *Phycomycetes*, as well as *Ascomycetes* and *Fungi Imperfecti* were represented and chromatographic evidence indicated that clavine alkaloids were common products. Rugulovasines A and B (XII)[10], two new isomeric alkaloids isolated from *Pen-*

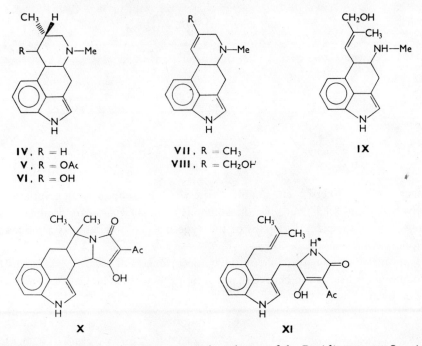

IV, R = H
V, R = OAc
VI, R = OH

VII, R = CH₃
VIII, R = CH₂OH

IX

X

XI

cillium concavo-rugulosum, were also present in cultures of the *Basidiomycetes Lenzites trabea, Corticium caeruleum* and *Pellicularia filamentosa*. It is difficult in a screening programme of this size to select culture conditions which will favour alkaloid formation in all species tested, or use detection methods which will reveal unusual derivatives, and these results must certainly underestimate the proportion of fungi which can produce ergoline derivatives. In view of the report by SALLAM and co-workers[11] that cultures of *Rhizopus nigricans, Aspergillus flavus* and *Geotrichum candidum* produce peptide derivatives of lysergic acid, the wide distribution of alkaloid-producing fungi assumes more importance. It extends opportunities for producing useful drugs by fermentation, but it also emphasizes the possibility for mycotoxicosis from contaminated foods.

A high proportion of 4-isoprenyltryptophan derivatives exhibit pharmacologic activity, and the discovery of ergoline alkaloids in higher plants stemmed directly from the psychomimetic activity of lysergic acid amide (XIII)[12]. Examination of "ololiuqui", the seeds of *Rivea corymbosa* used in magic ceremonies by Mexican Indians, revealed not only lysergic acid amide but several clavine alkaloids which are frequently found in ergot. The range of compounds now isolated from species of the

Family *Convolvulaceae* includes chanoclavine, agroclavine, elymoclavine, ergometrine (XIV) and ergosine[13]. Ergoline alkaloids have not so far been reported in other higher plant Families, but the extent of their distribution has received little serious attention.

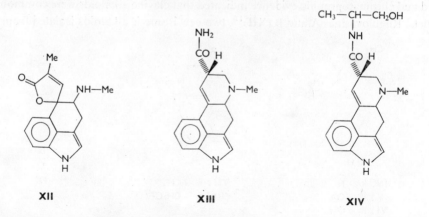

TABER and co-workers[14] were unable to detect indole compounds in a variety of non-convolvulaceous seeds, but since the alkaloids of *R. corymbosa* are produced in vegetative parts of the plant[15] a survey of this type may be unreliable as a guide to their taxonomic distribution.

Information now available on the co-occurrence and biosynthesis of this family of compounds leads us to believe that it is biogenetically homogeneous and based on the pathway shown:

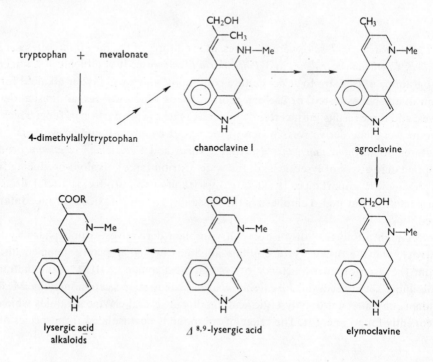

Much of our detailed knowledge of the biosynthetic pathway comes from radiotracer studies with *C. fusiformis*, but where comparable experiments have been carried out with *C. purpurea* or *C. paspali*, the results are in harmony[1,16,17]. Radiotracer experiments with *Convolvulaceae* species have established that ergoline alkaloid biosynthesis in higher plants proceeds from the same precursors as in *Claviceps*[15], and the co-occurrence of clavine and lysergic acid amide indicates that the two pathways are identical. However, this conclusion must be accepted cautiously until more complete information is obtained in view of differences between plants and fungi in the routes used to biosynthesize some aromatic metabolites.

ORIGIN OF THE PATHWAY

With these reservations in mind we can construct a family tree representing biochemical relationships among the 4-isoprenyltryptophan group of natural products (Fig. 1). It is branched at almost every biosynthetic intermediate on the route from tryptophan to lysergic acid depsipeptides. The position of the highest branch point in any organism seems to correspond with the extent to which the main pathway has developed. Terminal products from branching reactions represent alternative metabolic forms of the branching intermediate, and although prior intermediates may accumulate and be metabolized by parallel branching reactions, none of the later intermediates are found.

If this view of the elaboration of ergoline alkaloid biosynthesis is correct, we might relate it to biochemical evolution of the pathway in one of three ways:

(a) The pathway may have evolved slowly, in step with other adaptive variations and equally responsive to selective pressure. The branching reactions would reflect separation and independent development of related evolutionary lines. The weight of evidence is against such a simple relationship.

(b) The pathway may have developed independently in more than one evolutionary line. Whether this development was adaptive or an expression of biochemical predestination without adaptive advantage it implies parallel evolution, in such widely separated species as *Claviceps purpurea* and *Ipomoea argyrophylla*, of a complex series of reactions which generate identical lysergic acid depsipeptides.

(c) The enzyme catalyzing 4-isoprenylation of L-tryptophan may have existed as a primitive protein, and the pathway has evolved under positive selective pressure in a common ancestral line. What we now see randomly distributed in lower and higher plant species are surviving fragments and modifications of a primordial biochemical sequence which became redundant and was eliminated from most organisms. Unless some mechanism exists, or once existed, for transferring genetic information between lower and higher plant forms this appears to offer the simplest explanation.

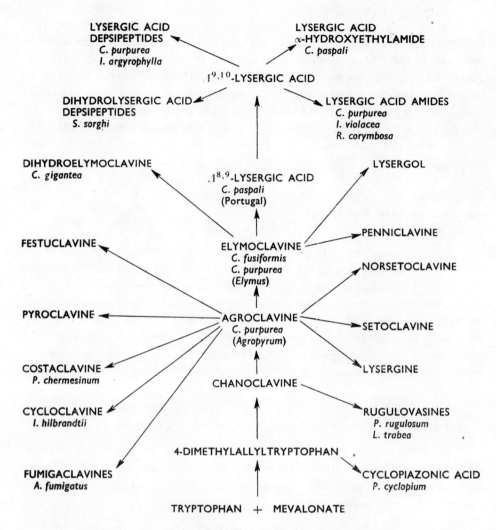

Fig. 1. Biogenetic relationships in the 4-isoprenyltryptophan pathway.

PHYSIOLOGY OF ALKALOID PRODUCTION

Alkaloid production in batch saprophytic cultures of *Claviceps* species can follow either of two patterns commonly seen in secondary-metabolite fermentations (Fig. 2). In the first and more familiar pattern 1 distinct metabolic phases are evident. A trophophase[18] in which rapid growth occurs without alkaloid production is followed by an idiophase[18] during which the growth rate falls and alkaloid accumulates. In cultures producing alkaloid by pattern 2 this metabolic phasing is absent. Alkaloid accumulates at a rate which matches the rate of mycelial growth.

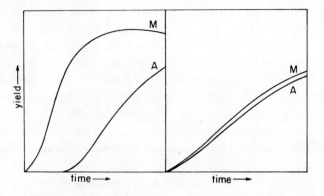

Fig. 2. Patterns of growth and alkaloid production in *Claviceps* cultures. *Left:* pattern 1; *right:* pattern 2; M mycelium, A alkaloid.

PATTERN 1

A large body of experimental data relating medium composition to alkaloid production indicates that pattern 1 is associated with exhaustion of an essential nutrient. For optimum yields of alkaloid conditions must initially favour balanced[19] (trophophase) growth and maximum accumulation of biomass. All essential nutrients must be available and easily assimilable. Terminating balanced growth by arranging for early exhaustion of phosphate in the medium appears to be the most effective method of promoting alkaloid synthesis, although suitable idiophase conditions have been obtained by limiting nitrogen, sulphate or oxygen[20].

One advantage of phosphate as a limiting nutrient may lie in the ability of *Claviceps* to store polyphosphate reserves and make them available, at a growth-limiting rate, after phosphate is exhausted from the medium. Although the literature on *Claviceps* is not conclusive, the behaviour of cultures grown on limiting phosphate probably parallels that of *Gibberella fujikuroi*. On phosphate exhaustion with other nutrients in excess the culture enters a transition phase[19], characterized by slow mycelial proliferation and intracellular accumulation of polyhydric alcohols and trehalose[21]. Under some circumstances sterols and lipid reserves may also accumulate in this period[22], but rapid increase in acetate-derived reserves normally occurs only after nitrogen has been depleted.

For a maximum yield of alkaloids the transition phase should be prolonged. Exhaustion of a second nutrient initiates the maintenance phase[19] in which the rate of alkaloid production declines. If this second nutrient is nitrogen a sharp increase in lipid accumulation competes with alkaloid synthesis, but the latter continues temporarily at the expense of nitrogenous reserves. If carbon is the second nutrient exhausted the organism draws upon carbohydrate reserves for secondary metabolism as well as maintenance, but eventually both processes cease and sporulation or lysis follow in the terminal phase[19]. In certain circumstances sequential accumulation of primary (carbo-hydrate) and secondary (alkaloid) "shunt" products may be observed[23] but this is not mandatory and is no doubt related to medium composition. Onset of alkaloid pro-duction with excess nitrogen in the medium is not associated with any marked change in total protein content of the mycelium, but the relatively high rate of protein syn-thesis in the idiophase implies rapid turnover[24,25]. The total amount of ribonucleic acid also shows only minor change under these conditions[21].

If carbon or nitrogen source is the limiting nutrient alkaloid production is wholly dependent upon endogenous reserves and predictably low. The type of carbon-limited fermentation used for penicillin, in which trophophase growth on a rich carbon source is followed by shift down to slower growth on a limited carbon ration, does not sup-port high alkaloid production. Possibly this is related to the mixed biosynthetic origin of ergoline derivatives since isoprenoid metabolites do not seem to accumulate in car-bon-limited conditions. On the other hand, maximum alkaloid yields cannot be achieved under phosphate limitation with glucose as the carbon source. The choice of carbon source depends on the characteristic of the *Claviceps* strain employed, but man-nitol is generally suitable. TABER[26] has suggested that it may have the advantage of suppressing endogenous polyol accumulation which competes with alkaloid synthesis. One strain of *C. purpurea* performed best on galactose, supplemented with a little gluc-ose to promote faster trophophase growth[27], and sucrose might offer an advantage with invertase producers which can convert the disaccharide to glucose and less assim-ilable oligosaccharides.

In a recent study of penicillin production in glucose-limited continuous cultures of *Penicillium chrysogenum*, PIRT and RIGHELATO[28] showed that penicillin synthesis per unit of biomass was independent of growth rate over a wide range, so that the production rate of a culture was proportional to total biomass. If growth was stopped by supplying only a maintenance ration of limiting nutrient the rate of penicillin syn-thesis decayed. The decay rate was inversely proportional to the previous growth rate and could be reversed by again supplying glucose in excess of the maintenance ration. The general conclusions seem to fit many of the known characteristics of alkal-oid production, provided that phosphate rather than glucose is considered as the growth-limiting nutrient, and application of continuous culture techniques to alkaloid production by *Claviceps* cultures should provide much valuable information.

PATTERN 2

Cultures in which alkaloid production follows pattern 2 display some features of carbon-limited fermentations. Maximum production is achieved with a slowly metabolized carbohydrate carbon source which limits the rate of growth with other nutrients in excess. However, these cultures also require high concentrations of an organic acid[29,30] which provides a supplementary source of carbon that is metabolized only slightly less rapidly than the carbohydrate. The acid, usually one of the tricarboxylic-acid-cycle intermediates, is used as an ammonium salt and the large surplus assits in maintaining pH control.

Recent work by TABER[31] on succinate uptake by *C. purpurea* suggests that the acid enters the mycelium by diffusion at a rate determined by the concentration gradient between undissociated acid in the medium and the cytoplasm. Succinate is metabolized and can support growth as the sole carbon source[32]. The relative rates at which the two available carbon sources are metabolized may be important and since uptake is passive a high external concentration of organic acid would be necessary to maintain the flow into amphibolic channels. The result might be a relatively high concentration of intermediates in and around the tricarboxylic-acid cycle, affecting the pool size of a critical intermediate or cofactor which regulates alkaloid biosynthesis. Arsenate and another uncoupling agent, 2,4-dinitrophenol, which stimulate alkaloid production in *C. paspali*[33], may have a similar effect. The influence of organic acids on alkaloid production is not unique, and increased yields have frequently been reported in fermentations where they have been included as a supplementary carbon source. A noteworthy example is the enhanced production of chloramphenicol by pattern 2 when high concentrations of lactate are included in the medium[34]. Expansion of pools intimately related to tricarboxylic cycle operation is also believed to trigger the formation of oligoketide metabolites in *Penicillium urticae*[18] and *P. islandicum*[35].

DISTRIBUTION OF PATTERNS

Culture conditions which allow alkaloid production by pattern 1 are suitable for *C. purpurea, C. fusiformis*, and *Sphacelia sorghi*[36], whereas *C. paspali* usually produces best under conditions giving pattern 2. Very little has been done with other fungi. There is some evidence that a medium supporting pattern 2 production is more generally effective in eliciting alkaloid synthesis[37], but with at least one strain of *C. paspali* the composition of the medium can be manipulated to give high production by either pattern[30]. In some *C. purpurea* fermentations with a moderate concentration of citric acid, alkaloid synthesis follows an intermediate course[22], and we may safely conclude that the production pattern *per se* is not an inherent property of the organism. It merely reflects the physiological conditions used to promote secondary metabolism in the culture. At times the basic pattern may be obscured by subsidiary factors such as the presence of surface-active agents. Certain detergents[38] and glycols[39] which stimulate alkaloid biosynthesis increase nutrient uptake, presumably by altering cell permeability, and it has been suggested that surface-active metabolites produced by the culture might have a similar role[38].

METABOLIC CONTROL OF ALKALOID BIOSYNTHESIS

All of the evidence from physiological studies is compatible with close metabolic control of alkaloid biosynthesis. Maximum, and usually complete, suppression is maintained during "balanced" cell proliferation on readily assimilated nutrients. Under conditions imposing slow or unbalanced growth this control is released. In cultures showing a trophophase—idiophase pattern, release of control in the idiophase starts protein turnover and several new enzyme activities appear along with those catalyzing alkaloid synthesis. However trophophase suppression is probably not maintained by a single general mechanism since alkaloids and carbohydrate reserves accumulate during phosphate-limited growth, whereas lipids, steroids, and oligoketide metabolites normally do not accumulate until nitrogen is exhausted. These control systems may be modified in heterokaryotic mycelium when nuclear interaction promotes a general release of metabolite repression in the transition phase. In the growth-linked pattern of alkaloid production synthesis is continuously switched on and it seems likely that the size of critical endogenous pools imposed by the nutrient balance in the medium allows the organism to grow without suppressing the alkaloid synthetase system.

The most important factors determining the yield of alkaloids, and probably of most other secondary metabolites as well, are not those which ensure an abundant supply of primary precursors, but those which promote a high level of enzymes catalyzing the synthesis. When high enzyme levels develop, the supply of precursors is not a limiting factor under most circumstances, and the flow from normal metabolic networks can meet the maximum demands of the system. However, the initial accumulation of a primary precursor may well influence the yield of product by inducing a high level of the required enzymes; there is now considerable evidence that tryptophan acts in this way to stimulate alkaloid production in *Claviceps* cultures[40-42]. Not all cultures respond to supplements of the amino acid, but in some strains of *C. fusiformis* producing alkaloids by pattern 1 the yield increases several-fold. The response is more likely to be related to their ability to accumulate large amounts of tryptophan from the medium[42] than to the absence of katabolic reactions which would prevent a large increase in pool size, since enzymes catalyzing the latter reactions are probably repressed during the trophophase[25].

For the maximum increase in alkaloid production tryptophan must be added to cultures of *C. fusiformis* at the beginning of trophophase growth. The effect is concentration dependent up to a saturation value which corresponds to the amount the culture is able to take up during the trophophase; the presence of tryptophan in the medium during the idiophase has negligible influence on yield, although the amino acid is readily incorporated into the alkaloid structure[42]. It is particularly noteworthy that the time at which alkaloid synthesis begins is not appreciably affected by the tryptophan supplement (Fig. 3). The inducing effect is not apparent until metabolic suppression is released at the end of trophophase growth and active enzyme appears in the mycelium. To explain the time lag between the two events we can construct numerous intriguing hypotheses, involving regulation at the level of transcription, translation, or enzyme activity, *e.g.*:

(a) Tryptophan stabilizes and thus allows the accumulation of an activator required for transcription of alkaloid synthetase genes after trophophase suppression is released. This model requires positive control of transcription as proposed for the L-arabinose system of *E. coli*[44].

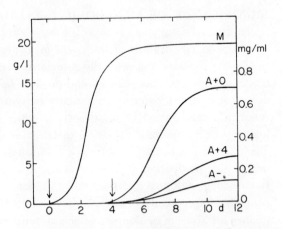

Fig. 3. Effect of tryptophan on alkaloid biosynthesis in *Claviceps fusiformis*. M mycelium (dry weight, g/l), A alkaloid (mg/ml); tryptophan added (+) at 0 (0) or 4 (4) days; — no tryptophan.

(b) Messenger ribonucleic acid is synthesized in a stable and inactive form during the trophophase and unmasked during the idiophase when translation occurs. Transcription under inducing conditions might generate a messenger ribonucleoprotein particle or informosome[45] which would pass from the nucleus and remain in the cytoplasm until the messenger ribonucleic acid was released. This might occur through the action of proteases formed at the end of trophophase growth or through a more specific form of translation control. Alternatively, translation of a "ticketed" messenger might have to await the formation of a particular class of ribosomes[46].

(c) The synthetase complex, or perhaps only the initial enzyme, is produced in an inactive form during the trophophase and is converted to an active form during the idiophase. Activation might require a second enzyme, as with glycogen and glutamine synthetases[47] which catalyze accumulation of storage compounds. If only the initial enzyme of the biosynthetic pathway is induced by tryptophan the remainder may be sequentially induced as their substrates appear.

All of these alternatives meet the observed requirement for protein synthesis to accompany the appearance of active alkaloid synthetase and there does not yet seem to be any evidence which would select between them. It is of interest that DEMAIN[48] observed marked stimulation of cephalosporin biosynthesis in a *Cephalosporium* species when methionine was added at the beginning of trophophase growth, but not if addition was delayed until the idiophase when cephalosporin was formed. The increased yield was attributed to a sparing effect on cysteine and might well be due to delayed enzyme induction by an enlarged cysteine pool. The existence of this type of metabolic control in secondary-metabolite fermentations would provide an attractive rationale for mutation programmes aimed at increasing yields. Eliminating feedback controls and other mechanisms limiting endogenous accumulation of the primary biosynthetic precursors should promote synthesis of the required enzyme systems.

SIGNIFICANCE OF METABOLIC CONTROLS

In primary metabolic pathways of microorganisms regulation by induction and metabolite repression is generally associated with katabolic reactions, while biosynthetic systems are controlled by end-product feedback. However, there is little evidence for feedback control of secondary biosynthetic pathways, and its apparent absence in alkaloid production suggests that the organism gains no advantage from limiting the accumulation of these metabolites. The control system which regulates alkaloid formation in the trophophase—idiophase pattern would be expected to prevent unnecessary dissipation of resources when conditions are favourable for rapid growth. In particular, a double requirement for induction and activation separated by a time delay should protect the organism from competitive growth disadvantage due to secondary metabolism in a nutrient-rich environment.

Groups of idiophase reactions may be initiated by a general control mechanism, possibly triggered by changes in pool size of an intermediate such as adenosine triphosphate or glutamate, but the induction of alkaloid synthetase by tryptophan appears to be aimed precisely at one secondary metabolic pathway. In *C. fusiformis*, the type as well as the timing of secondary metabolism are regulated, and the type is influenced by conditions encountered during the preceding growth phase. A regulatory system of this kind is well adapted for controlling biochemical processes which occur when trophophase growth is no longer possible, and is similar in many respects to those controlling reserve synthesis, bacterial sporulation[49], and cytodifferentiation[50]. Cultures in which alkaloid synthesis is linked to growth present an anomaly, but this pattern may, in fact, constitute an "unnatural" manipulation of the metabolic control system. It would be useful to know whether, under these conditions, activation of alkaloid biosynthesis is accompanied by other processes usually associated with idiophase metabolism, but the information does not yet seem to be available.

ALKALOIDS AND PARASITIC GROWTH

The physiological responses of *Claviceps* throughout its life cycle are so efficiently adapted to the demands of the environment that it would be remarkable if alkaloid production were an exception. In the parasitic sphacaelial stage nutrient supplied by the host supports mycelial proliferation and conidiation but little or no alkaloid is produced. KYBAL[51] has shown that sclerotial tissue originates from germinating coni- dia and his results indicate that phosphate exhaustion in the honeydew initiates de- velopment of the plectenchymatic growth form characteristic of the sclerotium[52]. At the same time, carbon and nitrogen nutrients remain plentiful, providing suitable con- ditions for idiophase growth and production of the large amounts of lipid and carbo- hydrate reserves which nourish the sclerotium during dormancy and outgrowth. AMICI and co-workers[53] have recently shown that heterokaryosis enhances the ability of *C. purpurea* to assimilate nutrients and to accumulate reserves. Since at least 78% of *C. purpurea* sclerotia contain heterokaryotic hyphae[54], active idiophase metabolism is probably of value under natural conditions.

Besides alkaloids a variety of unusual substances such as ricinoleic acid and pig- ments accumulate in sclerotial tissue and it is by no means apparent that all of these idiophase metabolites perform a useful function in the life cycle of the fungus. We might suspect that some of them are adventitious products of a set of conditions which, for other reasons, increase the ecological fitness of the organism. However, the potential for synthesizing a useless product, even though not expressed during tropho- phase growth, should confer a small selective disadvantage[55]. For this reason, and because specific regulatory systems exist to control idiophase metabolism, we believe alkaloid biosynthesis has persisted in *Claviceps* because it is of value to the fungus.

CONCLUSIONS

The 4-isoprenyltryptophan pathway of secondary metabolism occurs in more species of plants and fungi than was hitherto realized. In order to account for the apparently identical reaction sequence in such widely separated taxa as *Claviceps pur- purea* and *Ipomoea argyrophylla* we suggest that the pathway may have existed and served a useful function in a primitive common ancestor. Although superseded and eliminated from most progeny lines, in certain species it has been retained or modified because it serves a useful function.

As a component of idiophase metabolism the pathway is suppressed when con- ditions favour rapid mycelial proliferation. Thus it confers little competitive growth disadvantage, and enjoys the biochemical economy of being activated by the same general mechanism which relieves metabolic suppression of endogenous reserve syn- thesis. However, there is some evidence of a superimposed specific control in *C. fusi- formis* where the amount of alkaloid synthetase formed in the idiophase is predetermi- ned by the inducing action of a precursor accumulated during trophophase growth. If the existence of this regulatory system is confirmed, it lends some support to the view that the pathway yielding ergot alkaloids has persisted because it is needed. A hetero- karyotic phase in the life cycle could account for the survival of nonproducing strains.

REFERENCES

1. KELLEHER W. J.: Adv. Appl. Microbiol. 11, 211 (1970).
2. BRADY L. R.: Lloydia 25, 1 (1952).
3. SPILSBURY J. F., WILKINSON S.: J. Chem. Soc. 2085 (1961).
4. YAMANO T., KISHINO K., YAMATODANI S., ABE M.: Ann. Report Takeda Res. Lab. 21, 95 (1962).
5. TABER W. A., VINING L. C.: Canad. J. Microbiol. 4, 611 (1958).
6. AGURELL S. L.: Experientia 20, 25 (1964).
7. ABE M., YAMATODANI T., KOZU Y., YAMADA S.: Agric. Chem. Soc. Japan 41, 68 (1967).
8. HOLZAPFEL C. W., WILKINS D. C.: 5th Internat. Symp. Chemistry of Natural Products, London 1968. Abstracts, p. 169.
9. TABER W. A., VINING L. C., *unpublished results*.
10. YAMATODANI S., ASAHI Y., MATSUKURA A., OHMOMO S., ABE M.: Agric. Biol. Chem. 34, 485 (1970).
11. SALLAM L., EL-RAFAI A.-M., NAIM N.: Jap. J. Microbiol. 13, 218 (1969).
12. HOFMANN A., TSCHERTER H.: Experientia 16, 414 (1960).
13. STAUFFACHER D., TSCHERTER H., HOFMANN A.: Helv. Chim. Acta 48, 1379 (1965).
14. TABER W. A., VINING L. C., HEACOCK R. A.: Phytochemistry 2, 65 (1963).
15. GRÖGER D., MOTHES K., FLOSS H.-G., WEYGAND F.: Z. Naturforsch. 18b, 1123 (1963).
16. RAMSTAD E.: Lloydia 31, 327 (1968).
17. FLOSS H.-G.: Ber. Deut. Botan. Ges. 80, 705 (1968).
18. BU'LOCK J. D., POWELL A. J.: Experientia 21, 55 (1965).
19. BORROW A., JEFFERYS E. G., KESSELL J., LLOYD E. C., LLOYD P. B., NIXON I. S.: Canad. J. Microbiol. 7, 227 (1961).
20. WINDISCH S., BRONN W.: U. S. Pat. 2,936,226 (1960).
21. TABER W. A., VINING L. C.: Canad. J. Microbiol. 9, 1 (1963).
22. AMICI A. M., MINGHETTI A., SCOTTI T., SPALLA C., TOGNOLI L.: Appl. Microbiol. 18, 464 (1969).
23. TABER W. A.: Appl. Microbiol. 12, 321 (1964).
24. ROTHE U., FRITSCHE W.: Arch. Mikrobiol. 58, 77 (1967).
25. KAPLAN H., HORNEMANN U., KELLEY K. M., FLOSS H. G.: Lloydia 32, 489 (1969).
26. TABER W. A.: Lloydia 30, 39 (1967).
27. TABER W. A., VINING L. C.: Canad. J. Microbiol. 4, 612 (1958).
28. PIRT S. J., RIGHELATO R. C.: Appl. Microbiol. 15, 1284 (1967).
29. ARCAMONE F., CHAIN E. B., FERRETTI A., MINGHETTI A., PENNELLA P., TONOLO A., VERO L.: Proc. Roy. Soc. 155b, 26 (1961).
30. BRAR S. S., GIAM C. S., TABER W. A.: Mycologia 60, 806 (1968).
31. TABER W. A.: Bacteriol. Proc. 14 (1970).
32. TABER W. A., SIEPMANN R.: Appl. Microbiol. 14, 320 (1966).
33. KIM B. K., KELLEHER W. J., SCHWARTING A. E.: Lloydia 31, 422 (1968).
34. GOTTLIEB D., DIAMOND L.: Bull. Torrey Bot. Club 78, 56 (1951).
35. GATENBECK S., SJOLAND S.: Biochim. Biophys. Acta 93, 246 (1964).
36. MANTLE P. G.: Ann. Appl. Biol. 62, 443 (1968).
37. VINING L. C., NAIR P. M.: Canad. J. Microbiol. 12, 915 (1966).
38. KELLEHER W. J., KIM B. K., SCHWARTING A. E.: Lloydia 32, 327 (1969).
39. MIZRAHI A., MILLER G.: J. Bacteriol. 97, 1155 (1969).
40. FLOSS H.-G., MOTHES U.: Arch. Mikrobiol. 48, 213 (1964).
41. BU'LOCK J. D., BARR J. G.: Lloydia 31, 342 (1968).
42. VINING L. C.: Canad. J. Microbiol. 16, 473 (1970).

43. TEUSCHER E.: Flora **155**, 80 (1964).
44. ENGLESBERG E., IRR J., POWER J., LEE N.: J. Bacteriol. **90**, 946 (1965).
45. BELITSINA N. V., AJTKHOZHIN M. A., GAVRILOVA L. P., SPIRIN A. S.: Biokhimiya **29**, 363 (1964).
46. SUSSMAN M.: Nature **225**, 1243 (1970).
47. HOLZER H.: Adv. Enzymol. **32**, 297 (1969).
48. DEMAIN A. L.: Trans. N. Y. Acad. Sci. **25**, 731 (1963).
49. MANDELSTAM J., p. 377 in P. Meadow, S. J. Pirt (Eds.): 19th Symp. Soc. Gen. Microbiol. Cambridge University Press 1969.
50. SUSSMAN M., SUSSMAN R., p. 403 in P. Meadow, S. J. Pirt (Eds.): 19th Symp. Soc. Gen. Microbiol. Cambridge University Press 1969.
51. KYBAL J.: Planta Medica **12**, 166 (1964).
52. KYBAL J., MAJER J., KOMERSOVÁ I., WANI W. D.: Phytopathology **58**, 647 (1968).
53. AMICI A. M., MINGHETTI A., SCOTTI T., SPALLA C., TOGNOLI L.: Appl. Microbiol. **15**, 597 (1967).
54. SPALLA C., AMICI A. M., SCOTTI T., TOGNOLI L., p. 611 in D. Perlman (Ed.): *Fermentation Advances.* Academic Press, New York 1969.
55. ZAMENHOF S., EICHHORN H. H.: Nature **216**, 456 (1967).

Qualitative Changes in the Alkaloid Spectrum
of *Claviceps purpurea* after Mutation

H. Kobel and J.J. Sanglier

SANDOZ Ltd. Basle, Switzerland

Dedicated to Professor Dr. K. Mothes on the occasion of his 70th birthday.

INTRODUCTION

The ergot fungus *Claviceps purpurea* not only divides into numerous biogroups with various spectra of hosts but also into many chemical races with different alkaloid spectra. From the results of several investigators[1-7] it can be concluded directly or indirectly that the composition of the alkaloid complex is due to genetically fixed characteristics of the races. Within certain limits, the composition can vary by changing the culture conditions. *Claviceps paspali* displays relationships similar to those seen in *C. purpurea*[8].

For the alkaloid production it would be desirable to use strains with a minimum of secondary alkaloids. However, it seems to be very difficult to influence the alkaloid spectrum by mutagenic treatment. Accordingly, the literature concerning this type of research is rather sparse, but Gröger[1,9] and Strnadová[10-12] have reported interesting results.

Strnadová succeeded in obtaining numerous amino-acid auxotrophic mutants by UV-irradiation. Interestingly, together with a dependence on arginine she found a shifting from ergotamine to ergotoxine type alkaloids. This shift remained even after auxotrophic reversion.

In contrast, the employment of mutagenic agents for increasing alkaloid production is less difficult. In a previous publication[4] we reported on a mutant which yielded 45% more alkaloids than the initial wild strain.

With the goal of increasing alkaloid production in submerged cultures we set up a mutation program. The strain of *C. purpurea* we used produced mostly ergocornine and ergokryptine. In contrast to other groups[13-15] we employed a strain which produced a rich growth of spores on agar instead of nonsporulating strains. More than 99% of the conidia were mononuclear so that it became simple to isolate homokaryotic lines. Moreover, the mutagenic treatment of conidia is much less difficult than that of mycelial fragments.

METHOD

The initial strain used in our investigations was collected from *Festuca arundinacea* grown in Switzerland. It infested rye quite well and submerged cultures yielded peptide alkaloids. As culture media we employed a sucrose-casamino-acids Knop agar and a previously described culture medium[4] in a slightly modified form.

The agents used for mutation purposes were UV-light, N-methyl-N'-nitro-N-nitrosoguani-dine (NTG), ethyl methanesulfonate (EMS), methyl-*bis*-(chloroethyl)amine (MCA) and sodium nitrite. The conidia were exposed for a length of time sufficient to kill 90 to 99%.

The treated conidia were plate-cultured and the resulting colonies were transferred to agar slants for further growth. Then conidial suspensions made from these slants were used to inoculate shake cultures in 500 ml Erlenmeyer flasks which contained 50 ml of medium. These were placed on a shaking table and allowed to incubate at 24 °C. The cultures were harvested after 12 days. The alkaloid content, as well as the composition of the alkaloid complex were then determined. The best mutant of each series was then employed as the starting point for further mutagenic treatment.

RESULTS

In four successive mutation series the yield of peptide alkaloids could be increased stepwise six-fold. About 90% of all peptide alkaloids were found in the mycelium while about 10% passed over to the culture filtrate. Qualitative analyses were carried out on the mycelial alkaloids. Table I lists the per cent content of the alkaloids found in the initial wild strain as well as those of a quantitative (F III/237) and a qualitative (F III/405) mutant.

Mutant F III/405 had about the same content of total alkaloids as did the initial strain. Surprisingly, however, it did not contain ergokryptine, ergocornine and ergosine as did the initial strain. Instead, it contained only the peptide alkaloids ergocris-

TABLE I

Per Cent Content of Individual Alkaloids Comprising the Total Complex Found in the Mycelia of the Initial Strain and Two Mutants

Alkaloid or alkaloid pair	Initial strain W	Mutant	
		F III/237	F III/405
Ergokryptine/-kryptinine	53	53	—
Ergocornine/-corninine	33	33	—
Ergosine	6	8	—
Ergocristine/-cristinine	—	—	89
Ergotamine	—	—	1
Ergometrine	6	2	4
Clavines and others	2	4	6

tine and ergotamine. This strain had already shown a divergent morphology of the colonies on the plated agar as may be seen in Fig. 1.

The types of colonies found on plated agar may be characterized as follows:

Exy type. The initial strain and most quantitative mutants have pink colonies with deep radial and concentric folds. The marginal zone is small, white, flat and not pigmented.

Ec type. The ergocristine mutants display a central violet protuberance but other‑ wise an unpigmented, weakly folded, white mycelium with concentric zones of a somewhat dark pigmentation.

Further types. At least four further morphological types were seen occasionally — all with the ergokryptine/-cornine strain. They need not be described here since they are of little importance to the present problem. Furthermore, we recently repor‑ ted on deviating morphological types found while selecting strains of *C. paspali* with varying quantitative yields of alkaloids[16].

Frequency of formation of qualitative mutants. By lending our attention to these special morphological mutants we could isolate a total of twenty ergocristine strains. On plated agar, these always formed colonies of the Ec type. On the other hand, type Ec always yielded ergocristine strains and never those with alkaloid complexes iden‑ tical to those found in the initial strain. On various other agar media we could always differentiate the two types although this was not always plainly evident.

The amount of material at our disposal for the computation of mutation rates is rather small. In a series without mutagenic treatment we found one Ec mutant in 1000 colonies. But in a series with EMS treatment we found ten mutants in 1000 colonies — not an insignificant figure. The mutation rate was between 0.1 and 1% after treatment with UV, NTG and MCA. Up to now no qualitative mutants have been found in the series treated with nitrite.

Stability of the qualitative mutants. After several passages through agar and after various single-spore isolations the mutants remained constant as far as the composi‑ tion of their alkaloid spectra was concerned. We then inoculated rye in the field with an ergocornine/-kryptine strain and an ergocristine mutant. Both strains proved to be virulent. Their respective sclerotia and submerged cultures displayed indentical quali‑ tative alkaloid spectra.

Morphological differentiation of the sclerotia. There was a difference in the mor‑ phology of the sclerotia yielded by the two strains, as can be seen in Fig. 2. The scle‑ rotia of the ergokryptine/-cornine strain were more or less club-shaped with one end thicker than the other. The surface was frequently bark-like in appearance. The ergo‑ cristine mutant had cylindrical sclerotia equally thick at both ends. The surface was smooth, usually with small longitudinal furrows.

Auxotrophic investigations. In media devoid of amino acids both the initial wild strain and the Ec mutants displayed good rates of growth.

Back mutants. Mutagenic treatment of conidia from a strain of the Ec type producing ergocristine led in several cases to strains of the original ergokryptine/-cornine-producing Exy type.

DISCUSSION

The alkaloids influenced by mutation differ from one another in the composition of their peptide portions, as may be seen in Table II. Proline is common to all alkaloids. Both strains build onto their respective lysergic acid positions mostly valine and, to a much lesser extent, alanine. At the other end of the tripeptide, the initial strain incorporates leucine or valine whereas the mutant builds up phenylalanine.

TABLE II

Composition of the Peptide Portion of Alkaloids Influenced by Mutation

Alkaloid	Peptide portion
Ergokryptine	valine — proline — leucine
Ergocornine	valine — proline — valine
Ergosine	alanine — proline — leucine
Ergocristine	valine — proline — phenylalanine
Ergotamine	alanine — proline — phenylalanine

Our mutant seems to result from an interference with the mechanism catalyzing the coupling of two amino acids. Since, at the same time, morphological characteristics also are affected it seems that this enzyme also interferes with the synthesis of compounds other than alkaloids.

Our results are a good example of how chemical races develop. Since normal sexuality in *Claviceps* is disturbed and since it reproduces only vegetatively or by pseudogamic processes their mutant strains remain stable in free nature once they have been formed.

CONCLUSIONS

A homokaryotic strain of *C. purpurea* which in vitro yielded mostly ergokryptine and ergocornine with minor amounts of ergosine was subjected to mutagenic treat-

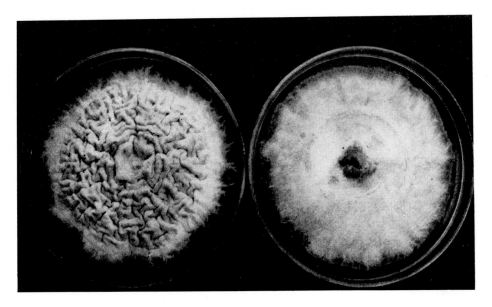

Fig. 1. Agar colonies of the ergokryptine/-cornine strain (Exy type; *left*) and an ergocristine-producing mutant (Ec type; *right*).

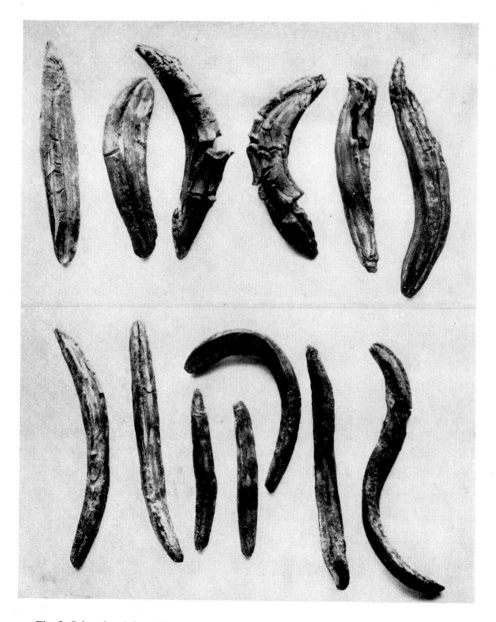

Fig. 2. Sclerotia of the wild type (Exy; *upper*) and the mutant (Ec; *lower*) grown on rye.

ment. Whereas the quantity of alkaloids produced was easily influenced by mutation the qualitative composition of the alkaloid complex proved to be quite stable.

In some cases, however, we could find mutants which did not yield the usual alkaloids but instead yielded ergocristine as the main alkaloid and ergotamine as a secondary alkaloid. At the same time, the agar colonies showed distinct morphological differences which facilitated selection.

The mutants proved to be stable even though it was possible to obtain back mutants leading to the initial wild strain. The chemical and morphological changes were also manifested in the sclerotia.

REFERENCES

1. GRÖGER D.: Kulturpflanze 4, Beiheft 1, 226 (1956).
2. HRONEŠ J., KYBAL J.: Z. Pflzücht. 48, 73 (1962).
3. KLEINEROVÁ E., KYBAL J.: Folia Microbiol. 14, 602 (1969).
4. KOBEL H., BRUNNER R., BRACK A.: Experientia 18, 140 (1962).
5. KYBAL J., BREJCHA V.: Pharmazie 10, 752 (1955).
6. VOIGT R., KEIPERT S.: Pharmazie 20, 785 (1965).
7. VOIGT R., KEIPERT S.: Pharmazie 22, 329 (1967).
8. KOBEL H., SCHREIER E., RUTSCHMANN J.: Helv. Chim. Acta 47, 1052 (1964).
9. GRÖGER D.: Abh. deut. Akad. Wiss. Berlin 7, 243 (1956).
10. STRNADOVÁ K.: Z. Pflzücht. 51, 167 (1964).
11. STRNADOVÁ K.: Planta Medica 12, 521 (1964).
12. STRNADOVÁ K.: Flora, Abt. A 157, 517 (1967).
13. AMICI A. M., MINGHETTI A., SCOTTI T., SPALLA C., TOGNOLI L.: Appl. Microbiol. 18, 464 (1969).
14. MANTLE P. G.: Trans. Brit. Mycol. Soc. 52, 381 (1969).
15. TONOLO A.: Nature 209, 1134 (1966).
16. KOBEL H.: Pathol. Microbiol. 34, 249 (1969).

Ergot Alkaloid Formation in Relation to the Cell Pools of Tryptophan and Adenosine 5'-Triphosphate

Z. Řeháček, J. Kozová, P. Sajdl, J. Kašlík

Department of Biogenesis of Natural Substances, Institute of Microbiology, Czechoslovak Academy of Sciences, Prague

In recent years considerable interest has centered around the ergot-alkaloid production by saprophytic strains. However, among a number of reports that have accumulated, the physiology of alkaloid formation has not been given sufficient attention. In view of this fact little is known about the somewhat complex role of cell-pool tryptophan and to the best of our knowledge there is no work on the relationship between the amount of biologically useful energy[1] and the biosynthesis of ergot alkaloids.

Here we record some observations on the intracellular tryptophan and adenosine 5'-triphosphate (ATP) in submerged cultures of two strains, *i.e. Claviceps paspali* (Stevens and Hall) labelled as MG-6, and *Claviceps purpurea* (Fr.) Tull designated Pla-4. Fermentation experiments were carried out on a simple defined medium with sorbitol and ammonium succinate but without tryptophan[2,3]. The flasks (300 ml) were incubated at 24 ± 1 °C on a rotary shaker operating at 240 r.p.m. with a 2.3 inch stroke. Each shaken flask with medium (100 ml) was inoculated with $7-10$ mg dry weight equivalent of washed vegetative cell inoculum blended for 15 seconds. The inoculum was obtained from 5-day shaken cultures grown in a broth containing 45.0 g malt extract Difco per litre of distilled water.

Submerged cultures of *C. paspali* MG-6 were found to produce alkaloids during the slow increase in mycelial dry weight. Total biomass synthesis and alkaloid formation paralleled one another. Maximum alkaloid production (200 mg/l) was reached within seven days. Lysergate α-hydroxyethylamide was the predominant component of the alkaloid mixture extracted from the culture broth. The alkaloid content of washed mycelium was negligible. As shown in Fig. 1 the rate of protein synthesis changed during the fermentation and a difference was ascertained between strain MG-6 and Pla-4 in that respect. However, in both strains a decrease in protein synthesis corresponded to the relative increase in the alkaloid formation intensity. The reduction of protein synthesis should make available the simple nitrogenous precursors for synthesis of lysergic acid derivatives[5].

The investigation undertaken in view of the fact that tryptophan serves as one of the precursors of ergot alkaloids[6] manifested the ability of both strains to grow in the absence of added tryptophan and to accumulate extra endogenous tryptophan.

This is illustrated in Fig. 2. In strain MG-6 with intensive alkaloid synthesis the accumulation of an internal metabolic pool of tryptophan, as well as the level of mycelial protein tryptophan, were much lower as compared with the data belonging to the low-producing strain Pla-4. Fig. 3 shows that alkaloid production took place while the free-tryptophan level was decreasing and it continued at a greater rate when free-tryptophan level was increasing. Alkaloid production commenced while tryptophan synthase activity was growing. In the course of alkaloid formation, cell-pool tryptophan showed a single minimum. The change in protein-tryptophan level was insignificant. The drop in the free tryptophan was accompanied by intensive alkaloid synthesis and was inversely proportional to the increase in tryptophan synthase activity. In the subsequent phase, tryptophan synthase activity markedly dropped. At this stage, tryptophan oxygenase (formerly called tryptophan pyrrolase) manifested its activity, alkaloid formation was impaired and excreted imidazolylglycerol was found in the fer-

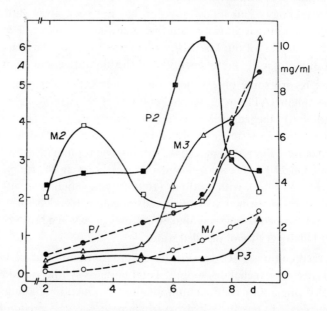

Fig. 1. Protein synthesis during alkaloid fermentation by *C. paspali* MG-6 (M) and *C. purpurea* Pla-4 (P). *1* dry weight, mg/ml, *2* incorporation of U-^{14}C-L-leucine into the trichloroacetic acid -insoluble fraction of the mycelium, c.p.m. $\times 10^{-3}$ (A), *3* total alkaloids determined colorimetrically[4] and expressed as ergometrine equivalents, mg/l $\times$ 0.1 (A). — For measurement of ^{14}C-leucine incorporation 50 mg of chilled mycelium was blended with 10 ml of supernatant at low speed for 30 s, then 5 ml of the blended mycelium was dispensed in 25 ml conical flasks containing 5.0 µCi of U-^{14}C-L-leucine and 2.5 mM of DL-leucine and incubated at 24 °C for 30 min. Leucine incorporation was stopped by adding cold 10% trichloroacetic acid (TCA). The TCA-treated samples were filtered on membrane filtres (pore size 0.6 µm). The washed and dried layers of TCA-insoluble mycelial fractions were placed into vials with 5 ml of scintillation fluid containing 2,5-diphenyloxazole 4.0 g, 1,4-di-2-(5-phenyloxazolyl)benzene 0.1 g, toluene 1000.0 ml. The radioactivity measurements were made in a liquid-scintillation computer (Mark I, Nuclear Chicago Corp.).

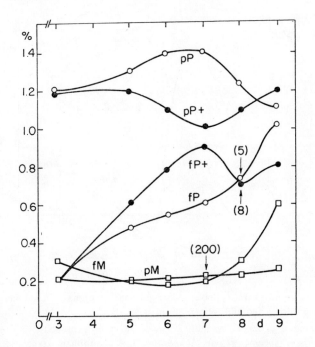

Fig. 2. Mycelial free (f) and protein (p) tryptophan (% of dry mycelium) in *Claviceps* strains MG-6 and Pla-4. M MG-6, P Pla-4, P+ Pla-4 with amitrole added (6 mM). Arrows indicate the time of maximal alkaloid yield presented by data in brackets (mg/l calculated as ergometrine). Tryptophan was determined according to TEUSCHER[7].

mentation broth. The relative rate of induction of tryptophan oxygenase was lower than that of tryptophan synthase, *i. e.* the pattern of induction was sequential in nature. Each of the enzymes was probably regulated by a different level of tryptophan. Whereas the increase in the tryptophan synthase activity was inversely proportional to the fall of cell-pool tryptophan, tryptophan oxygenase was able to manifest its activity when cell-pool tryptophan was enhanced. In view of some previous observations[10,12], tryptophan might protect tryptophan oxygenase from various forms of denaturation and proteolytic attack.

We conclude that the delayed production of alkaloids did not seem to be due to the time required to accumulate an excess of tryptophan as evidenced by the negative influence of a higher level of endogenous tryptophan on alkaloid yield. However, a certain concentration of cell-pool tryptophan was needed to promote alkaloid synthesis. These observations provide support for the hypothesis[13-15] suggesting endogenous free tryptophan as a possible inducer of the alkaloid-synthesizing enzyme system. In this connection, studies[16] on an inducible amidase of *Pseudomonas aeruginosa* can be mentioned. When the concentration of repressing metabolites is high, changes in inducer concentration have little effect on amidase synthesis. Only when the concentration of such metabolites falls is enzyme synthesis derepressed and depends on

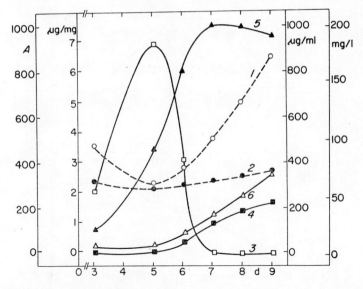

Fig. 3. Tryptophan, tryptophan synthase and tryptophan oxygenase in submerged mycelium of *C. paspali* MG-6 during alkaloid fermentation.

1 cell-pool tryptophan[7], μg/mg dry weight, *2* mycelial protein tryptophan[7], μg/mg dry weight, *3* tryptophan synthase specific activity[8] (A, $\times 10^{-3}$), *4* tryptophan oxygenase specific activity[9,10] (in cell-free extracts obtained by French press treatment; A), *5* total alkaloids[4], mg/l, *6* imidazolylglycerol[11], μg/ml.

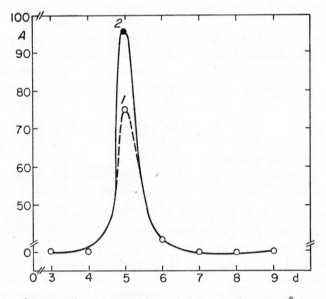

Fig. 4. Influence of ergometrine on tryptophan synthase specific activity[8] (A, $\times 10^{-2}$) in cell-free preparations of *C. purpurea* Pla-4 of various age. *1* control, *2* ergometrine (final concentration 1 mM) added.

the concentration of the inducer. Increasing levels of cell-pool tryptophan in the exponential phase of *Claviceps* culture growth indicated that tryptophan biosynthesis was
controlled less tightly than usual[17]. On the other hand, we assume a positive link between rapid metabolism of endogenous tryptophan and the alkaloid-producing capacity. Here we may point out the low activity of tryptophan synthase in the low producing strain Pla-4, *i.e.* only 8% of that in the alkaloid-producing strain MG-6. Fur-

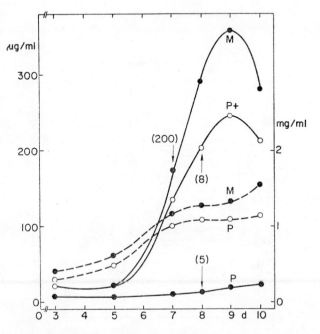

Fig. 5. Occurrence of imidazolylglycerol during submerged alkaloid synthesis by *C. paspali*
MG-6 (M) and *C. purpurea* Pla-4 (P). Arrows indicate the time of maximal alkaloid yield presented by data in brackets (mg/l calculated as ergometrine). *Solid lines:* imidazolylglycerol,
µg/ml; *dashed lines:* dry weight, mg/ml; P+ Pla-4 with amitrole added (6 mM) to the 3-day cultures.

thermore, Fig. 4 demonstrates that added ergometrine increased (27%) tryptophan
synthase activity of the Pla-4 strain. This gave rise to the hypothesis that alkaloid synthesis may reflect a regulatory device in tryptophan metabolism.

   As distinct from the alkaloid-forming strain MG-6, the Pla-4 strain formed only
an insignificant amount of imidazolylglycerol (Fig. 5). The accumulation of imidazolylglycerol was substantially increased after treatment of the Pla-4 culture with amitrole (3-amino-1,2,4-triazole). The mechanism of amitrole action resulted from competitive inhibition of imidazolylglycerol-phosphate dehydratase which was involved
in the histidine pathway[18]. As a consequence of amitrole inhibition imidazolylglycerol
accumulated. Simultaneously, the proportion between mycelial protein tryptophan and cell-pool tryptophan was altered, as illustrated in Fig. 2. Free tryptophan

was favoured over protein tryptophan. It seems reasonable to expect that the tryptophan—histidine cross-pathway regulation[19] could serve a useful function in alkaloid
formation. When the histidine pathway is inhibited, as indicated by accumulation of
imidazolylglycerol, phosphoribosylpyrophosphate which is a common intermediate
for both pathways becomes more available for the tryptophan pathway. This pattern
is favourable for alkaloid biosynthesis. Particular importance was also attached to the

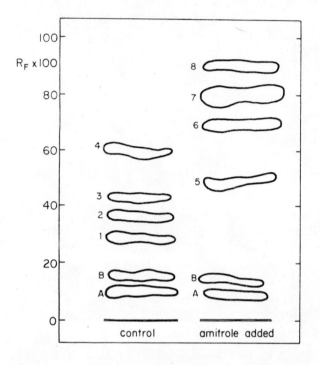

Fig. 6. Change of alkaloid spectrum of *C. purpurea* Pla-4 by adding amitrole (6 mм) to the 3-day
cultures. 10-day fermentation broths were made alkaline with ammonia, extracted with an equal
volume of ether (in three lots) and analyzed by thin-layer chromatography on silica gel G (Merck,
Darmstadt) using chloroform—methanol (4 : 1; *v/v*). A, B unidentified blue fluorescent spots,
1 ergometrine, 2 ergine, 3 ergometrinine, 4 isosetoclavine, 5 setoclavine, 6 ergotamine, 7 ergocristine, 8 ergocristinine.

apparent change of alkaloid spectrum due to the added amitrole. This is shown in
Fig. 6. Whereas the control sample contained ergine, ergometrine, ergometrinine and
isosetoclavine, the amitrole-treated culture produced setoclavine, ergotamine, ergocristine and ergocristinine. These data provide further support for the opinion[20] that
alkaloids do not represent a really decisive characteristic of the producing strain.

Next, our attention has been focused on intracellular ATP because of its role as
the primary energy mediator in metabolism. The evidence of adenylate kinase activity

in cell-free extracts of the investigated *Claviceps* cultures suggested that endogenous ATP formation was in part due to the following adenylate kinase reaction:

$$2\ ADP \rightleftarrows AMP + ATP$$

Fig. 7 demonstrates the different shapes of adenylate kinase curves of both *Claviceps* strains. It is noteworthy that during alkaloid biosynthesis the mentioned nonoxidative

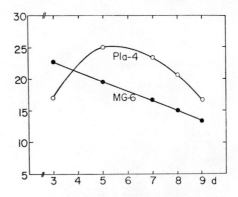

Fig. 7. Adenylate kinase activity (nmol ATP/mg protein per min) during submerged cultivation of *Claviceps* strains MG-6 and Pla-4. (Ascertained[21] in cell-free extracts obtained by French press treatment.)

ATP formation decreased. Cell-free extracts of *Claviceps* cultures were found to possess ATPase activity. The enzyme, the rough characteristics of which are given in Tab. I, appears to be a true ATPase (EC 3.6.1.3) which splits only the terminal phosphate from ATP. The ATPase activity in the alkaloid-producing strain MG-6 was inhibited by ergometrine (Fig. 8). This observation may indicate the possible role of alkaloid formation in the physiology of the culture.

In comparison with other microorganisms[22,23] the amount of ATP per known mycelial mass, *i.e.* the ATP pool, was relatively low in *Claviceps* cells (Tab. II). In the

TABLE I

Properties of ATPase in Subcellular Preparations of Submerged *Claviceps* Cultures

Conditions	*C. paspali* MG-6	*C. purpurea* Pla-4
Mg^{2+} activation	present	present
Ca^{2+} activation	absent	absent
ATP to Mg^{2+} ratio	0.1 or 0.05	0.1 or 0.05
K^+ stimulation[a]	105% at 0.2 M	62% at 0.2 M
Digestion product	ADP	ADP
Optimal pH	7.5	7.5

[a] Increase in activity in addition to that caused by Mg^{2+}.

Pla-4 strain producing insignificant amount of alkaloids as culture growth proceeded the ATP concentration fell and was inversely proportional to the increase in the ATPase activity (Fig. 9 *above*). The ATP pool and ATP utilization were proportional to the growth yield. On the other hand, in the alkaloid-producing strain MG-6 the amount of intracellular ATP was high and was accompanied by a decrease in ATPase activity (Fig. 9 *below*). The marked rise in the ATP pool might have been due to a low-

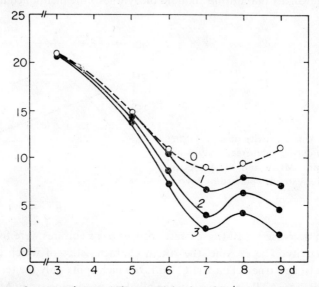

Fig. 8. Influence of ergometrine on ATPase activity (nmol P_i/mg protein per min) of *C. paspali* MG-6 during alkaloid fermentation. Final concentrations of ergometrine added (mM): *1* 0.25, *2* 0.5, *3* 1.0, *0* control.

er rate of growth. The slow growth was probably a rate-limiting step in biosynthesis that was not energy-requiring. It is reasonable to suggest that this might have been the case of alkaloid formation.

TABLE II

Comparison of ATP Pools in Several Microorganisms

Organism	ATP mM
Claviceps purpurea Pla-4	0.0004 — 0.0008
Claviceps paspali MG-6	0.0006 — 0.0016
Yeast	0.8 — 2.0
Streptomyces aureofaciens 8425	0.01 — 0.1
Streptomyces aureofaciens RIA 57	0.02 — 0.5
Escherichia coli	0.0003 — 0.007

Fig. 10 shows a schematic illustration of the role proposed for ATP in regulatory mechanisms[27-29] of the investigated *Claviceps* strains. When the ATP concentration is high the AMP level is necessarily low. Under these conditions, which are typical of the alkaloid-producing strain MG-6, the expected result is that a low AMP concentration will cause fatty acid biosynthesis to compete more markedly for acetylCoA with the citric-acid cycle because of the high ATP concentration. This suggestion is supported by our previous data[30] indicating an inverse correlation between alkaloid formation and attainment of the maximum cell-pool level of citric and succinic acids in particular and other organic acids in general. In this connection, a positive correlation

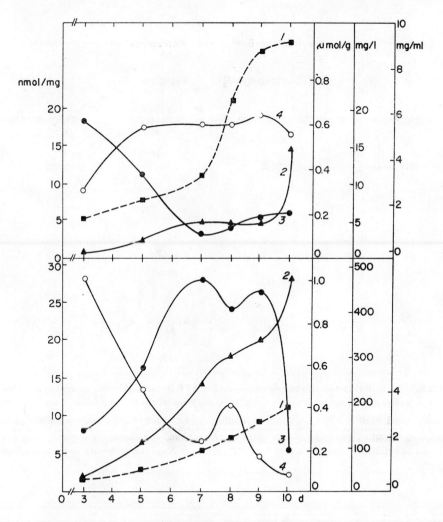

Fig. 9. Changes in ATP concentration and ATPase activity in *Claviceps* strains Pla-4 (*above*) and MG-6 (*below*) during alkaloid fermentation. *1* dry weight, mg/ml, *2* total alkaloids[4], mg/l *3* ATP[23], μmol/g of dry cells, *4* ATPase[25,26], nmol P_i/mg protein per min.

between lipids and sterols and alkaloid biosynthesis[31,32] can be mentioned. Furthermore, in the MG-6 strain as distinct from the low-producing strain Pla-4 the intracellular level of ATP might have been sufficient for inhibition of both phosphoenolpyruvate carboxylase[33] which equilibrates the level of oxalacetic acid in the citricacid cycle[34] and citrate synthase[35] which catalyzes the formation of citrate from acetylCoA and oxalacetate.

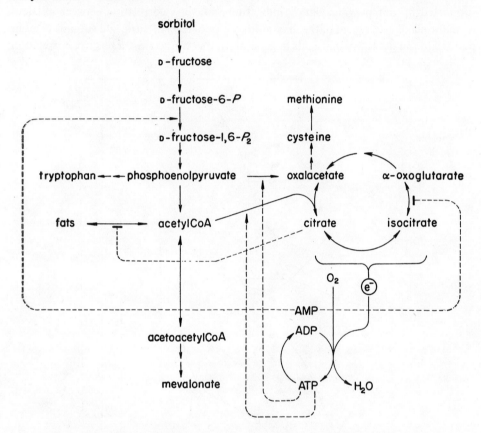

Fig. 10. Schematic illustration of the role proposed for ATP in the regulation of energy metabolism. *Broken lines* connect effector compounds to the enzymes which they modulate: – – – –→ negative effect, – – –▮ positive effect. ⓔ indicates supply of electrons (from oxidative reactions) to the electron transport phosphorylation system. Tryptophan, mevalonate and methionine are mentioned as building stones of the ergoline ring[27]. Intermediates not necessary to the discussion are omitted.

CONCLUSIONS

The investigated *Claviceps* strains grown in the absence of added tryptophan accumulated extra endogenous tryptophan. The delayed production of alkaloids was not due to the time required to accumulate excess tryptophan. However, a certain concentration of tryptophan was needed to promote alkaloid synthesis. The pattern of induction of tryptophan synthase and tryptophan oxygenase was sequential and each of the enzymes was probably regulated by a different level of tryptophan. A positive link between rapid metabolism of endogenous tryptophan and the alkaloid-producing capacity is suggested. The tryptophan—histidine cross-pathway regulation is expected to serve a useful function in the biosynthesis of alkaloids. The result gave rise to the hypothesis that alkaloid formation represents a regulatory device in tryptophan metabolism of the *Claviceps* cultures. Another physiological role of alkaloids was demonstrated by an inhibition of ATPase activity and stimulation of tryptophan synthase activity by ergometrine. The evidence that a rapid increase in the ATP pool and reduced ATP utilization were favourable for alkaloid production leads to the suggestion that energy production was not limiting in alkaloid biosynthesis. The results of ATP experiments and the level of organic acids in the cell pool during fermentation indicate that alkaloid formation was inversely proportional to the rate of competition of the citric-acid cycle for acetylCoA. The change of alkaloid spectrum caused by adding amitrole provides support for the opinion that alkaloids do not represent a really decisive characteristic of the producing strain.

This investigation was supported by funds provided be the International Atomic Energy Agency, research contract No 845/RB.

The authors wish to express their thanks to Prof. H. Rochelmeyer (Claviceps paspali MG-6) and Dr. J. Kybal (Claviceps purpurea Pla-4) for kindly supplying the cultures indicated. Thanks are also due to Prof. D. Gröger, Dr. A. Hofmann, Prof. C. Spalla, Dr. A. Tonolo (alkaloids) and Dr. T. Cleary (amitrole) for generous help in sending reference samples.

REFERENCES

1. BAUCHOP F., ELSDEN S. R.: J. Gen. Microbiol. **23**, 457 (1960).
2. KOBEL H., SCHREIER E., RUTSCHMANN J.: Helv. Chim. Acta **47**, 1052 (1964).
3. ŘEHÁČEK Z., KOZOVÁ J., ŘIČICOVÁ A., KAŠLÍK J., SAJDL P., ŠVARC S., BASAPPA S. C.: Fol. Microbiol. **16**, 35 (1971).
4. ROEDER K., MUTSCHLER E., ROCHELMEYER H.: Pharm. Acta Helv. **42**, 407 (1967).
5. ARCAMONE F., CASSINELLI G., FERMI G., PENNELLA P., POL C.: Paper at 3rd Internat. Ferment. Symp., New Brunswick 1968.
6. WEYGAND F., FLOSS H. G.: Angew. Chem., Internat. Ed. **2**, 243 (1963).
7. TEUSCHER E.: Pharmazie **16**, 570 (1962).
8. YANOVSKY C., p. 233 in S. P. Colowick, N. O. Kaplan (Eds.): *Methods in Enzymology*, Vol. II. Academic Press, New York 1955.
9. TREMBLEY G. C., GOTTLIEB J. A., KNOX W. E.: J. Bacteriol. **93**, 168 (1967).

10. KNOX W. E., MEHLER A. H.: Science **113**, 237 (1951).
11. CASSELTON P. J.: Physiol. Plantarum **19**, 411 (1966).
12. FEIGELSON P., GREENGARD O.: J. Biol. Chem. **237**, 1908 (1962).
13. FLOSS H. G., MOTHES K.: Arch. Mikrobiol. **48**, 213 (1964).
14. BU'LOCK J. D., BARR J. G.: Lloydia **31**, 342 (1966).
15. VINING L. C.: Can. J. Microbiol. **16**, 473 (1970).
16. CLARKE P. H., HOULDSWORTH M. A., LILLY M. D.: J. Gen. Microbiol. **51**, 225 (1968).
17. LINGENS F., GOEBEL W., UESSELER H.: Naturwiss. **54**, 141 (1967).
18. HILTON J. L., p. 71 in *Isotopes in Weed Research.* Internat. Atomic Energy Agency, Vienna 1966.
19. JENSEN R. A.: J. Biol. Chem. **244**, 2816 (1969).
20. MOTHES K.: Lloydia **29**, 156 (1966).
21. Boehringer & Söhne GmbH, Mannheim: *Biochemical Test Combination.* Adenosine-5'-triphosphate, UV-method TC-J, Cat. N° 15379 TAAC.
22. POLAKIS E. S., BARTLEY W.: Biochem. J. **99**, 521 (1966).
23. COLE H. A., WIMPENNY J. W. T., HUGHES D. E.: Biochim. Biophys. Acta **143**, 445 (1967).
24. COLE C. V., ROSS C.: Anal. Biochem. **17**, 526 (1966).
25. MICKELSON M. N.: J. Bacteriol. **100**, 895 (1969).
26. AGURELL S.: Acta Pharm. Suecica **3**, 71 (1966).
27. HATHAWAY J. A., ATKINSON D. E.: J. Biol. Chem. **238**, 2875 (1963).
28. RAMAIAH A., HATHAWAY J. A., ATKINSON D. E.: J. Biol. Chem. **239**, 3619 (1964).
29. ATKINSON D. E.: Science **150**, 851 (1965).
30. ŘEHÁČEK Z., BASAPPA S. C.: Fol. Microbiol. **16**, 110 (1971).
31. AMICI A. M., MINGHETTI A., SCOTTI T., SPALLA C., TOGNOLI L.: Appl. Microbiol. **15**, 597 (1966).
32. AMICI A. M., MINGHETTI A., SCOTTI T., SPALLA C., TOGNOLI L.: Appl. Microbiol. **18**, 464 (1969).
33. VOŘÍŠEK J., POWELL A. J., VANĚK Z.: Fol. Microbiol. **14**, 398 (1969).
34. CANOVAS J. L., KORNBERG H. L.: Proc. Roy. Soc. B **165**, 189 (1966).
35. HOŠŤÁLEK Z., RYABUSHKO T. A., CUDLÍN J., VANĚK Z.: Fol. Microbiol. **14**, 121 (1969).

AUTHOR INDEX

Numbers in *italics* are page numbers of the author's own chapter

GENETIC INDEX

1. Chromosomal

2. Episomes and plasmids

MICROORGANISM INDEX

COMPOUND INDEX

ENZYME INDEX

SUBJECT INDEX

Aberrations, chromosome 87
Acceptor, nuclear 76
Accumulation *see* Production
Acetylation 224
 —, of homoserine 151
N-Acetylation, of D-tryptophan 116
Actinophage 30, 224
 — —host interactions 30
Activation, of anthranilate synthase 119
 — of prephenate dehydrogenase 113
Activity, auxinic 400
 —, enzyme 101, 171, 247, 365, 415
 —, of ATPase 433
 —, of methionyl-tRNA synthetase 169
 —, psychomimetic 407
Acylation, of tRNA^{Met} 173
Adaptation, ecological 406
Aeration 356
Aggregate, enzyme 114
 —, protein 107
Aldol condensation 286, 300
Alkaloid 116, 215
 — biosynthesis 373, 414, 436
 —, clavine 406
 —, peptide 393, 422
 — production 109, 120, 375, 393, 405, 421, 427
Alleles 39, 91, 257
 —, *argA1/arg$^+$* 126
 —, *proA1/pro$^+$* 126
 —, *trp/trp$^+$* 126
 —, chromosomal 24, 40
 — frequencies 54, 55, 65
 —, mutant 60, 82, 228
 —, resistance 160, 167
 —, wild-type 153, 303
Allelic complementation 98, 104, 128
 — mutations 181
Allosteric control 195
 — effector 195, 210
 — feedback inhibition 159
 — properties 102
Altered enzyme properties 98, 101
Amber mutation 12
 — suppressor 177